AF479123

Philosophy and Medicine

Volume 128

Marta Soniewicka

Editor

The Ethics of Reproductive Genetics

Between Utility, Principles, and Virtues

 Springer

Editor
Marta Soniewicka
Department of Philosophy of Law
and Legal Ethics
Jagiellonian University
Krakow, Poland

ISSN 0376-7418 ISSN 2215-0080 (electronic)
Philosophy and Medicine
ISBN 978-3-319-60683-5 ISBN 978-3-319-60684-2 (eBook)
https://doi.org/10.1007/978-3-319-60684-2

Library of Congress Control Number: 2017959720

Printed on acid-free paper

This Springer imprint is published by the registered company Springer International Publishing AG part of Springer Nature.
The registered company address is: Gewerbestrasse 11, 6330 Cham, Switzerland

Acknowledgments

I owe special thanks to Professor Robert Audi for his extraordinary support and for our philosophical discussions in Krakow which had an impact on my work. I would like to also extend my thanks to Professor Thaddeus Metz for his encouragement, as well as to Professor Roberto Andorno for his remarks on the general ideas which framed the book. I would like to use this opportunity to extend my sincere thanks to Professor George Annas for the inspiring time I had as a Fulbright scholar in Boston when he suggested me to work on the ethics of reproductive genetics under his supervision.

I am also grateful to Dr. Aeddan Shaw for his help in the linguistic edition of the volume, as well as to Filip Bardziński for his technical editorial work.

The preparation of this volume was financed by the Polish National Science Centre (Dec-2013/10/E/HS5/00157).

Contents

**Part II   The Moral, Legal and Social Challenges
                  of Reproductive Genetics**

Contributors

Roberto Andorno is an associate professor at the Faculty of Law and research fellow at the Institute of Biomedical Ethics, University of Zurich, Switzerland. He is a former member of the UNESCO International Bioethics Committee. Dr. Andorno has published extensively on various issues at the intersection of law and bioethics, notably *Bioética y Dignidad de la Persona* (Madrid 2012) and *Principles of International Biolaw* (Brussels 2014).

Robert Audi is John A. O'Brien professor of philosophy at the University of Notre Dame (USA) and a past president of the American Philosophical Association and the Society of Christian Philosophers. He is internationally known for contributions to epistemology, ethics (especially on intuitionism and in political church–state issues), and the theory of action. His books include *The Structure of Justification* (Cambridge University Press, 1993); *Religious Commitment and Secular Reason* (Cambridge University Press, 2000); *The Architecture of Reason: The Structure and Substance of Rationality* (Oxford University Press, 2001); *The Good in the Right: A Theory of Intuition and Intrinsic Value* (Princeton University Press, 2004); and *Means, Ends, and Persons: The Meaning and Psychological Dimensions of Kant's Humanity Formula* (Oxford University Press, 2016).

Ewa Baum is an assistant professor at the Department of Philosophy of Medicine and Bioethics, Faculty of Health Sciences, Poznan University of Medical Sciences, Poland. She authored the book *Stem Cells as a Bioethical Problem of Modern Medicine* and authored or coauthored (with M. Musielak and K. Pawlaczyk) dozens of articles on ethical problems of the development of medicine, including "Utopian trends in modern bioethics. Ethical aspects of research on and using stem cells" (Bull. Trans. Univ. Brasov New Ser. Ser. B2 2005: 12(47) p 221–224), "Who is right about legalization of organ trade?" (Prz. Filoz. Nowa Seria, 2012: R. 21, nr 2 (82), s. 141–150), and "Levels of hepatocyte growth factor in serum correlate with quality of life in hemodialysis patients" *(Int. J. Clin. Exp. Pathol.* 2015:Vol. 8, nr 10, s. 13477–13,482).

Sylvester C. Chima is an associate professor and head of the Programme of Bio & Research Ethics and Medical Law, School of Public Health, Nelson R Mandela School of Medicine, University of KwaZulu-Natal in Durban, South Africa. His research interests concentrate on medical law, bioethics, human rights, and neuro-pathology. He has authored numerous articles in international journals such as *BMJ*, *BMC Medical Ethics*, *Journal of General Virology*, *Human Biology*, *Nigerian Journal of Clinical Practice*, etc. He is the author/coauthor of three books on medical law, bioethics, and African health issues. His latest book is entitled *A Primer on Medical Law, Bioethics and Human Rights for African Scholars* (Chimason Educational Books, Durban, 2011).

Barbara Chyrowicz is a professor and chair of the Department of Applied Ethics, Faculty of Philosophy, the John Paul II Catholic University in Lublin, Poland, and the founding member of the Polish Bioethics Society. In 2013, she received the Officer's Cross of the Order of Polonia Restituta. She published numerous articles and books on ethics and bioethics, including *Bioetyka i ryzyko. Argument "równi pochyłej" w dyskusji wokół osiągnięć współczesnej genetyki* (Lublin: Towarzystwo Naukowe KUL, 2000), *O sytuacjach bez wyjścia w etyce. Dylematy moralne, ich natura, rodzaje i sposoby rozstrzygania* (Kraków: Znak, 2008), and *Bioetyka. Anatomia Sporu* (Kraków: Znak 2015).

Iñigo De Miguel Beriain is a senior research fellow at the Department of Public Law, University of the Basque Country, Bilbao, Spain. His research has addressed issues related to the philosophy of law, ethics of economy, bioethics, and biolaw. He is the author of several articles and books, the latter including *Matrix: La Humanidad ante la Encrucijada* (València: Tirant lo Blanch 2004) and *El Embrión y la Biotecnología: Un Análisis Etico-jurídico* (Granada: Editorial Comares 2004). His paper entitled "Bioethics and new Biotechnologies in Human Health" was awarded with the International Prize by the "House of Representatives of the Principality of Asturias-International Society of Bioethics (SIBI)."

Ana Maria Marcos del Cano is an accredited professor of philosophy of law at the Faculty of Law of the UNED and the director of the Department of Legal Philosophy. She authored numerous articles, chapters, and books on philosophy of law, including *La eutanasia: estudio filosófico-jurídico* (Madrid: Librerías Marcial Pons 1999).

Jan Domaradzki is an assistant professor of sociology at the Poznan University of Medical Sciences. His current research includes social implications of new genetics, geneticization of public discourse, and medicalization. He is the author of two books in the field of the sociology of religion and several articles.

Małgorzata Karbarz is an assistant professor at the Institute of Biotechnology and Basic Sciences, University of Rzeszow, Poland. Her research interests include bioethics and social aspects of biotechnology and plant genetics. She authored and coauthored several articles, including "Science, safety, and trust: The case of transgenic food" (*Croatian Medical Journal* 2013: 54(1), p 91–96) and "Gene transfer and modulation for the production of food with enhanced quali-quantitative values: potentials, promises and achievements" (*Food Technology And Biotechnology* 2013: 51(2), p 208–223).

Wojciech Lewandowski is an assistant professor at the Department of Applied Ethics, Faculty of Philosophy, the John Paul II Catholic University in Lublin, Poland. His research interests concentrate on the problem of justification of partiality and impartiality in contemporary ethical theories and the problem of practical reasoning in ethics. He recently published *Przyszłość i odpowiedzialność. Problem uzasadnienia odpowiedzialności za przyszłe pokolenia we współczesnej etyce* (Lublin: Wydawnictwo KUL, 2015).

Thaddeus Metz is research professor at the University of Johannesburg, where he is also distinguished professor affiliated with the new Institute for Pan-African Thought and Conversation (2015–2019). He has had more than 150 journal articles and book chapters in ethical, political, and legal philosophy published, about half of which are in the field of African philosophy. His book titled *A Relational Moral Theory: Africa's Contribution to Global Ethics* is under contract with Oxford University Press in the UK (2017), which also published his previous book, *Meaning in Life: An Analytic Study* (2013).

Jakub Pawlikowski is an associate professor at the Department of Ethics and Human Philosophy, Medical University of Lublin, Poland, and a visiting associate professor at Harvard T.H. Chan School of Public Health, Harvard University, USA. He is a member of the Transparency Council of the Agency for Health Technology Assessment and Tariff System and former member of the team of experts on Molecular Genetic Research and Biobanking at the Ministry of Science and Higher Education of Poland. He coedited (with P. Stanisz and M. Ordon) *Sprzeciw sumienia w praktyce medycznej: aspekty etyczne i prawne* (Lublin: Wydawnictwo KUL, 2014).

Henning Rosenau is a professor at the Martin Luther University of Halle-Wittenberg and executive director of the Interdisciplinary Centre for Medicine–Ethic–Law. He is a member of the Ständige Kommission Organtransplantation der Bundesärztekammer. He authored and coauthored several books, including *Schriften zum Bio-, Gesundheits- und Medizinrecht* (Baden-Baden: Nomos Verlag, 2008), *Ein zeitgemäßes Fortpflanzungsmedizingesetz für Deutschland* (Baden-Baden: Nomos Verlag, 2012), and *Fortpflanzungsmedizingesetz* (AME-FMedG) (Tübingen: Mohr Siebeck Verlag, 2013), and is coeditor of *Medstra* (medical criminal law; Heidelberg: C.F. Müller Verlag).

Nete Schwennesen is a postdoctoral researcher at the Department of Anthropology/Department of Public Health, University of Copenhagen. She is concentrating her research on the topics of health, digital technology, care, aging, and reproductive technologies. She is the author and coeditor of several articles and books, including "At the Margins of Life: Making Fetal Life Matter in Trajectories of First Trimester Prenatal Risk Assessment" (in *Bio-objects: Life in the 21st Century*, eds. N. Vermeulen, S. Tamminen & A. Webster, 2012, Farnham: Ashgate, p 117–131) and "Representing and intervening: Doing good care in first trimester prenatal risk assessment (FTPRA)" (*Sociology of Health and Illness* 34(2), p 283–298).

Aeddan Shaw is an assistant professor at the Jesuit University of Philosophy and Education in Krakow and senior lecturer at the Jagiellonian University. His broad academic interests range from the philosophy of language, teaching methodology, travel writing, and the history of ideas. He has written chapters for a number of publications on normativity, is the author of a series of popular science articles in *The Teacher*, and is currently preparing a book for publication on the misconceptions frequently associated with Noam Chomsky entitled *Furious Purple Herrings: An Examination of the Logical, Bio-logical and Bio-Cultural Accounts of Language Through the Prism of Normativity*.

Marta Soniewicka is an assistant professor at the Department of Philosophy of Law and Legal Ethics at the Jagiellonian University in Krakow, Poland. Her research interest concentrates on the philosophy of law, philosophy of politics, and ethics. She authored and coauthored several articles, chapters, and books including "Failures of Imagination: Disability and the Ethics of Selective Reproduction" (*Bioethics*, 29: 557–563), *Paradoksy bioetyki prawniczej* (Warszawa: Wolters Kluwer Polska 2010; along with J. Stelmach & B. Brożek & W. Załuski), and *After God. The Normative Power of the Will from the Nietzschean Perspective* (*Studies in Philosophy and Social Sciences*, Dia-Logos, Frankfurt a. Main: Peter Lang Edition, 2017).

Peter Sýkora is a professor of philosophy at the University of Sts. Cyril and Methodius (UCM) in Trnava, Slovak Republic. He is a founder and director of the Center for Bioethics at UCM. He has written on evolutionary biology, ontology, and bioethics. He authored and coauthored numerous articles, chapters, and books including "Macroevolution of plasmids: A model for plasmid speciation" (*J. Theor. Biology* 159: 53), *A Tale of Two Countries: Czech and Slovak Stem Cell Biopolicies* (in *Contested Cells*, BJ Capps & AV Campbell, eds, Imperial College Press, London 2010), and *Altruism Reconsidered: Exploring New Approaches to Property in Human Tissue* (with M. Steinmann and U. Wiesing, eds, Aldershot, Hampshire: Ashgate. 2009).

Adriana Warmbier is an assistant professor at the Institute of Philosophy, Jagiellonian University in Krakow, Poland. Her research interests include virtue ethics of Aristotle and its interpretations, moral philosophy, theory of action, philosophy of the subject, and post-humanism. She has authored several articles on Aristotelian moral philosophy and human perfection, such as "Moral perfection and the demand for human enhancement" (*Ethics in Progress* 2016 6(1)), as well as edited several books including *The Idea of Excellence and Human Enhancement: Reconsidering the Debate on Transhumanism in Light of Moral Philosophy and Science* (Frankfurt a. Main: Peter Lang Edition, forthcoming 2017).

Introduction

Revolutionary advances in biotechnology have significantly transformed medicine, yet, at the same time, these technological improvements (e.g., CRISPR, mitochondrial gene transfer, therapeutic cloning, PGD, PND, ART) have brought new challenges to our societies and to ethics and law in particular. The Promethean dream of contemporary medicine poses questions about the ends and boundaries of medicine. In an attempt to answer this question, one should turn to the philosophy of medicine.

This book is aimed at analyzing *the foundations of the ethics of reproductive genetics* by considering different moral theories and their implications for judgments in clinical practice and policy-making. One usually claims that there are some basic principles (non-maleficence, beneficence, confidentiality, autonomy, and justice) which constitute the foundations of bioethics and medical ethics (Beauchamp and Childress 1989; Engelhardt 1996). Yet these principles conflict with one another, and one needs some criteria to solve these conflicts and to specify the scope of application of these principles. Thus, some argue that one should integrate these principles into a broader theoretical framework (Brody 1988; Dworkin 2011), while others advocate the superiority of case-based reasoning (casuistry) over theory in solving moral dilemmas (Kamm 2013). In turn, some justify moral decisions in medicine with reference to principles (rights and duties), while others argue that cost-and-benefit analysis is the best way to address healthcare issues, in particular the distribution of healthcare resources (Singer et al. 1995; Singer 2011). Still others emphasize the role of the virtues (fidelity, compassion, *phronesis*, justice, fortitude, temperance, integrity, self-effacement) in constituting the ethos of the medical profession, as well as in addressing the special relationship between doctor and patient which should be based on trust, responsibility, and the caring bond (Pellegrino 1993). Exploring these miscellaneous ethical approaches as introduced to biomedicine, particularly *to reproductive genetics*, the book shall elucidate their different assumptions concerning human nature, the meaning of medical practice, and the relations between healthcare providers, recipients, and other affected parties (e.g., progeny, relatives, other patients, society). It is worth stressing that these approaches answer the question of how medical ethics fits into our moral life in general differently. The book attempts to answer the question of whether the ten-

sion between these ethical doctrines would generate conflict in the field of biomedicine alone or if these competing approaches could complement each other in some
way.

The book consists of two parts. In the first part, the contributions review a number of moral theories such as consequentialism (in particular the utilitarian
approach), principle-based ethics (in particular Kantian deontology), virtue ethics,
and commonsense intuitionism in order to identify the justification of arguments
given in biomedical debates with regard to intricate moral dilemmas, with a special
emphasis on the aspects of reproductive genetics which give rise to the most heated
debate. In the first chapter, Robert Audi proposes a wide-ranging pluralistic theory
for approaching moral questions, with an eye on the applicability of ethical standards to concrete biomedical decisions, including clinical decisions and the determination of research programs, especially those involving genetic engineering and
fetal tissue use. In the second chapter, Roberto Andorno elaborates upon the theme
of whether it is possible to answer morally complex questions without adhering to
certain moral principles or to explain one's moral views through rational argumentation alone. In the third chapter, Marta Soniewicka argues that the neo-Aristotelian
virtue approach to ethics could significantly enhance the understanding of medical
ethics and its application in the clinical context of reproductive genetics. In the
fourth chapter, Thaddeus Metz presents a novel bioethics of communion, inspired
by relational ideals from the African philosophical tradition, applicable to bioethical
issues in general and human procreation in particular, putting special emphasis on
people's capacity to commune with respect, where communing is a matter of identifying with others and exhibiting solidarity with them. In the fifth chapter, Wojciech
Lewandowski claims that in the ethics of reproductive genetics, the category of the
special obligations of the parents based on relationship-centered reasons is the best
way to guarantee the continuity of reasons before and after a child was brought into
existence. In the sixth chapter, Adriana Warmbier continues the discussion about the
morality of reproductive choices and objection to the principle of procreative beneficence by appealing to Aristotelian virtue ethics and its concepts of practical wisdom and eudaimonia which may provide us with a deeper understanding of the
good life and flourishing. The seventh chapter by Jakub Pawlikowski tackles the
problem of the conscientious objection of healthcare workers in the context of
genetic testing, claiming that it results mainly from situations which violate the
principle of respect for human life and the principle of non-maleficence and therefore cannot be reduced to individual moral or religious beliefs. Aeddan Shaw
addresses in the eighth chapter the problem of flourishing from a different angle,
introducing environmental ethics such as Attfield's practice-consequentialism with
its emphasis on the impact of our deontic decisions upon the environment and future
generations. Finally, in the ninth chapter and the last of this section, Barbara
Chyrowicz discusses the issue of resolving moral dilemmas which seems reasonable within a theory and not outside it since, as she claims, dilemmas do not appear
on the theoretical level but in practice and recognizing them as dilemmas is connected with the fact that the decision-maker who has norms accepted in the domain

of a given theory cannot cope "here and now" with deciding which of the rival options should be given priority.

The second part of the book consists of the contributions which encompass the crucial moral, legal, and social challenges of reproductive genetics. Legal issues are the subject of the next chapter in which Hennig Rosenau ponders the problem of reproductive and therapeutic cloning and addresses the question of the normative status of an embryo, arguing for an urgent reform of the current German regulations which are no longer up to date and therefore insufficient to cover these issues in a satisfactory way. The eleventh chapter by Peter Sykora discusses the prohibition of germline gene interventions in humans which was recently challenged by the legalization of mitochondrial replacements in the UK and by the edition of germline genes in nonviable human embryos by CRISPR/Cas9 technology in China. CRISPR gene edition is the subject of Chap. 12, written by Iñigo de Miguel Beriain and Ana María Marcos del Cano, which is concerned with such issues as: the necessity of this technology and the risk issues involved, the embryo loss involved, the alteration of the human genome and human identity, and the enhancement/eugenics issue. In Chap. 13, Ewa Baum and Jan Domaradzki present some of the major dilemmas that accompany the development of genetics in the context of medical and reproductive ethics and discuss them from the perspective of the basic principles of ethics and medical law. Małgorzata Karbarz examines in Chap. 14 the increased technical capabilities as well as technical limitations of the diagnosis in PND and PGD which may raise ethical questions concerning reproductive decisions. In the next chapter, Nete Schennesen analyzes the process of prenatal decision-making in the context of first trimester prenatal risk assessment (FTPRA) in Denmark, claiming that the problems and solutions about prenatal testing should be framed as processes of knowledge production. Sylvester Chima devotes the final chapter to the legal and ethical issues of the termination of pregnancy in African countries, arguing for the improvement of women's reproductive rights.

Some of these issues are new and require insightful ethical consideration; some have already been discussed, but the arguments can still be interpreted from a fresh perspective in the debate, thus providing new insights into old problems.

Jagiellonian UniversityMarta Soniewicka
Krakow, Poland

Part I
Moral Theories Applied to Biomedicine

Chapter 1
Ethical Theory and Moral Intuitions in Biomedical Decision-Making

Robert Audi

Abstract One of the most important problems of contemporary life is how to bring ethics to bear on the development and applications of technology. This problem is nowhere more acute than in the biomedical sphere. The ethical guidance of biomedical research and medical practice depends on adequately clear sound ethical standards, but it also depends on the internalization of these standards in both research and therapeutic practice. This paper focuses on the problem of determining adequate ethical standards and proposes a wide-ranging pluralistic theory for approaching moral questions. It does this in broad outline, but also with an eye to the applicability of ethical standards to concrete biomedical decisions, including clinical decisions and determination of research programs, especially those involving genetic engineering and fetal tissue use.

Keywords Virtue ethics • Kantian ethics • Utilitarianism • Intuitionism • Genetic engineering • Integrative ethical theory

Four Major Kinds of Ethical Theory

Both moral philosophy and ethics as studied outside philosophy have centered on four kinds of ethical theory. Often the term 'approach' is more apt than 'theory' for cases in which decision-makers have an ethical orientation grounded in one or another kind of theory, but do not self-consciously hold a theory as such. My first task here is to review these and note some strengths and weaknesses. In the light of that review, I will suggest how we might proceed toward a decision framework that draws on the best elements in each of the widely known approaches.[1]

---

[1] This section and the next draw heavily on, though they also refine and somewhat revise, points made I have made in (2007a), especially Chap. 1.

R. Audi (✉)
University of Notre Dame, Notre Dame, IN, USA
e-mail: audi.1@nd.edu

© Springer International Publishing AG, part of Springer Nature 2018
M. Soniewicka (ed.), *The Ethics of Reproductive Genetics*,
Philosophy and Medicine 128, https://doi.org/10.1007/978-3-319-60684-2_1

Virtue Ethics

In both Western and Eastern philosophy, virtue ethics has long appealed to many thinkers. Plato, Aristotle, and Confucius developed moral views of this sort. Aristotle – whose influence in bioethics since the 1970s has been major – described (for instance) just acts as the kind that a just person would perform; he also said that a just person is not to be defined as one who characteristically performs just acts. As this wording suggests, he apparently took moral traits of character to be ethically more basic than moral acts. He said, for instance, regarding the types of acts that are right, "Actions are called just or temperate when they are the sort that a just or temperate person would do" (Nicomachean Ethics 1105b5ff). It is virtues, such as justice and temperance, rather than acts, that are ethically central for Aristotle. Among his overall descriptions of it importance is that "virtue makes us aim at the right target, and practical wisdom makes us use the right means" (1144a).

For a virtue ethics, then, agents and their traits, as opposed to rules of action, are morally basic. The idea is that we are to understand what it is to behave justly through studying the nature and tendencies of the just person, not the other way around. We do not, for instance, first construct a notion of just deeds as those that, say, treat people equally, and then characterize a just person as one who characteristically does deeds of this sort. Thus, for adults as well as for children, and in ordinary life as in the professions, role models – who should exhibit good character as well as right actions – are absolutely crucial for moral learning. The person of practical wisdom is the chief role model in ethics; such people exemplify all of the moral virtues and also tend to be good advisors in ethical decisions.

Aristotle understood the virtues in the context of his theory of the good for human beings. For him, happiness (flourishing in some translations) is central. Happiness is our final unifying end: we may seek other things for their own sake, but only when "through them" we can achieve happiness. Happiness is not, however, a passive state; it requires a life in which "actions and activities (…) that involve reason" (our distinctive characteristic) is central; the "human good," then, proves to be activity of the soul in accord with virtue" (1098b214-17). Here virtue is understood as excellence: the happy life must, in some predominating way, be one in which we achieve excellence in our activities. Aristotle said a great deal about what constitutes excellence, and – on the plausible assumption that excellence is a human good – this notion is of major importance for biomedical contexts in which we must determine what is good for, in the interest of, or a detriment to, a person, including ourselves.

If, however, we take traits as ethically more basic than acts, we face a problem particularly acute in practical ethics: how does a virtue theory enable us to determine what to do? Ethics largely concerns conduct. How do we figure out what counts as, for instance, being generous or honorable? Virtue ethics has resources for answering this, including the appeal to practical wisdom as applied to the context of decision. A person of practical wisdom is a paradigm of one having virtue, and in a famous passage Aristotle calls virtue "a state that decides, consisting of a mean, the mean relative to us, which is defined by reason … It is a mean between two vices, one of excess one of deficiency" (1097a1-4). Consider beneficence. If I am selfish and ignore others' needs, this is a deficiency; if I give so much at once that I am prevented from significantly helping others later, I am excessive. Good ethical decisions may be seen in the light of such comparisons.

Both in and outside philosophy, there is great appreciation for the wisdom of the best virtue-ethical approaches. But it is widely agreed that these approaches are insufficiently determinate in implications for practical decision and need supplementation by standards of a different kind. The main contrasting approaches, not surprisingly, are rule theories – so called because they makes rules of action central and call for specific kinds of deeds. Let us briefly consider three categories of rule ethics.

Kantian Ethics

Among the most famous moral principles is Immanuel Kant's master principle, the Categorical Imperative. It says, in one formulation,

> Act as if the maxim of your action [roughly the principle underlying it] were to become through your will a universal law of nature. (Kant 2002, sec. 422)

Thus, I should not leave someone to bleed to death on the roadside if I could not rationally will the universality of the practice – say, even where I am the victim. More specifically, we would not want to universalize, and thus live by, the callous principle: One should stop for someone bleeding to death provided it requires no self-sacrifice. Similarly, I should not make a lying promise to repay money if I could not rationally universalize the underlying principle, say that when I myself can get money only by making a lying promise to repay it, I will do this. One way to see why the Imperative disallows this is to note that I count on sincere promises from others and cannot rationally endorse the universality of a deceitful practice that would victimize me.

Kant also gave a less abstract formulation of the Categorical Imperative, the Humanity formula (sec. 429):

> Act in such a way that you always treat humanity, whether in your own person or in the person of any other, never merely as a means, but always at the same time as an end.

Always treat people as ends in themselves, never merely as means. The Imperative seems to say: Never use people, as in manipulatively lying to them; instead, respect them. Treating people as ends clearly requires caring about their good. They matter as persons, and one must to some extent act for their sake whether or not one benefits from doing so.[2]

---

[2] The notions of treating persons as ends and of treating them merely as means can be clarified even independently of Kant's ethical writings. For an indication of how and references to literature on Kantian ethics see chapter 3 of my (2004).

This formulation, more obviously than the other one quoted, applies to oneself as well as others; it requires a kind of respect for persons, and this includes self-respect – an attitude that bears on how the voluntary sale of organs is to be viewed.[3] If we take Kant's two formulations together (and he considered them equivalent), then apparently we must not only treat persons as ends but – as the rational universalizability of our principles would suggest – treat them equally so. Everyone matters, and matters equally.

Utilitarianism

A very different kind of rule theory is suggested by a question likely to figure influentially in biomedical decision-making contexts: what good are rules unless they contribute to our well-being – unless (above all) following them enhances human happiness and reduces human suffering? This kind of concern leads to utilitarianism, the position of Jeremy Bentham and John Stuart Mill. For Mill, the master principle is roughly this: choose that act from among your options which is best from the twin points of view of increasing human happiness and reducing human suffering. In Mill's words:

> The creed which accepts as the foundation of morals "utility" … holds that actions are right in proportion as they tend to promote happiness, wrong as they tend to produce the reverse of happiness. By happiness is intended pleasure, and the absence of pain. (Mill 1957, 10)

If one act produces more happiness than another, it is preferable, other things equal. If the first also produces suffering, other things are not equal. We have to weigh good consequences of projected acts against any bad ones and subtract the negative value from the positive. Ideally, we would at the same time produce pleasure and reduce suffering.

The ethical aim for action is to find options second to none in total value understood in terms of happiness.[4] For instance, lying causes suffering, at least in the long run; truthfulness contributes, over time, to our well-being – roughly, how well off we are from the point of view of happiness as the positive element and suffering as the negative one. Mill argued similarly in support of other morally required conduct, such as fairness and non-interference with other people's conduct.

---

[3] I have discussed a number of questions concerning this issue in (1996).

[4] Mill's quoted formulation is less clear than the formulation I have given in the preceding text; that represents a major kind of utilitarianism–though not the only kind found in Mill – as a sort of ethics by cost-benefit analysis: for each of our options, such as giving a donation to A vs. giving it to B, we assign probabilities of relevant outcomes and, for each of those outcomes we assign values; we multiply the probabilities by the positive or negative numbers representing the good and bad outcomes respectively; and we then rank our options accordingly. Right acts are those that maximize the good. What makes this ethics rather than a kind of economics is that it makes goodness, not profit, the standard of conduct.

Utilitarianism is commonly formulated as the position that for an act to be morally right is for it to produce "the greatest good for the greatest number."[5] This misrepresents the view. Utilitarians are concerned above all to maximize the good, understood non-morally (since no moral notion is to be presupposed in the principle that is to determine what is the morally required thing to do). Some ways to produce it, say by providing education for all children, are no doubt better than others because – other things equal – they positively affect more people; but the idea that doing (or producing) good for more people rather than fewer as a basic concern is not appropriate to defining the position. For instance, if providing public libraries only in highly educated communities would produce more good (say, in stimulating innovations and productivity) than providing them equally to a whole population, the former would be preferred.[6]

Common-Sense Intuitionism

Suppose one agrees with virtue theorists that there are as many different dimensions of morality as there are moral virtues, and with rule theorists in holding that morality demands that we have and act on principles. This may lead to the kind of common-sense ethical theory set out by the 20th-century English moral philosopher, W. D. Ross.[7] His approach – a pluralistic multiple-rule view – is to categorize our basic duties (moral obligations). He did this by considering the kinds of grounds on which moral obligations rest; for instance, making a promise to help you weed your garden is a ground of an obligation to do it; injuring someone in rushing to a class is a ground of an obligation to make reparations; and seeing someone bleeding by the wayside, as the Good Samaritan did, is a ground of an obligation to help, even if not necessarily a predominating ground. For Ross, the basic obligations are

---

[5] Joseph DesJardins, e.g., in his work, says, "Utilitarianism is typically identified with the policy of 'maximizing the overall good' or, in a slightly different version, of producing 'the greatest good for the greatest number'" (2005, 30). He does not discuss the difference (which is far from slight), and he discusses utilitarianism in relation to both characterizations. Bentham may be the main source of the greatest number formulation. His *Introductory View of the Rationale of Evidence* has, "Of legislation the proper end may, it is hoped, be stated as being–not but that there are those who will deny it–in every community, the creation and preservation of the greatest happiness of the greatest number." (1843, 6) He does not, however, present this as equivalent to his principle of utility: "that principle which approves or disapproves of every action whatsoever, according to the tendency which it appears to have to augment or diminish the happiness of the party whose interest is in question." (1996, 12).

[6] This assumes that the narrow distribution of libraries would not create a degree of resentment that would cause suffering so great as to outweigh the benefits of favoring the educated. Utilitarians always seek to consider the total effect of a possible action; the point here is that inequality of distribution is not automatically or in itself to be avoided. The overall good is the sole standard of conduct.

[7] The reference is to the much-discussed Ross (1930, 29–30). Some of Ross's principles are clarified or qualified in chapter 5 of my (2004).

to (1) keep promises, (2) act justly, (3) do good deeds towards others, and (4) express gratitude for services rendered, as well as to (5) avoid injuring others, (6) make reparations for wrong-doing, (7) avoid lying, and (positively) (8) improve oneself. He took it to be intuitively clear and indeed self-evident that we have these obligations: you can see this by engaging in sufficiently clear and deep reflection – a kind of intuitive thinking – on the moral concepts in question. Hence the name 'intuitionism' for the position that morality is to be conceived in terms of such principles as those expressing these commonly recognized obligations.[8]

Ross knew that prima facie obligations can conflict. Consider the Good Samaritan, who went to great lengths to help a wounded stranger (Luke 10:30–35). Suppose he had promised to help his daughter weed her garden and was unable to do this given the delay caused by ministering to the stranger. Ross thought that where two or more duties (his term for obligations) conflict, we often need practical wisdom (wisdom in human affairs) to determine which duty is final, that is, which duty is, all things considered, the one we ought to fulfill, as opposed to our "prima facie duty," our duty relative to the moral grounds in the situation, here a wounded stranger's need for one's help and a promise to one's daughter. Our final duty is what we ought to do "in the end," and it will be the same as our prima facie duty if no other such duty of equal weight conflicts with that. If I promise to write you and have no conflicting duty, writing you is what I ought to do.

There is an Aristotelian element in Ross's common-sense ethics. Practical wisdom is what Aristotle took to be essential in determining what acts express virtue; and Ross thought, as Aristotle may have, that sometimes it is intuitive, or even obvious, which duty takes precedence where two conflict. Saving an injured person may be quite obviously a stronger duty than keeping a promise to weed a garden. By contrast, selecting one good candidate over another good one to fill an important position may rarely be obviously right. Here morality counsels both humility and the practice of a retrospective self-scrutiny that helps us to rectify past mistakes and to avoid future errors.

Similarities, Differences, and Deficiencies Among the Four Approaches

Reflection reveals both differences among the four kinds of ethical theory just described and some degree of complementarity. Let us first consider some differences and, in that light, explore the possibility of a unified view that captures the best of each.

To see some respects in which these basic kinds of ethical views differ, consider a case in which your grandfather (who has outlived your parents) puts you in charge of directing his medical treatment if he becomes incompetent. You have promised to

---

[8] For a detailed account of Ross's intuitionism and a defense of a more comprehensive view that incorporates major elements of it, see my (2004), especially chapters 2 and 3.

let him die with dignity if he is suffering, unable to communicate, and clearly terminally ill. His lung disease prevents normal breathing, and putting him on a respirator is suggested. He suffers when conscious, cannot communicate or even understand what is said to him, and is being fed through tubes. Many facts that such a case presents cannot be filled in here, but we can even at this point see some differences between the approaches. Take common-sense intuitionism first, since it views our promissory obligations as a morally basic kind. Unless we find some conflicting obligation of equal weight, we must do as we promised and decline to allow a respirator. Imagine, however, that other grandchildren have asked to come to him one last time and need a day to make the trip. Here one might have an obligation of beneficence – to do something good for them – that would favor a respirator if he would otherwise die too quickly. Suppose one could confine its use for this short-term purpose. Allowing its use might then be consistent with the original promise.

A virtue ethics could (though it need not) lead one to a similar decision. The virtue of fidelity is the one most relevant here. Fidelity to one's word is central, but the virtue is broader and encompasses a wider loyalty to others. There is a virtue of beneficence as well, and this would incline one much as the Rossian duty of beneficence would. One's central focus, however, would be on what kind of person to be in the situation; this conception is supposed to lead one to the right deed. The procedure is not to consider types of action and bring rules to bear on them. It is crucial to see that as different as these approaches are, they may, like different ways of building a bridge, take one to the same destination.

The Kantian and utilitarian accounts both differ strikingly from the intuitionist and virtue ethical views. They are each what might be called master principle theories of right action, whereas the former are highly pluralistic.[9] A virtue approach embodies a plurality of virtues central for ethical thinking; intuitionism embraces a plurality of rules. This is not to imply that you must make different decisions depending on which of the master principle views is your guide. Indeed, Kant's Categorical Imperative is commonly taken to imply a subsidiary moral principle expressing a strong obligation to keep promises, as well as a principle of beneficence calling for good deeds. This characteristic makes Kantian ethics like intuitionism in a certain kind of application, and indeed utilitarians may also formulate principles that are far more specific than the master principle quoted from Mill and (they may argue) derivable from it. But a Kantian would likely arrive at a decision differently than someone guided by one of the other views. On a Kantian view, we should try to think of a principle for the case that could be used by anyone in the same situation, for instance the principle that when a promised release from suffering and indignity (say by administering additional morphine) can be carried out with just a slight delay by accommodating relatives with a deep and loving concern

---

[9] This contrast is not sharp (and deserves analysis not possible here). Even supposing Kant's formulations of the Categorical Imperative are all equivalent, he appeals (in its intrinsic end formulation) to a plurality of moral considerations, e.g. an obligation to avoid treating people merely as means and an obligation (not entailed by that) to treat them as ends. For Mill, too there is at least the plurality that comes from taking value to have both negative and positive dimensions.

to be present, the delay is warranted. We might also ask whether the grandfather is being treated as an end and not merely as a means. If we were sure he would not have agreed to a delay in such a case, we would likely not think we are treating him as an end (roughly, as mattering for his own sake); but apart from such an unlikely factor, we could reasonably think he might have wanted to have his other grandchildren present. We can then see the delay as treating him as an end.

On the kind of utilitarian view sketched, our focus must be on the (non-moral) good to be done by making one decision rather than another. We might now focus on how much suffering he will endure in the extra day on the respirator and might compare that with the suffering of the grandchildren if they cannot get to him before he dies. We might also think about the effects of the example we set if we delay (or if we do not); being seen as breaking a promise can have very bad consequences, for instance in reducing trust regarding promising. Even the pressure for hospital space and the costs of the extra medical care will be relevant. None of these things need be irrelevant on the other views; but for utilitarianism, facts are relevant on the basis of their bearing on the consequences of our options for the happiness of all affected, not on the basis of their bearing on whether we are keeping a promise, being virtuous, or following a rule that is universalizable in the way Kant intended. This makes a great difference in approach; and even if one often reaches the same moral destination, one may not do so in just any causally effective way. We could be influenced by the monetary costs much more than on the other views, perhaps thinking of how much good could be done with the savings. For intuitionism, by contrast, the obligation of beneficence – which is the overarching obligation for utilitarianism – is only one important moral consideration; the promise also has moral weight, and even a duty of gratitude toward the grandfather may add to the grounds for adhering to the original promise.

In the kind of broad terminology required by a single paper of this kind, we might identify main sources of dissatisfaction with the four kinds of theory as follows – and here I choose the terms of description with bioethical decision foremost in mind. Virtue ethics is insufficiently determinate in its implications for many practical decisions. Kantian ethics is caught between, on one interpretation, excessive formality and serious vagueness and, given the easiest remedy for these (apart from recourse to moral intuition), rigorism in formulating acceptable rules. Utilitarianism seems often in danger of compromising justice in the quest for maximizing non-moral value. Intuitionism appears to lack a procedure for resolving conflicts of obligation, which Ross and other proponents grant are a challenge for the theory.

Prospects for an Integrated Theory Incorporating of the Four Leading Views

Two or more of these ethical views can be fruitfully combined; for instance, you might hold that we are to maximize happiness, as Mill requires, but only within the limits of never treating people merely as means, as Kant demands. On this combined view, in, for example, a military hospital with many battlefield victims, you cannot sacrifice an innocent person to harvest organs that will save six others, even if the total resulting happiness would be greater (in which case a purely utilitarian approach would call for the sacrifice).

Many who reflect on ethics find something of value in virtue ethics, in Kantianism, and in utilitarianism, and all three are standardly given attention in courses in bioethics. Might a single wide principle capture much of their content? If a single principle is needed, an approach I find promising is to combine elements in these three historically most influential theories in ethics. In suggesting this, I here assume that there are at least three conceptually independent factors that a good ethical theory should take into account: happiness – roughly, welfare, conceived in terms of pleasure, pain, and suffering; justice, conceived largely as requiring equal treatment of persons; and freedom. These are all reflected on the Rossian list of basic obligations, but for simplicity I leave out the others, which may in any case be justifiably affirmed on the basis of these.[10]

On this approach – call it pluralist universalism – our broadest moral principle would require optimizing happiness so far as possible without either producing injustice or curtailing freedom (including one's own); and this principle is to be internalized – roughly, automatically presupposed and normally also a strong motivator – in a way that yields moral virtue. Each value becomes, then, a guiding standard, and mature moral agents will develop a sense of how to act when the values pull in different directions. This sense need not always call for explicit consideration of each element; practiced moral agents may often discern in an intuitive way the overall decision the elements collectively favor.[11]

It might be pointed out here that no highly specific, single standard, however, can be our sole moral guide. This is especially so in the case of principles that appeal to

---

[10] The interpretation of justice, freedom, and happiness (and especially the first two) is treated in detail in my (2004, ch. 5), which also introduces duties of manner as a distinct category. These are obligations concerning how we should do what we ought to do, e.g. respectfully or generously as opposed to resentfully. The entire set of "Rossian" common-sense obligations may also be integrated under an interpretation of Kant's Categorical Imperative; this Kantian intuitionism, as I call it, is not as easily explained as the pluralist universalism formulated in the text, but below I will describe it in some detail.

[11] I speak of optimizing rather than maximizing happiness because, for one thing, a maximization standard, even with the limitations the principle expresses, may be too demanding. I also agree with Mill (and Aristotle as I read him) that the quality as well as quantity of happiness is relevant, which makes talk of maximizing it at best misleading. I have dealt with this kind of demandingness problem in ethics in, e.g., (2004, ch. 4). See my (2004, ch. 5) for rationale for taking freedom to be morally important independently of the other Rossian obligations.

different and potentially conflicting elements. How should we balance these elements in the pluralist universalism? A priority rule for achieving a balance among the three values is this. Considerations of justice and freedom take priority over considerations of happiness; justice and freedom (presumably) do not conflict because justice requires protecting the highest level of freedom possible within the limits of peaceful coexistence, and this is as much freedom as any reasonable ideal of liberty demands. Thus, a drug that gives pleasure but reduces freedom would be condemned by the triple-barreled principle (here the freedom element is predominant); a social program that make a multitude happy but is unfair to a minority would be rejected as unjust (here the justice element predominates). Moreover, although one may voluntarily devote one's life to enhancing the happiness (if only by reducing the suffering) of humanity this is not obligatory. Thus, coercive force may not be used to produce such beneficence, nor is there an (overall) obligation on the part of each of us as individuals to maximize the good, but, within the limits of liberty and justice, beneficence may predominate in this way.[12]

The pluralist universalism just sketched is compatible with, and I think clarified by, a wider integration of major elements in the four kinds of ethical theory we have considered. We should begin by noting that some of the moral virtues stressed by Aristotle and other virtue ethicists are reflected in Kantian categorical imperatives with a small 'c', for instance, the imperatives to avoid lying and to do good deeds – imperatives of veracity and beneficence, which each represent moral virtues. Less noticed are the "secondary rules" stressed by Mill at the end of Chapter Two in *Utilitarianism*. As to intuitionism, surely the intuitive common-sense standards articulated by Ross (and briefly noted above) are common coin among the major theories and many major religions:

1. Justice: including the positive obligation to prevent and rectify injustice, and the negative obligation to avoid commission of injustice;
2. Non-injury: roughly, the obligation to avoid harming others;
3. Fidelity: the obligation to keep one's promises
4. Veracity: avoidance of lying–this obligation, like that of fidelity (under which Ross subsumed it) is a kind of fidelity to our word;
5. Reparation: the obligation to make amends for wrong-doing;
6. Beneficence: the obligation to contribute to virtue, knowledge, or pleasure in others;
7. Self-improvement: the obligation to better oneself; and
8. Gratitude: the obligation to respond in an appropriately appreciative way to those who do good deeds toward us.

One further comment is needed here.[13] Justice should be taken to entail not only treating people in accord with their merit (as Ross put it) but also equally in some

---

[12] The problem of deciding just how much one ought to do for others is difficult on any plausible ethical view, and especially for utilitarianism, which makes maximization of the good the central obligation. I have discussed in detail how this problem may be dealt with in my (2004, chs. 3 & 4).

[13] Detailed discussion of Ross's principles, together with a rationale for my two added ones, is provided in my (2004, ch. 5).

(doubtless related) proportionate sense.[14] Equal treatment is something to which people are sensitive even early in life. Take small children of the same age who are given different privileges in playing with toys. They will tend to compare the toys each may play with and become upset if theirs is visibly less elaborate. Resentment of preferentially unequal treatment may well trace to just such cases.

There are, on my view, other Rossian elements: roughly, principles of the same a priori status and also apparently basic in guiding intuition and inference:

  9. Liberty: the obligation to preserve and enhance liberty;
10. Respectfulness: the obligations of manner – to treat persons in respectful ways.

Obligations of manner concern the way we do what is obligatory as opposed to what we must do. Clearly one can do the right thing for the right reason but in a morally unacceptable way – say, telling a patient that painful chemotherapy is the only hope for remission yet in a way that projects annoyance and no perceptible concern for the suffering to occur.[15]

Is there any prospect of a unified pluralistic theory that is more specific than pluralist universalism and aids us in biomedical decision? I have argued in previous work that there is: a Kantian intuitionism. The rough idea is that in making moral or morally significant decisions – which includes most decisions in bioethics – we should do what accords with the Rossian duties if there is no conflict among them and, if there is, resolve that conflict with the help of the Humanity Formula interpreted in the light of a conception of what it is to treat persons as means and, by contrast, as ends. I have provided a detailed account of the Humanity Formula in descriptive terms (overlapping Kant's but also quite different), so that its application can be guided without presupposing moral judgments.[16] I have also proposed a set of adjunctive principles that supplement the treatment of conflicting obligations which Kantian intuitionism makes possible.

What, then, is the overarching standard of obligation? It is to fulfill one's obligation under the intuitive categories just presented, with any conflicts between or among them to be resolved at least in part by determining whether the decision treats all concerned as ends in themselves and, especially, avoids treating anyone merely as a means. One might like to eliminate 'at least in part', but here I think we must grant that in certain cases practical wisdom is needed to supplement, perhaps through moral imagination, our reflections in accordance with principles. I take such wisdom to yield intuitions that are not idiosyncratic and can be defended, and here I would argue that the universalizability formulation of the Categorical

---

[14] For a fine-grained and illuminating theory of equality, see Temkin (1993).

[15] Detailed discussion of what constitutes a manner of action and how it is morally important is provided in my (2016).

[16] No disrespect for Kant is implied by taking his formula to have force that is independent of his own interpretation of it. His greatness surely consists partly in his using universally applicable categories with meaning that transcends even his own interpretation of the formula. His interpretation is not represented even by him as complete, and (as the literature of Kantian ethics shows) he had views, such as some of the rigoristic ones, that are at best difficult to square with the best interpretation of the Humanity Formula. See, e.g., (Wood 2008).

Imperative is useful. Practical decision will have a basis in the facts of the case, and these – when adequately considered – will often make it obvious what to do given the primacy of one of the intuitive principles or the overall preferability of one option. Kantian intuitionism, taken together with sufficiently described facts of a case, also enables us to frame a generalization for like cases. If a generalization that accounts for our projected decision is not rationally universalizable within the framework of values implicit in Kantian intuitionism – which, like the ten Rossian principles, takes persons as ends in the sense demanding respect – then it is morally unacceptable. (In the light of these clarifications, Kantian intuitionism may be viewed as explicating the force of the pluralist universalism sketched above.)

Some readers will have noticed that Kantian intuitionism is not an ordinary right-based theory: it does not simply indicate what kinds of acts are right without any dependence on commitments regarding intrinsic value. We cannot understand even the common-sense obligation of beneficence without a notion of what is good for persons, and we cannot understand what it is to treat persons as ends apart from a clear sense of what is for their good. But this is not to say that the right is, as it is for utilitarianism, is derived from the good in the sense that right action is simply a matter of maximizing the good, understood non-morally, say hedonically or in terms of preference as psychological properties of persons. This entails a major difference from utilitarianism, which is a good-based theory. Granted that Kantian intuitionism countenances non-moral goods such as reduction of suffering as entailing normative reasons for action, it also countenances distinctively moral values, as naturalistic versions of utilitarianism do not. These elements allow the view to overcome the dichotomy between starkly right-based ethics and starkly naturalistic consequentialism, which is good-based. The view will become clearer in the context of biomedical decisions.

Clinical Decisions

A major and widely respected ethical standard in bioethics is preservation of patient autonomy. Autonomy is a matter both of being free to determine one's fate (within certain limits) and of being free in doing the things one chooses. The former is crucial for having autonomy, the latter for exercising it. The liberty obligation (ninth among the common-sense principles listed above) addresses this directly. One might think that we could fully account for this obligation within a utilitarian framework, but that is doubtful. Consider a case in which you are legally empowered to decide for a patient whether to have a hysterectomy or simply remove ovarian cysts. Her well-being might be equally affected either way, assuming that she is past childbearing. But should one not have a clear moral preference that she make the decision herself? Is there not a value in the self-determination in question? Indeed, suppose the probability of less overall suffering is higher on one option and other things are equal. If the decision is given to me to make, should I take this fact as a reason to proceed so rather than ask her to make the decision herself, given that I am quite sure she would decide differently? I believe not.

There is no need to deny that an irrational exercise of autonomy may be blocked by an appropriate party (such as a parent), but that is not my case, which concerns the moral weight of considerations of freedom independently of considerations of utility. One might now argue that, for utilitarian reasons, we need a policy of respecting autonomy to prevent fear and thereby suffering. This of course raises the question just how the policy is to be understood and whether it makes room for exceptions that, on reflection, we can agree are morally unwarranted. Suppose, with Mill and other naturalistic utilitarians, our theory of value is hedonistic (as Mill's clearly is at the opening of Chapter Two of Utilitarianism). Then it is at best difficult to see how someone holding the theory of value underlying utilitarianism could resist the need to maximize intrinsic value – especially once it is seen that the need to do this makes biomedical and other decisions non-personal in the sense that the consequences of the various options for persons other than patients or experimental subjects must be relevant and may be overriding.

Example like those I have used also bring out a defect in preference utilitarianism as conceived by economists and others considering public policy. The roots of this view are in some version of Humean instrumentalism. On this view, reasons for action derive entirely from (non-instrumental) desire – "passion" for Hume, "preference" in decision-theoretic terms. This theory is unsound, as argued in detail by a number of philosophers.[17] Note how this theory bears on making exceptions to the autonomy obligation. We normally prevent ill-considered suicides if there is adequate evidence of irrationality. But suppose a person has non-instrumental desires that point, if not to committing suicide, then to making ill-considered drug-induced experiences. We may withhold morphine under such conditions on the basis of a conception of what is in the interest of the patient as well as on the basis of how others are affected. But what ground can we have, on a desire-based account of reasons and practical rationality, for withholding drugs from not only patients but the population at large? Mill sought to deal with this kind of problem by distinguishing qualities of pleasure and giving much greater weight to the quality of pleasure over its quantity. But reflection shows that his majoritarian attempt (in Chapter 2 of Utilitarianism) to deal naturalistically with the problem and the related difficulty of weighing quality against quantity is a failure.

When it comes to distinguishing between rational and irrational desires, one point to be stressed is that scarcely anyone thinks that it is not rational to desire (for its own sake, hence non-instrumentally) any of the elements of value implicit in the common-sense framework described here. Imagine someone saying, for instance, "I see that you want to avoid lying, but why is there any reason to do that or want others to?" As a skeptical theoretical question, we can understand this; but suppose it is a request for a rationale, not a serious challenge to the reason-giving status of the veracity standard. These are very different – so different that, on the skeptical count,

---

[17] For detailed discussion of preference and desire as bases of reasons for action (such as moral reasons), see, e.g., Parfit (2011) and my (2002).

one might ask the same of the principle that a clear, steadfast visual impression of a car rushing toward one is a reason to believe there is a car approaching. In practice, virtually everyone abides by and is committed to accepting this principle.

Utilitarianism in all its major forms and instrumentalism in the theory of practical reason share a property that sets them apart from all the other major approaches to ethics and practical reason. They are content-neutral. This is obvious for instrumentalism: there are no specific kinds of things that must be objects of desire or preference in order for such basic motivational states to determine our reasons for action. This is why Hume said that "reason is and ought only to be the slave of passion." As to utilitarianism, even in its hedonistic versions it does not specify any specific act-types as right or wrong: any type of act whose instantiation maximizes hedonic value is right – in principle, even if this is by cruel sadistic acts whose consequences embody pleasure outweighing any pain they cause. Theorists in these traditions tend to agree with the common-sense judgment that some act-types are intuitively wrong and to argue that in practice such judgments can be accommodated. Here I can only say that these arguments seem to me to fail and I believe that the kind of alternative view I propose is superior.

One may think that the Kantian intuitionist framework is committed to "the dignity of persons." There is much to be said for that idea as an overarching guide in practical ethics; but Kantian intuitionism does not depend on it. Indeed, the theory can serve as a way to understand what we must do and value if we are to honor that ideal.[18] Treating persons as ends, for instance, is achievable to a substantial extent by doing things for their good where this is as concrete as reducing their suffering, enhancing their capacities for pleasure in exercising their rational faculties, treating them equally with others, and avoiding even approach to treating them merely as a means. This is the kind pattern supportable both by at least major elements in the major religions of the world and even by purely secular governments and institutions. The standards in question do not depend on support from religion or political theory, but this does not imply that such support is insignificant.

Genetic Engineering as an Illustrative Case

So far, my main examples have come from individual relationships. But many biomedical decisions concern research – what we should support and how we should do it. Here again, I will draw some contrasts between the common-sense intuitive moral framework I propose and the kind of largely utilitarian view that so often seems worthy of determining policy. Genetic engineering is a good case for discussion.

Writers in bioethics commonly distinguish therapeutic from non-therapeutic cases. Consider, e.g., altering genes to prevent certain physical defects or diseases

---

[18] Helpful discussions of dignity are provided by Waldron (2012) and, especially in relation to bioethics, Bostrom (2008).

vs. producing "designer children."[19] To be sure, since genetically determined deficiencies in intelligence are considered defects, it is easy to think that the permissibility of preventing them implies the permissibility of enhancing intelligence beyond the norm. But that does not follow. To think so invites such reasoning as that, since being underweight to the point of emaciation is unhealthy, we may endorse producing weight gains far beyond some standard norm. Moreover, we already know that intellectual superiority over "ordinary" people can facilitate to domination of a kind that is uncontroversially viewed as something to be avoided. One element in the Kantian intuitionist framework readily comes to mind here: the wrongness of treating people merely as means. The wrongness of such merely instrumental treatment of persons is a constraint on any enhancements that would make it easy to treat them so.

A similar kind of question comes to the fore if we contrast genetic engineering aimed at increased longevity – a kind of engineering for which we may expect support from the many persons (including many of influence) who would like to live longer than they currently expect to – as opposed to genetic engineering aimed at avoiding the foreshortened lives prematurely ended owing to disease. Here an element comes into play that is not as clear for utilitarianism: the greater urgency of preventing suffering than of simply increasing pleasure. Intuitionism has always taken non-injury to be by and large a more compelling moral demand than positive beneficence, and doing so also corresponds to the typical stronger obligation to avoid treating persons merely as means as opposed to treating them positively as ends. (Cf. "Do no harm" as a physicians' injunction.) Mill could justify respecting this prima facie preferability only by appeal to qualities of experience, a strategy for which hedonism as such does not have an adequate rationale.

It remains true, however, that Kantian intuitionism does not imply that increasing longevity is, by itself, a morally inadmissible project. Suppose, however, that it reduces the well-being of a great many people to spread the world's resources to a larger population with many aged persons. Then Kantian intuitionism would have a rationale for avoiding that. It would, for instance, not only do potential harm to those losing resources but (in democratic societies, at least) fail in commitments to their well-being made by their governments and fail as well to treat them as ends in making them merely part of a utility calculation. For utilitarians, resisting it would apparently be outweighed by the greater aggregate pleasure that might occur, even if average quality of life would be diminished.

If climate change is going to have as serious consequences as some anticipate, the question arises whether some genetic engineering might be justified as preventive, say as altering human biology to equip people to deal better with warmer temperatures, unstable climate, and rising waters. As with therapy and enhancement, the Kantian intuitionist framework is more nuanced. A major difference in policy, beyond weighting prevention and elimination of suffering more heavily than positive increases in well-being, is that equal treatment of persons matters in itself as a

---

[19] For recent discussion of the distinction between the therapeutic and non-therapeutic, as well as exploration of other aspects of biomedical enhancement research, see Buchanan (2011).

matter of justice. This is an important variable given that, at least in its early stages, genetic engineering for enhancement purposes would likely be more readily available to the affluent than to the general population.

Under utilitarianism, by contrast, distribution principles have no basic normative status: they must earn their normative authority in terms of the aggregate effects of people's adhering to them. This point is easily missed if policy makers conceive utilitarianism in terms of the popular formula, "the greatest good for the greatest number." To be sure, this influential formula for right actions (as those that optimally promote this distributive good) is not relied on by any major utilitarian. But its rhetorical flair makes the formula hard to forget once it is heard, as it often is in applied ethics, and it obscures the point that, for utilitarianism, how we should distribute benefits and burdens depends on what distribution principles will maximize the "ratio" of the (non-moral) good to the (non-moral) bad.[20]

There is no simple formula for applying Kantian intuitionism to genetic engineering. But it should now be clear that this does not imply that it has only meager resources. Rather, the point indicates complexities that must be faced and, when considered properly, yield ethically better results than those obtainable by applying such superficially simpler views as utilitarianism.

Medical Research Using Fetal Tissue

Another testing ground for Kantian intuitionism in the biomedical realm is the issue of how fetal tissue is to be used. Here it must be said at the outset that any ethical theory that takes the moral status of persons seriously must first decide whether it is attained in the human species during fetal development.

This question already indicates a difference yet stressed between Kantian intuitionism and utilitarianism. The latter has no principled way of giving preference to the well-being of persons over that of other living things that have pleasure and pain (I assume a broad hedonism here). Animals as well as persons must matter, though how much they matter is, for utilitarianism, undecidable apart from such criteria as quantities and qualities of pleasure and pain in the animal kingdom. But more to the point, none of us or any other beings matter as persons, say, but only in relation to our potential to contribute to the ratio of the (non-morally) good to the (non-morally) bad. This is why there is such a serious demandingness problem for utilitarianism: in a world like this, are we really ethical in, say, taking a long vacation when we could be serving suffering people in sub-Saharan Africa? And if a fetus brought

---

[20] There is even an ambiguity in the formula that, unnoticed (as is common) affects its application. Are we to give priority to maximizing aggregate good and then seek (if there remain options) to distribute it as widely as we can, or are we to prioritize wide distribution, doing some good for the maximum number of people our options permit us to help and only then seek to maximize the aggregate good we produce? These aggregative vs. distribute readings make a vast difference in policy. Detailed discussion of this famous formula is provided in my (2007b).

to term and treated by good pediatric medicine will still have a pained life overall, what difference can its personhood make in itself? By contrast, persons in themselves matter for Kantian intuitionism, and the humanity formula bears directly on demandingness: we can at least say that morality does not require our treating ourselves merely as means to maximally enhancing the hedonic ratio. It does not require even our approaching that.

To be sure, Kantian intuitionism does not presuppose highly specific criteria for personhood. This is partly an ontological question and certainly not one for ethics alone. What can be said here regarding policies concerning fetal research is this. First, let us assume (as is widely accepted by democratic societies) law-making in democratic societies should be determined by standards that are defensible on the basis of natural reason and thus do not depend on theology or religion. Thus, if arguments that appeal to natural reason do not settle this matter, then laws should be structured so as to take account of the unmistakable personhood of pregnant women in contrast to the rationally contestable personhood of the fetus, which would appear to indicate according higher moral status to the former. But second, it is appropriate to the Kantian intuitionist framework to take potential personhood as conferring some kind of moral status. This is significant, though its policy implications are indefinite and must be determined in each cultural context. It should be added here that no leading ethical theory provides any clearly adequate way to determine the moral status of the conceptus or even the well-developed fetus. This remains a difficult issue on which reasonable disagreements appear likely to persist.

A further implication of Kantian intuitionism – and indeed of any moral theory that affirms equal liberty rights of persons irrespective of religious convictions – is that, for establishing constitutional or other legal standards affecting the extent of liberty protected by law, governments should, so far as possible, use criteria of personhood that are, on the one hand, neutral among religious understandings of personhood and, on the other hand, supportive of the most extensive freedom of religion possible within the basic moral rights of all. This principle requires much elaboration and many qualifications,[21] but the underlying idea is to protect citizens from domination by any religious group while also supporting the maximum liberty possible within a plausible account of human rights.

If, when fetal tissue is sought for some biomedically legitimate purpose, the only plausible arguments for the personhood of the fetus depend on theology or pronouncements by clergy, government should not be required to outlaw its use, on pain of violation of the rights of persons; but government also should not force religious people who accept those pronouncements to agree to allow use of fetal tissue drawn from their own biology (as with a miscarriage). These two principles leave open what "secular" criteria for personhood or, more broadly, for moral status, should guide fetal research. They do, however, suggest a framework for bioethical decisions that does not depend on theology or religion, while allowing that, as practical wisdom counsels, some theological and religious considerations may be relevant to these decisions. They

---

[21] I have explicated, qualified, and defended the broad idea presented here in a number of works. For detailed discussion and extensive references to relevant literature see my (2000) and (2011).

may clarify secular arguments, if only by stating and exemplifying an opposing case; they may indicate rights that might otherwise be ignored; and they may articulate ideals that can enhance common ground among the religious and non-religious.

If the fetus has any moral status, as is plausible on more than one reasonable ethical view, then uses of fetal tissue should be constrained by a respect for it. Indeed, the fetus might have moral significance even apart from having moral status or might have the former importance at any stage and the latter status later in development. Just what its moral importance and moral status might be, and how these may change with development toward live birth (or with the analogous development in an artificial setting) must be determined by inquiry informed by moral philosophy, biology, and (in my view) metaphysics. The point here is that once personhood is a serious focus of moral thinking, as it is for Kantian intuitionism but not for at least classical utilitarianism, there is a prima facie case for maintaining that fetal tissue cannot be used "merely as a means."

Conclusion

No historically influential position in ethics is by itself adequate to guide biomedical decisions, and the best response to this is to seek to formulate a view that captures the best elements in each. Doing this reduces dependence on appeals to practical wisdom and moral intuition, but it is naïve to think that these appeals can be entirely eliminated. This was certainly the view of W. D. Ross, whose pluralistic common-sense intuitionism represents an increasingly appreciated contribution to the history of ethics. My own contribution derives from integrating the intuitive moral principles formulated by Ross and clarified and supplemented by principles, distinctions, and constructs drawn from my own reflections. The result is a moral theory that facilitates everyday decisions by starting with basic obligations – the intuitive kind that guide much of ordinary life and much of moral education, while providing a theoretical framework, using notions Kant rightly made central, for dealing with difficult decisions in which such obligations conflict. The problem is in good part one of incommensurable normative considerations. But incommensurability does not preclude rational comparisons or rational judgment regarding action.[22] The results of such comparisons, moreover, may be intuitively tested in terms of comparison with the plurality of obligations and values implicit in taking persons as possessing the moral status of ends in themselves. They may also be tested by reflection on the generalizations that arise from analysis of the basis of individual decisions. This approach does not yield easy decisions in bioethics, but those are not in general possible in any major domain of human life, and, as illustrated by our examples, the Kantian

---

[22] For defense of this view see Regan (1997). For critical discussion of the view, especially as represented by Chang (1997), see Boot (2009) and (2016), which contains detailed discussion of the issues.

intuitionist theory appears to be a better guide to practical decision than competing views in the field. It accommodates insights from virtue ethics. It takes consequences seriously as major elements in evaluating options. It calls for rational universalizability of principles that rise from generalizing on the basis of particular judgments. And it provides for explanation and justification of moral decisions in relation to a framework that takes account of universally respected principles and values.[23]

References

Audi R (1996) The Ethics of Organ Transplantation. Utilitas 8(2):141–158

Audi R (2000) Religious Commitment and Secular Reason Cambridge University Press, Cambridge

Audi R (2002) Prospects for a Naturalization of Practical Reason: Humean Instrumentalism and the Normative Authority of Desire. Int J Phil Stud 10(3):235–263

Audi R (2004) The Good in the Right: A Theory of Intuition and Intrinsic Value. Princeton University Press, Princeton

Audi R (2007a) Moral Value and Human Diversity. Oxford University Press, New York & Oxford

Audi R (2007b) Can Utilitarianism Be Distributive? Maximization and Distribution as Standards in Managerial Decisions. Bus Eth Quart 17(4):593–611

Audi R (2011) Democratic Authority and the Separation of Church and State. Oxford University Press, New York & Oxford

Audi R (2016) Means, Ends, and Persons: The Meaning and Psychological Dimensions of Kant's Humanity Formula. Oxford University Press, New York & Oxford

Bentham J (1843) Introductory View of the Rationale of Evidence. In: Bowring J (ed.) The Works of Jeremy Bentham, Vol. 6. Simpkin, Marshall & Co., Edinburgh

Bentham J (1996) An Introduction to the Principles of Morals and Legislation. In: Burns JH, Hart HLA The Collected Works of Jeremy Bentham: An Introduction to the Principles of Morals and Legislation. Clarendon Press, Oxford

Boot M (2009) Parity, Incomparability and Justified Choice. Phil Stud 146:75–92

Boot M (2016) Incommensurability of Values. Rowman & Littlefield, New York & Toronto (in press)

Bostrom N (2008) Dignity and Enhancement. In: Human Dignity and Bioethics: Essays Commissioned by the President's Council on Bioethics. Washington, D.C.

Buchanan A (2011) Beyond Humanity? Oxford University Press, Oxford

Chang R (1997) Incommensurability, Incomparability, and Practical Reason. Harvard University Press, Cambridge, MA

DesJardins J (2005) Introduction to Business Ethics. McGraw-Hill, New York

Kant I (2002) Groundwork of the Metaphysics of Morals (trans. A.W. Wood). Yale University Press, New Haven

Mill JS (1957) Utilitarianism. Macmillan, New York

Parfit D (2011) On What Matters Vol. 1. Oxford University Press, Oxford

Regan D (1997) Value, Comparability, and Choice. In: Chang R (ed.) Incommensurability, Incomparability, and Practical Reason. Harvard University Press, Cambridge

Ross WD (1930) The Right and the Good. Oxford University Press, Oxford

Temkin L (1993) Inequality. Oxford University Press, New York & Oxford

Waldron J (2012) Dignity, Rank, and Rights. Oxford University Press, Oxford

Wood AW (2008) Kantian Ethics. Cambridge University Press, Cambridge

---

[23] For valuable discussion of many of the topics in this paper I thank Marta Soniewicka and Adriana Warmbier.

Chapter 2
Do Our Moral Judgements Need to Be Guided by Principles?

Roberto Andorno

2.1 Introduction

"At Home" is a short story by Anton Chekhov about a lawyer, Eugene Bikovsky, who tries to explain to his 7 year-old son, Seriozha, why he did wrong by taking a cigarette from his desk and by smoking it (Chekhov 1996, 16–24). The story begins with Bikovsky coming home from a session at the court, and with the governess telling him about his son's wrongdoing. Seriozha is a frail and innocent child; he admires and loves his father, and he had been entrusted to the care of a governess since the loss of his mother.

Bikovsky wonders what to tell his son, but before he had time to think of anything to say, Seriozha has already entered the study. "'Good evening, papa,' he says in a gentle voice, climbing on to his father's knee and swiftly kissing his neck. 'Did you send for me?'" When they began to talk, Bikovsky finds that his son is not at all aware that he has committed any fault. The lawyer first appeals to pure rationality; he tries to explain to the child the conceptual distinction between *meum* and *tuum* in property law, and why it is wrong to take things that belong to others. But Seriozha has a world of his own in his mind and pays little attention to his father's abstract explanations.

Next Bikovsky tries to convey his disapproval of his son's behavior by arguing that smoking itself is wrong, because "tobacco is very bad for the health, and men who smoke die sooner than they should". But the result is equally fruitless, not least because the father is unable to justify why he smokes himself. The lawyer is now frustrated by his inability to get through to Seriozha, but then he suddenly realizes

This article was first published as: Roberto Andorno, "Do Our Moral Judgments Need to Be Guided by Principles?", Cambridge Quarterly of Healthcare Ethics, Volume 21(4), pp. 457–465 (2012), © Cambridge University Press, reproduced with permission.

R. Andorno (✉)
University of Zurich, Zurich, Switzerland
e-mail: roberto.andorno@uzh.ch

© Springer International Publishing AG, part of Springer Nature 2018

M. Soniewicka (ed.), *The Ethics of Reproductive Genetics*,
Philosophy and Medicine 128, https://doi.org/10.1007/978-3-319-60684-2_2

that to communicate effectively with his son, he must appeal to another strategy. Instead of using a purely objective logic, he has to enter the subjective world of his son in a manner that will enable him to convey the message about the perils of smoking for a child. Bikovsky improvises a story of a king who had a long, grey beard and lived in a palace of crystal surrounded by a wonderful garden with oranges and bergamot pears and wild cherry trees.

> The old king had only one son, who was heir to the kingdom, a little boy, just as little as you are. He was a good boy; he was never capricious, and he went to bed early, and never touched anything on his father's table…. He had only one failing – he smoked…. Because he smoked, the king's son fell ill of consumption and died when he was twenty years old. The old man, decrepit and ill, was left without anyone to take care of him, and there was no one to govern the kingdom or to protect the palace. Foes came and killed the old man and destroyed the palace, and now there are no wild cherry trees left in the garden, and no birds and no bells. (Chekhov 1996, 23)

The tale makes a deep impression on Seriozha. His eyes become full of sadness and, after a minute of reflection, he says in a low voice: "I won't smoke any more." Chekhov's story concludes with the father wondering why it is so hard to present morality as the result of purely abstract logic, and why always become more acceptable when it is accompanied by examples, parables, or stories that directly relate to the listener's (or the reader's) personal experience, interests, and concerns.

This chapter's purpose is neither to discuss the value of narrative ethics nor to evaluate the role that stories might have in moral education. Rather, it intends to focus on another issue conveyed by Chekhov's story: do our moral decisions need to be guided by principles? My intention here is to argue that, although principles play a key role in our moral judgments, these latter cannot be reduced to a result of purely deductive reasoning, because they previously require another kind of rationality. Instead of being purely deductive, our moral decisions appear to be the result of a combined inductive-deductive process. This claim is developed in two parts. The first part briefly presents some of the criticisms leveled in recent decades against purely deductive moral theories. The second part argues, appealing to Aristotle's account of the knowledge's process, that an inductive-deductive model provides a more realistic account of how sound moral judgments are actually made.

2.2 The Criticism of Deductive Moral Theories

During recent decades there has been extensive criticism of the dominance and adequacy of principles to guide ethical decisions. Some scholars even speak of an "empirical turn" in this field and suggest that we are entering into a new phase in the history of ethics, which is characterized by an increasing emphasis on context sensitivity. (Musschenga 2005) This discussion was especially intense in the field of medical ethics, in which critics of the so-called principlism as developed by

Beauchamp and Childress (1989)[1] have faulted this theory for being "too abstract, too rationalistic, and too far removed from the psychological milieu in which moral choices are actually made." (Pellegrino 1993; see also: Clouser and Gert 1990; Ten Have 1994) But this debate is obviously not limited to the field of medical ethics. Rather, it takes place in the broader philosophical context of dissatisfaction with the very idea of appealing to abstract principles for making moral decisions. The doubts take various forms but include, among others, the question of whether moral judgements can be codified or captured by any theoretical structure, and therefore, whether our moral lives can be reduced to the legalistic application of a set of principles. (Andorno 2011).

One of the first key essays to raise doubts about deductive ethical theories was Elizabeth Anscombe's "Modern Moral Philosophy", originally published in 1958. In this paper, the British philosopher criticized what she characterized as a "law conception of ethics". Appealing to Aristotle, she called for a return to concepts such as character, virtue and human flourishing, that is, for relying on persons rather than on norms. From this perspective, the key question of ethics is not so much "what I should do", as Kant claimed, but rather "how can I become a good person?" The point is that morality does not merely consist in doing certain kinds of actions; instead it is about being a particular kind of person. Of course, to move towards this latter objective, the moral agent needs to do (or to abstain from) certain actions, and this inevitably means to comply with some moral norms that command (or disapprove) them. But in this approach, formalized moral principles are not regarded as end in themselves; they are rather means that aim to contribute to the flourishing of oneself, of others, and of society at large. In addition, virtue ethicists emphasize that equally or even more important than externally doing certain things is to internally adhere to the goods which are pursued with those particular actions, and this latter condition is impossible to meet without personal virtues.

Another seminal paper in this line was Michael Stocker's "The Schizophrenia of Modern Ethical Theory," published in 1976. Stocker argues that deontological and utilitarian moral theories create a serious dichotomy between the principles they advocate and the motives that inspire moral agents in real life. This situation leads practitioners of such theories to suffer from a "moral schizophrenia" because they will necessarily have a gap between their values and their real motives for action. Stocker calls such a gap a "malady of spirit" and suggests that, because these theories result in this malady, they are seriously flawed. To illustrate this, he gives several examples of persons who are possible candidates for morally schizophrenic utilitarians and deontologists. In one of them, the fictitious Smith, who is committed to Kantianism, comes to visit a sick friend at the hospital, and cannot admit to himself that he is doing so because he enjoys his friend's company or wants to cheer him up (these would constitute "heteronomous" motives). He feels obliged to think that he is visiting his friend solely out of a sense of duty. Therefore, he will experience an internal conflict between his theoretical principles and his real motives, which

---

[1] In later editions of this work, the authors have made significant efforts to incorporate the criticisms to their principlist approach.

makes his moral life schizophrenic. According to Stocker, this and other similar examples show well that what, in the end, is lacking in deductive moral theories is simply love for the other person, which is an essential feature of the most significant human relationships and constitutive of a human life worth living.

In this same line of thinking, MacIntyre has advocated for a radical change in the way we think about morality and for a return to a virtue-centred ethics. His hypothesis is that modern moral theories (namely, deontology and utilitarianism) have failed because they have rejected Aristotle's claim that human beings have an intrinsic good or end (telos) to aim for, and have ignored the fact that we cannot reach this natural end without proper preparation, which consists in an adequate education and in personal effort in the practice of virtues. (MacIntyre 1984, 54–55).

Thus, the key point raised by the above-mentioned scholars is that moral life does not consist in merely learning some rules and then making sure that each of our actions lives up to those rules. The idea that knowing moral theories is enough for making sound moral decisions is as naïve as expecting that just by reading a book about how to swim one will be able to swim. As an Aristotelian would say, moral life only becomes possible by the effort aimed at developing good habits of character, and never by normative argument as such. In reality, when the individual is shaped by certain habits of virtuous conduct, recourse to strict arguments is rather superfluous. This does not mean that the adequacy of one's judgments is measured by pure inner conviction, but only that one's capacity to distinguish between good and bad judgments cannot be reduced to pure science. (Beiner 1999, 43).

From a different philosophical standpoint, Bernard Williams addressed similar criticisms to both utilitarianism and deontology. In his view, both theories have represented a flight from reality, and have failed to understand the complexity of moral choices. (Nussbaum 2009, 213) Williams insisted on the need for what he called "internal reasons for action," which relate to our genuine reasons to act and are connected with things that we really care about. In his view, mere "external" arguments – for example, the proposition that X is morally good – cannot really move us to act. We will only have a reason to act if there is something contingently about us (our personal education, our psychological states, our feelings, etc.) that motivates us to behave in a particular way. (Williams 1985) In addition, Williams has stressed the crucial importance of the historical context for any account of morality. On the ground that it is impossible to provide "a general test for the correctness of basic ethical beliefs and principles," (Williams 1985, 72) he rejected both Rawls' contractualism and Hare's utilitarianism, as they erroneously assume a reflective agent capable of distancing himself from the life and character he is examining. (Williams 1985, 78–92) In contrast to both philosophers, Williams envisions a nontheoretical process beginning and ending with socially and historically conditioned ethical intuitions.

Also the proponents of casuistry and moral particularism are severe critics of principle-based approaches in ethics. Advocates of casuistry claim that absolute moral principles are "tyrannical" and do not play any substantial role in justifying particular moral judgements. In their view, we start from particular cases in which we are confident of our judgements and then reason by analogy by comparing each new situation with others and with paradigm cases. (Jonsen and Toulmin 1988) But casuists accept, at least, the generalization that cases can be sufficiently similar so that we should judge them similarly. (Dworkin 2006) In this respect, moral particularists are even more radical in their rejection of generalizations. For instance, Jonathan Dancy maintains that what is a reason in one case may be no reason at all in another, or even a reason on the other side. In his view, a feature that makes one action better can make another one worse, and can make no difference at all to a third. Therefore, moral reasons are necessarily holistic, or context-specific. (Dancy 2004).

2.3 Inductive-Deductive Reasoning

This chapter argues that some of the above-mentioned shortcomings of purely deductive ethical theories could be better addressed by appealing to a more comprehensive, inductive-deductive understanding of moral reasoning. This model includes a first inductive step in which we abstract from our experience the specific normative criteria relevant to the case at hand, and a second step in which we deduce what to do by applying those criteria to that particular situation. Actually, this seems to be the way in which we make our moral judgments in everyday life, often without being aware of it.

2.3.1 Induction: Moving from Experience to General Principles

One of the first philosophers, if not the first, to develop a careful explanation of the inductive-deductive model of thinking was Aristotle. His understanding of the cognition process is considered as one of his most influential contributions to the philosophy of science. In this regard, it has been said that "current explanations of the scientific method feature Aristotle's iterative process as the central core." (Gauch 2004, 221) From specific observations, inductive reasoning provides general principles (bottom-up movement), and, with those principles serving as premises, deduction attempts to explain observed phenomena (top-down movement). This leads to successive cycles of observations and generalizations, back up again to observations to verify concepts and to obtain more accurate generalizations.

Aristotle's understanding of the cognition process, which can be found in the opening lines of the Metaphysics, is especially developed in the Posterior Analytics, in which he argues that our intellect (nous) grasps first principles through induction (*epagoge*). (Aristotle 1949 II, 19) He distinguishes five stages in this process: (1) Perception (*aisthesis*) discriminates among particulars. (2) Memory retains these perceptions. (3) Repeated memories develop experience of a universal (*katholou*). (4) Higher universals are inferred. (5) First principles are inferred. (Welch 2001).

Interestingly, Aristotle claims that the inductive-deductive process does not only apply to theoretical knowledge but also to practical reasoning. In the Nicomachean Ethics he maintains that the first principles (*archai*) of both science and morality find their source in inductive reasoning and that those principles inferred from experience constitute the starting point of deductive reasoning. One of the relevant passages is the following: "Induction supplies a first principle or universal, deduction works from universals; therefore, there are first principles from which deduction starts, which cannot be proved by deduction (*syllogismos*); therefore, they are reached by induction (*epagoge*)". (Aristotle 1982, VI.3.3.)

Because Aristotle assigns a foundational role to perception in his account of knowledge and concept-acquisition, it is not surprising that he is often thought of as the first empiricist. Even the well-known axiom *Nihil est. in intellectu quod non prius fuerit in sensu* (Nothing is in the intellect that was not first in the senses), which is often associated with the philosophical position of British empiricists, notably John Locke, has in reality its roots in Aristotle's thinking (Aristotle 1975, III, 8, 432a), which later inspired Thomas Aquinas. Indeed, the axiom *nihil est. in intellectu...* can be literally found in Thomas Aquinas's writings. (See Thomas Aquinas 1976, q. 2 a. 3 arg. 19) It is noteworthy that Locke used this argument in order to criticize Descartes' theory of innate ideas and to make the case that all our ideas have their origin in experience. But Aristotle's and Locke's views are in this regard radically different. In Aristotle, empirical data are just the starting point of knowledge, but knowledge is much more than a mere association of simple ideas, as Locke claims. (Locke 1995) For the Greek philosopher, our minds are so constituted as to be able to transcend the material realm and reach universal concepts by abstraction, and this is precisely what empiricists deny. Induction entails, according to Aristotle, a real process of abstraction. It is indeed a kind of rationality, and not a mere feeling, even if reason operates here in an implicit or informal way. How is the passage made from sensitive cognition to an intellective one? The Aristotelian explanation is that, once exposed to the power of our intellect, sensible objects lose their individualizing matter, and that what remains in our minds is the concept of each of them, which assumes the character of universality. As a matter of fact, without this first abstraction step, intellectual knowledge would be impossible.

In any case, experiences play a crucial role in this process. We cannot come to know the first premises of knowledge without such experience of particulars, in the same way that we cannot see colors without the presence of colored objects. (Smith 2010) In other words, induction is the transformation of sense perception into knowledge that goes beyond the limited data of experience. This is why induction involves a kind of "creation from nothing," because human experience is inevitably individual, whereas concepts are universal. (Groarke 2009, 331) Thus, for Aristotle, the first principles of moral reasoning are obtained by induction and cannot be demonstrated but only assumed. Contrary to what is often believed, he does not claim that we derive those principles from the concept of "nature" or "natural" (this is why he does not commit any "naturalistic fallacy"). In his view, only noninferentially justified first principles make moral knowledge possible without facing an infinite regress or a vicious circle; this is for him the only alternative to skepticism.

This recourse to first principles is often seen in contemporary discussions as a weak point of Aristotelian philosophy, because "nothing is more generally unacceptable in recent philosophy than any conception of a first principle". (MacIntyre 1998, 171) However the fact is that no philosophical system attempting to bring some substantive account of truth can avoid relying, at least implicitly, on some first, nondemonstrable principles. As Richard Hare writes, "many of the ethical theories which have been proposed in the past may without injustice be called 'Cartesian' in character: that is to say, they try to deduce particular duties from some self-evident first principle". (1961, 39) These principles mark a starting point of these theories (for instance, the Categorical Imperative in Kantian ethics or the principle of utility in Mill and Bentham). The advantage of Aristotelian ethics in this respect lies in its attempt to stay as close as possible to common sense, and in its effort to reflect the complexity of moral decisions in real life. This puts in evidence what a scholar calls the "tremendous modesty" that characterizes Aristotelian ethics. (Beiner 1999, 43).

But there is an additional clue supporting the Aristotelian claim that the basic moral principles are obtained by induction from experience, and that theoretical explanations come along afterwards to explain what was already implicitly known: the fact, well documented by anthropological studies, that basic moral standards are remarkably similar in all cultures, in spite of them having very different traditions and social and religious backgrounds. The best-known example of this is the Golden Rule, which embodies an ethics of reciprocity ("Do not do to others what you do not want them to do to you") and can be found in virtually the same wording in all major cultures and religions. But there are many examples of other more substantive norms that are strikingly similar among cultures, although expressed in different conceptual terms or with different emphasis (Beauchamp 2003, 2010). How can this phenomenon be explained? My hypothesis is that we all, as humans, share the use of practical reason (i.e., the use of reason concerning action) and are therefore able to identify from our experience the most basic human goods or interests for us and for the society in which we live. On this basis, we infer the principles that command (or forbid) certain behaviors, depending on whether or not they contribute or to those basic goods, or interests, or needs.

2.3.2 Deduction: Moving from General Principles to Particular Conclusions

According to Aristotle, the underlying structure of practical reasoning is always the same. We move up from individual experiences to a general moral principle and then down to a concrete application of the principle to the particular situation in which we are placed. For instance, as we are growing up, we come to understand, by experience and education, that generosity is a good thing. As an adult I am confronted with somebody in need (e.g., an aged and lonely neighbor who is ill in bed and has called me up to tell me that he needs to get a medicine from the pharmacy). After having induced from the circumstances of the case the morally relevant factors, and having (implicitly) in mind the first principle of morality (good is to be done and evil avoided), I reason as follows: Generosity is a good thing; helping this person is generosity; so I go to the pharmacy and buy the drug for him.

In this top-down movement from universal knowledge to a set of particulars, the central role is played by practical wisdom (phronesis, or prudentia). Phronesis is the "right reason in matters of conduct." (Aristotle 1982, VI.13.5.) It is concerned with how to act in particular situations; it makes us choose the right means to achieve good ends; it enables us to act in the right way, for the right reasons, and at the right time; it is the ability to determine what ought to be done in the concrete case. From this it is not difficult to see that practical wisdom cannot be achieved by a mere mechanical rule following. As MacIntyre writes "knowing which rule to apply in which situation and being able to apply that rule relevantly are not themselves rule-governed activities". (MacIntyre 1998, 189) In other words, the application of rules itself requires an exercise of judgment. This is why phronesis is, in Aristotle's terms, an intellectual, not a moral virtue. However, it still entails practical, not speculative reasoning: it is not exercised in order to know something, but in order to do something.

An important point to stress here is that, for Aristotle, practical reasoning is an approximate form of reasoning. It is not exact theory as mathematics or physics are. In his view we must be content if, in dealing with ethical subjects, we succeed in "presenting a broad outline of the truth." (Aristotle 1982, I.3.4) Since practical wisdom deals with contingent matters (i.e., with things that can be other than they are), it cannot be codified in advance in a very detailed fashion. The problem is not that there is no definite right and wrong but rather that reliable standards of right and wrong have to be applied to the variable conditions of human life.

But, of course, we also reason theoretically about morality matters. If a colleague asks me my view on, say, human cloning, the moral judgement that I make does not aim to do something, or to apply the general principles to a particular case, but simply to develop an argument on a specific topic. Here I engage in reasoning that is directed at the resolution of questions that are theoretical rather than practical. Knowledge is here pursued for its own sake, and without ulterior purpose or practical application. One is tempted to say that in such situations we use a purely deductive reasoning to come to a conclusion. But the fact is that induction is also here present, although in a less immediate way. As mentioned above, both theoretical and practical knowledge proceed using an inductive-deductive reasoning. When we are confronted with purely theoretical issues, the principles that we apply to come to a conclusion are not inferred from a concrete case, but have been induced from our experience of the world all throughout our lives.

2.4 Conclusion

In sum, do our moral judgements need to be guided by principles? The answer is: yes. Because how could it be otherwise? If morality is not a merely descriptive undertaking but has, at its core, a normative dimension, how could it avoid the recourse to some guiding standards? But they should not be simply imposed a priori; this is an excessively artificial, counterfactual, and inoperative way of conceiving them. Principles should be "empirically informed and less reductionistic than in current conceptions". (Louden 1992, 139) They must be the result of a process of induction by moral agents themselves, and only afterwards be conceptually structured in their own minds to help them decide what to do in a particular situation (practical reasoning), or what to think in moral terms about some general topic (theoretical reasoning).

The crucial point is that moral agents should, as far as possible, gain access to moral norms from the inside, and not have them imposed from the outside. This is not a merely academic debate, but has very practical consequences, also in the specific field of medical ethics. For instance, in recent years a number of scholars have stressed the importance of promoting dialogue, deliberation, and storytelling as starting points for a more fruitful decisionmaking process in clinical practice. (see for instance: Widdershoven and Abma 2007; Steinkamp and Gordijn 2003; Molewijk et al. 2008) The use of inductive-deductive reasoning is also encouraged today as the best way for adequately teaching medical ethics: rather than setting out a range of disparate and often conflicting theories at the beginning, it is recommended to start by examining particular moral problems and seek to build up to a unified theory from the answers given to the cases. (Saunders 2010) Similarly, it has been suggested that the promotion of empathy among medical students by confronting them to concrete cases should become a priority of ethics education. (Maxwell and Racine 2010).

References

Andorno R (2011) Anti-theory. In: Chadwick R, ten Have H, Meslin E (eds) Sage handbook of health care ethics. Sage, London, pp 31–38

Anscombe E (1958) Modern moral philosophy. Philosophy 33:1–19. Reprinted in: crisp R, Slote M (eds.) Virtue ethics. Oxford University Press, Oxford, pp 26–44

Aquinas T (1976) Quaestiones disputatae; De Veritate. In: Thomas Aquinas. Opera Omnia (Leonine edition), Vol 22. Editori di San Tommaso, Rome

Aristotle (1949) Posterior analytics (trans: Mure GRG). In: Ross WD (ed) Aristotle's prior and posterior analytics. Clarendon Press, Oxford

Aristotle (1975) On the soul (trans: Hett WS). Harvard University Press, Cambridge, MA

Aristotle (1982) Nicomachean ethics (transl: Rackham H). Harvard University Press, Cambridge, MA

Beauchamp TL (2003) A defense of the common morality. Kennedy Inst Ethics J 13(3):259–274

Beauchamp TL (2010) Universal principles and universal rights. In: den Exter A (ed) Biomedicine and human rights. Makku, Antwerpen, pp 49–66

Beauchamp TL, Childress JF (1989) Principles of biomedical ethics, 3rd edn. Oxford University Press, New York

Beiner R (1999) Do we need a philosophical ethics? In: Bartlett R, Collins S (eds) Action and contemplation. Studies in the moral and philosophical thought of Aristotle. State University of New York Press, New York, pp 37–52

Chekhov A (1996) Selected stories. Wordsworth, Hertfordshire

Clouser KD, Gert B (1990) A critique of principalism. J Med Phil 15:219–236

Dancy J (2004) Ethics without principles. Oxford University Press, New York

Dworkin G (2006) Theory and practice of moral reasoning. In: Copp D (ed) The Oxford handbook of ethical theory. Oxford University Press, New York, pp 624–644

Gauch H (2004) Scientific method in practice. Cambridge University Press, Cambridge

Groarke L (2009) An Aristotelian account of induction: creating something from nothing. McGill-Queen's University Press, Montreal

Hare RM (1961) The language of morals. Oxford University Press, Oxford

Jonsen A, Toulmin S (1988) The abuse of casuistry. A history of moral reasoning. University of California Press, Berkeley

Locke J (1995) An essay concerning human understanding. Book II. Prometheus Books, New York

Louden R (1992) Morality and moral theory: a reappraisal and reaffirmation. Oxford University Press, New York

MacIntyre A (1984) After virtue. A study in moral theory, 2nd edn. University of Notre Dame Press, Notre Dame

MacIntyre A (1998) First principles, final ends and contemporary philosophical issues. In: Knight K (ed) The MacIntyre reader. University of Notre Dame Press, Notre Dame

Maxwell B, Racine E (2010) Should empathic development be a priority in biomedical ethics teaching? A critical perspective. Camb Q Healthc Ethics 19(4):433–445

Molewijk B, van Zadelhoff E, Lendemeijer B et al (2008) Implementing moral case deliberation in Dutch health care; improving moral competency of professionals and the quality of care. Bioethica forum. Swiss J Biomed Ethics 1(1):57–65

Musschenga A (2005) Empirical ethics, context-sensitivity and contextualism. J Med Phil 30(5):467–490

Nussbaum M (2009) Bernard Williams: tragedies, hope, justice. In: Calcut D (ed) Reading Bernard Williams. Routledge, New York

Pellegrino E (1993) The metamorphosis of medical ethics: a 30 years perspective. JAMA 269:1158–1162

Saunders B (2010) How to teach moral theories in applied ethics. J Med Ethics 36(10):635–638

Smith R (2010) Aristotle's logic. In: Zalta EN (ed.) The Stanford Encyclopedia of philosophy. http://plato.stanford.edu/entries/aristotle-logic/. Accessed 14 Dec 2010

Steinkamp N, Gordijn B (2003) Ethical case deliberation on the ward. A comparison of four methods. Med Health Care Philos 6(3):235–246

Stocker M (1976) The schizophrenia of modern ethical theory. J Phil 73(14):453–466. Reprinted in: Crisp R, Slote M (eds) virtue ethics. Oxford University Press, Oxford, pp 66–78

Ten Have H (1994) The hyperreality of clinical ethics: a unitary theory and hermeneutics. Theor Med 15:113–131

Welch J (2001) Cleansing the doors of perception: Aristotle on induction. In: Boudouris K (ed) Greek Philosophy and Epistemology, vol. II. International Association for Greek Philosophy, Athens, pp 204–213

Widdershoven G, Abma T (2007) Hermeneutic ethics between practice and theory. In: Ashcroft R, Dawson A, Draper H (eds) Principles of health care ethics. Wiley, Chichester, pp 215–221

Williams B (1985) Ethics and the limits of philosophy. Fontana Press, London

Chapter 3
The Moral Philosophy of Genetic Counseling: Principles, Virtues and Utility Reconsidered

Marta Soniewicka

3.1 Introduction

The moral philosophy of medicine defines the ends of medicine and how it should be practiced (Pellegrino and Thomasma 1981, 1993); it encompasses medical ethics and bioethics that address the moral issues which have particular salience in healthcare and biotechnology.

There are various philosophical traditions in terms of which medical ethics can be considered. Considered from the historical perspective, medical ethics in Western culture was constructed on the basis of the Hippocratic Oath and has been developed within a broad scheme of Aristotelian virtue-ethics (Pellegrino and Thomasma 1993). Yet in modern, post-Enlightenment moral philosophy, there was a significant shift from ethics based on a character (on who we are and on who ought we to be) to ethics based on duties and principles (addressing primarily the question what ought we to do?). This shift from an integral moral philosophy based on an ideal of life to the rational ethics based on principles governing our actions, has strongly influenced modern Western medical ethics which was dominated on one hand by the four-principles approach developed by Tom Beauchamp and James Childress (Beauchamp and Childress 1989), and by utilitarian thinking defended by such philosophers as Peter Singer (Kuhse and Singer 2006; Singer 2011), on the other. We owe the revival of Aristotelian virtue ethics in contemporary moral thinking to such outstanding philosophers as Elisabeth Anscombe and Alasdair MacIntyre (Anscombe 1981; MacIntyre 2007). Virtue-based ethics also garnered attention in medical philosophy, being introduced to medical practice by such scholars as Edmund Pellegrino & David Thomasma (Pellegrino and Thomasma 1993; see also: Oakley and Cocking 2001; Toon 2014).

M. Soniewicka (✉)
Department of Philosophy of Law and Legal Ethics, Jagiellonian University, Krakow, Poland
e-mail: marta.soniewicka@uj.edu.pl

© Springer International Publishing AG, part of Springer Nature 2018
M. Soniewicka (ed.), *The Ethics of Reproductive Genetics*,
Philosophy and Medicine 128, https://doi.org/10.1007/978-3-319-60684-2_3

In this paper I will briefly present the principle-based ethics of genetic counseling (Sect. 3.2) and its application to morally controversial issues of genetic selection (Sect. 3.3). Then, I will discuss the shortcomings of the principlism in the context of the ethics of reproductive genetics (Sect. 3.4) and I will argue for enriching the principle-based ethics by the virtue approach (Sect. 3.5). I will claim that the virtue-based approach is particularly significant in medical ethics which is not only concerned with shaping the behavior of physicians by a set of rules, but also with developing the ethos of the medical profession based on certain values and ends.

3.2   The Ethics of Genetic Counseling: Aims and Principles

The main goals of genetic counseling are: (1) providing useful information (to deliver genetic information to the parents to help them make reproductive choices; to help them understand and personalize technical and probabilistic genetic information; to elucidate the consequences of their choice based on genetic information); (2) providing medical help (enhancing parental ability to adopt to the consequences of their choice including information about medical help and treatment); (3) providing education (exploring the meaning of the information in the light of personal values and beliefs of the parents promoting parental preferences and self-determination in exercising reproductive choice); (4) providing psychological assistance (helping to minimize psychological burden of the parents and to increase personal control of the parents); (Biesecker 2003; Murray 2003). Besides, one could name the goal (5) to provide assistance to the prospective parents in coping with the moral problem which may occur if the values on which their decision is made are in conflict (helping the parents to identify the moral problem and to understand it according to their intuitions).

Genetic counseling promotes the idea of the self-determination of patients, especially in their reproductive choices and is primarily aimed at providing neutral, useful, reliable and understandable information to make the decisions of prospective parents possible on the grounds of their own beliefs and sets of values. One may identify the following principles which were established to guarantee the realization of this aim: (1) non-directiveness (promoting autonomy); (2) nonmaleficence and beneficence; (3) confidentiality and protecting privacy; (4) veracity and truth-telling (Murray 2003; cf. Beauchamp and Childress 1989; Engelhardt 1996).

3.3   Genetic Selection as the Moral Challenge to Genetic Counseling

It seems that not only the principles presented in the previous section may come into conflict with each other, but also meeting the afore-mentioned goals of genetic counseling seems to be in some situations at odds with some principles, e.g. with the principle of non-directiveness. I shall elucidate this problem using the example of genetic selection.

Reproductive genetic testing is aimed at the increase of reproductive opportunities, in particular the opportunity to decide whether one wants to become a parent of a child with a certain genetic makeup or not. Reproductive genetics enables selective reproduction which consists of deciding about which children *will be born* (preimplantation genetic diagnosis, PGD) or which children *will not be born* (prenatal diagnosis, PND) on the basis of their genetic traits and conditions (in particular potential abnormalities) (Press and Ariail 2003). Thus, genetic testing which identifies the potential untreatable diseases of one's progeny may arouse moral controversy and skepticism which was expressed by Adrienne Asch:

> The tests do nothing to promote the health of the developing fetus or the health of the pregnant woman. Rather, they are offered so that people may decide against becoming a parent of a child with a particular characteristic that clinicians and policy makers understand to be detrimental to a satisfying life for the child or the family, or that may require outlays of societal resources (Asch 2003, 336–337).

Consider a couple with one child, a boy, who discovered that he had hemophilia when he was at the age of 3. After that discovery, both parents performed genetic testing which revealed that the mother was a carrier for this genetic disease and the father was healthy. The couple wanted to extend their family and to have more children. Until the 1960s, the lifespan of people with hemophilia was limited in a drastic way to around the age of 30, but nowadays, thanks to the rapid development of medicine, one can live with hemophilia up to the age of 70 and fully control the disease with medicines. Nevertheless, having the disease requires frequent medical check-ups, medical treatment and special caution. Thus, one may say that it still constitutes a burden, both mental and physical, on one's health condition, and it also generates costs (constituting a social burden if there is social healthcare system). The physician who provides genetic counseling should supply the parents with information which is needed to take a fully-informed reproductive decision. Yet the physician can put no pressure on the prospective parents by suggesting any particular view to them. The physician will of course inform the prospective parents about the risk of having another child with hemophilia – if the mother is a carrier and the father is healthy, there is a 25% risk of giving birth to a girl who would be a carrier of a disease and a 25% risk of giving a birth to a boy who would be sick (every son of the mother-carrier has a 50% risk to be born with disease and every daughter of a mother-carrier has a 50% risk of being a carrier too).[1] What is more, the physician may inform the prospective parents about the existing possibilities for preventing the birth of a sick boy – one of them is to opt for IVF and the use of PGD to select the sex of the child which is not linked to a disease, in this case female; another option is not having biological children and adopting a child instead. If the physi-

---

[1] The risk is different when one of the parents, the father or the mother (rare case), are sick or when both of them are sick. There are many combinations possible here – for instance the father sick and the mother being a carrier or not being a carrier etc., in each of them the risk of having a sick child differs.

cian suggests the parents restrain from natural reproduction and choose the IVF procedure, it would violate the principle of non-directiveness, even if the parents are confused as to what to do. According to this principle, the parents should be encouraged to take the decision themselves, having the information about the existing opportunities and their results, and not being influenced by the views of the physician.

There are some who may argue that giving birth to a child with a genetic disease which limits the quality of life is harmful if it could have been prevented and if the disease is serious enough (cf. Parfit 1987; Feinberg 1992; Savulescu 2001; Buchanan et. al. 2007; Savulescu & Kahane 2009; Kamm 2013). Assuming that hemophilia is considered a serious disease by the physician, would it be justified to suggest that the prospective parents use IVF and PGD to prevent the birth of a sick boy on the grounds of the priority of the nonmaleficence principle over the principle of non-directiveness in this particular case? Since no one can benefit from not being born, it seems unconvincing that the violation of the principle of non-directiveness would be justified.

Another set of questions would arise if we consider a case of a woman who is pregnant with her child and who performed PND and got to know that her child would probably be born with Down's syndrome. She is struck by the tragic news, she does not know much about this genetic disorder, she is frightened and has no idea what to do. Would it be sufficient if the physician provides the patient with the information about the risk and the right to terminate the pregnancy? Would the decision be fully informed given this information? It is worth emphasizing here, as Asch does, that most selective procreative decisions against disability are based on a misinterpretation of what disability is – on stereotypes of burdens and suffering (Parens and Asch 2003). Thus, to guarantee a truly informed reproductive choice, one has to improve education about life with disabilities. If parents who face the problem of giving birth to a child with disability have no experience in that matter, they should have the opportunity to not only get a medical description of the disability, but also to meet people with such disabilities and their families. They should be provided not only with the information about their reproductive rights but also about the legal rights and social capabilities of people with disabilities to be able to fully understand what life with a disability means and how it can be accommodated by our societies.[2]

One can also use PGD for positive selection, and instead of weeding out a certain medical condition, one may choose to opt for it. For instance, there are some parents who are born deaf and understand their deafness as a significant part of their identity and culture and want their children to share it with them, so they use sperm sorting or PGD to select for deafness (Davis 1997a, b; Davis 2010). Some may argue that in such cases some degree of persuasion of prospective parents by the genetic counselors is justified (Davis 1997a, b; Davis 2010), while the others argue that such

---

[2] Whatever we learn about disabilities, we would never be able to fully understand people with the most severe disabilities which may constitute the failure of imagination – I address the problem elsewhere (Soniewicka 2015).

reproductive choices should be restricted by the law as harmful (Buchanan et al. 2007). Assisting the parents in their idea of creating a disabled child must be a great moral challenge for genetic counseling deeply devoted to the model of non-directiveness and reproductive autonomy (Caplan 1993; Davis 1997a, b).

In the next section I will elucidate why the principle-based approach to the ethics of genetic counseling is not fully sufficient to cover the reproductive issues and to solve the afore-mentioned conflicts.

3.4 Principlism in Genetic Counseling and Its Shortcomings

There are different principle-based moral theories, such as Kantian deontology based on the categorical imperative and utilitarianism based on the principle of utility. These theories are based on rational reasoning and, however different they are, provide us with general or universal guides to action. Yet they have certain limitations, being: too abstract, too formalized, insensitive to the nuances of real personal life, neglecting context and diminishing the role of our relationships etc. They capture some important aspects of our moral thought, while neglecting others at the same time. They seem unable to cover the complexity of our moral life and can be described as incomplete. To elaborate this claim, I will apply principle-based theories to the cases introduced in the previous section.

In each of these cases there was a conflict of principles, yet it was not fully clear of which principles. The ethical theory based on principles has to provide us with an answer to the question of how such conflicts should be solved. It can be done in at least two ways: by giving a hierarchy of the principles, or by providing a meta-principle.

3.4.1 Kantian Ethics

In Kantian deontology, one may regard the categorical imperative as a kind of meta-principle. The categorical imperative was introduced in two formulations – the formula of universality and the formula of humanity (Audi 2016). Both formulations are based on the assumption of the moral capacities of human beings – the capacity of every rational being to subject its will to universal moral law which is associated with human dignity (Kant 2002). Human dignity is based on the notion of autonomy which may be understood as the moral self-determination of rational beings (O'Neill 2005). Only an act performed out of duty, not merely consistent with a duty, is morally valuable within Kantian ethics; moral worth is not derived from an end or consequences of an act, but from the principle which determines the will.[3] Ethics, as

---

[3] Kantian moral rigor may bring about counter-intuitive moral claims as the frequently truth-telling paradox exposes (i.e. the duty to tell the truth even if it would be harmful to others).

Kant argues, is not based on experience nor is it derived from the relationship, a moral judgment is aprioristic, the Ought precedes an act. Moral insights must come from reason, not from our senses or habits of the heart. A virtuous person for Kant is one who is capable and willing to obey the duty.

Applying Kantian deontology to reproductive decisions seems to overlook some important aspects of these choices. According the German philosopher, morally valuable acts must be free from inclinations, and thus acting from duty means overcoming inclinations. Parental love, in particular a mother's caring bond with the child, is a very strong inclination which may determine our actions but may not affect the moral value of the act or decision. Of course, there are plenty of examples in which people do morally wrong things out of affection but excluding this component from our moral thought seems to be counterintuitive and misleading. For instance, if we consider the motivation of the prospective parents to take the risk of giving birth to a disabled or chronically ill child (the hemophilia case or Down's syndrome case), we may strongly admire those who argue not to term the pregnancy or not to select against disability out of unconditioned love toward the future child. Yet for Kant, delivering a child with a high probability of disability would be considered as morally valuable only if such a decision would be taken out of moral duty to respect human life in every condition. Considering the parent-child relationship in terms of mere duties seems to be a completely confusing concept of this relationship. Of course, one may object to the existence of such a duty to respect human life before the birth if one rejects the assumption of the normative status of a fetus or an embryo. In the latter case, one would follow the duty to respect the autonomy of the parents to decide whether to give birth to such a child, leaving their decision beyond the moral realm.

Looking at the above described cases from the Kantian perspective, we can see a significant moral difference between the cases in which the prospective parents let the disabled (Down's syndrome case) or chronically ill (the hemophilia case) child be born and the case in which the parents intentionally create a disabled child (the deafness case). Bringing a child who is intentionally disabled into the world is not equivalent to harming the child but it may be interpreted as violating the formula of humanity – one may argue that the child is treated by the parents as a mere means to an end which is the satisfaction of their desires to have a child sharing their disability. The parents may argue, on the other hand, that they do not treat the child as a mere means to an end but also as an end in itself since they consider deafness as good for the child too. Yet it is hard to maintain that deafness may be considered as an objective good, it seems rather to be a subjective good from the perspective of those who have never had the possibility to hear and are proud of their difference.

3.4.2 *Utilitarianism*

On the grounds of utilitarian ethics, the meta-principle is the formula of welfare maximization. In contrast to Kantian ethics, the moral value of an act depends on its consequences (intentions are irrelevant) and consequences are evaluated from the perspective of an end, which is the maximization of utility (Mill 1871). There are many miscellaneous kinds of utilitarian ethics and it is not the purpose of this paper to discuss their nuances and differences here. My only aim is to capture the main characteristic of the principle-based utilitarian reasoning in order to show how it would work when applied to genetic counseling as the solution to moral conflicts. The utilitarian view on persons is significantly different from the Kantian view – persons are considered as bearers of utility functions; and it is utility, not persons as such, which matters in utilitarian thinking and therefore the difference between persons is ignored (Diamond 1990; Kymlicka 2001).[4]

Thus, from the utilitarian perspective there is no significant moral difference in all three cases introduced in the previous section. It is irrelevant from the utilitarian perspective whether the parents let the disabled or chronically ill child to be born, or if they intentionally created a disabled child. In all three cases giving birth to a disabled or chronically ill child, if it only could have been prevented, would be considered as equally wrong (Hare 2002; Singer 2011). In utilitarian ethics one may use the same arguments to claim for avoiding disability and to claim for not choosing to give birth to a disabled child when one could have chosen assisted reproductive technology (ART) to guarantee that a healthy child would be born instead. What is more, causing it to be that there will be a disabled person will be equally wrong to causing any existing person to have a disability (cf. Diamond 1990). Utilitarianism disregards the morally significant distinction between what we do to people through our action (doings) and what happens to them through our action (allowings-to-happen), neglects the difference between action and omission (Diamond 1990). The only thing which matters, from the consequentialist utilitarian point of view, is that there is more disability (and therefore more weakness, more suffering, more frustration and unhappiness) introduced to the world as it could be if the parents have prevented to have a disabled child and had a healthy one instead (Hare 2002; Singer 2011). Thus, it seems that the physician would be morally justified to suggest the prospective parents such a decision which would maximize the welfare and health and minimize the burden of disability in the world (cf. Kamm 2013; Buchanan et al. 2007).

From the utilitarian perspective, birth is considered as a mere fact, and disability as an abstract idea of a social and personal burden – and these facts seem to be possible to calculate according to one single measure of utility. This way of thinking may result in the conclusion that genetic fitness can be detached from persons and

---

[4] This leads to challenging the non-identity problem with the same-number-principle and allows utilitarian thinkers to consider giving birth to a disabled child as harmful though harm does not affect that particular child who was born (Parfit 1987; Buchanan et al. 2007).

can be considered in terms of rights (the right to be born healthy). There is something important missing in this picture. Most parents do not consider the birth of their child as a mere fact which constitutes rights, costs and interests. They rather think of the birth of a child as a significant, partly mysterious event which transforms their life, and they consider reproduction in terms of a relationship which is created by that fact. Disability or disease is considered by the prospective parents not as an abstract fact but as the tragic circumstance affecting particular persons – their progeny and themselves. From the real life perspective, both gaining and losing life has the same worth for every human being, irrespective of its genetic and health condition (Diamond 1990). Thus, the above presented reproductive decision may be considered as hard and tragic for the parents but definitely not as a mere welfare calculation.

3.4.3   *The Missing Element in Rational Moral Thought*

Of course, both of these ethical doctrines are far from providing us with the answers to all of the moral questions posed in the above-mentioned intricate reproductive cases. Yet it seems that the deontological view which captures the moral difference between dysgenic decisions and the decisions "to let nature decide", is more consistent with our basic moral intuitions. Nevertheless, both kinds of rational moral reasoning fail to fully address such key concepts as reproduction or disability. What we consider as good and evil, as right or wrong, is tightly connected to our understanding of ourselves, our social practices, our social roles and relationships, as well as to our understanding of such significant aspects of human life as procreation, birth or death which shape our idea of life and enter our moral thinking (cf. Diamond 1990). The pivotal difference in the moral doctrines regards what counts in moral judgments – consequences only, intentions only, or perhaps character, context and relationship too? My claim is that to take reproductive decisions one needs to first answer the question of how one understands reproduction and parenthood, what does the good life mean (and in most of the cases the answer to this question would be much more complex than the answer based on the quality of life measure offered by the utilitarian thinking) (Sandel 2007). The same concerns medical professionals providing genetic counseling, who in order to know how to assist the parents in these cases they need to know first how to understand their profession and the relationship between the physician and the patient on which this profession is based.

To sum up, in both kinds of reasoning something important is missing which is particularly vivid in reproductive choices that are rarely based on abstract principles, but rather on experience, on how the agents understand themselves and their roles, and on significant relationships which shape the interpretation of principles and duties and determine their application. The abstract and general principles are so formulated that they tend to conflict with each other frequently and the theories offer little help in their ranking and interpretation. My thesis, broadly inspired by Pellegrino and Thomasma (Pellegrino and Thomasma 1993), is that the principle-based doctrine can be enhanced, not replaced, by the virtue-based approach which makes our understanding of moral life more complete. This concerns medical ethics in particular which depends on the understanding of medical profession, its aims and values. It helps us interpret and apply the principles defined in the Sect. 3.2.

3.5 Virtue Ethics and Its Application to Genetic Counselling

Virtue, as defined by Aristotle in the Book II of *The Nicomachean Ethics*, is a settled and purposive disposition to act and feel (to respond to our feelings) in particular ways which enables achievement of the *telos* of a man (Aristotle 2011). In other words, virtue is a habitual disposition to act well; it encompasses both feeling what is good and the capacity to act according to it; it is not an automatic reflex, but rather a habit guided by reason. Virtues (which can be moral and intellectual) enable us to fulfill our potentials, they are discovered by reason, they can be teachable (they are acquired by practice, habituation, teaching, following good examples). Aristotle writes about living a good life and being good in what one does in our social lives – both are tightly interwoven. Leaving the problem of an individual ideal of the good life aside, let us ponder the latter question – what does it mean to be good in what one does and how it is linked with principles.

Justification of the principles may be expressed in such a way: "For every A, A ought to be B" is equivalent with the statement that "For every A, A which is not B, is a bad A" (Ossowska 2004, 247). Translating this reasoning into medical ethics, one can say that for instance: "The physician ought to provide medical help to the patient and not to harm the patient" (the principle of beneficence and nonmaleficence) which means that "every physician which does not provide help or which harms the patient is a bad physician". The reason to act in a certain way (following the rule) is based here on a certain internal value or virtues which define an ideal and an end of a profession. This approach requires an analysis of the meaning of the social practice such as the medical profession. One may object to this approach in a twofold manner, arguing that by identifying appropriate meaning and values we have to fall into either *essentialism* (if we argue that an appropriate meaning results from the essence of the good) or in *conventionalism* (if we argue that an appropriate meaning results from a social convention). The former assumes moral foundationalism and the latter moral relativism, both of which are hard to maintain. Yet one may challenge the objection and defend the meaning of medical practice on the grounds of our shared intuitions and moral insights, which are open to the permanent interpretation of those who are involved in this practice and those who participate in a society in which this practice takes place (Walzer 1985; Sandel 2010).

In defining the meaning of medical practice, the distinction made by MacIntyre between goods which are external and internal to a practice may be of help (MacIntyre 2007). External goods are "contingently attached" to a practice by "the accidents of social circumstance" (MacIntyre 2007, 188). Internal goods, on the other hand, can be only specified in terms of a particular practice and "can only be identified and recognized by the experience of participating in the practice in question." (MacIntyre 2007, 188). The internal good of the medical practice can be identified with healing (which includes: health, cure, assistance to the suffering, medical competence, medical excellence etc.), while among external goods one may name: money, reputation, fame, prestige or power. The physician may provide genetic counseling of good quality because she is aimed at healing the patient as the internal aim of her practice combined with her aspiration for excellence. One may also provide genetic counseling of high quality when only motivated by self-interest, for instance aiming at such external goods as financial benefit, caring about reputation, or being afraid of malpractice suits. Since external goods are not specific to that particular activity, a person interested in them may find other ways to achieve them. The physician interested in financial rewards only provides genetic advice of high quality to wealthy patients at his private practice and not at the public hospital where she provides care for the average patients paid by social insurance system. She might be tempted not to care that much about less educated patients or foreigners who are less likely to sue her for malpractice. If she were involved in very well paid clinical research, she might be also tempted to care more about the good of the researches than the good of her patients involved in them. Once achieved, external goods become some individual's possession, so they are the object of competition where there are losers and winners; while the achievement of internal goods, on the other hand, is good for the whole community who participate in the practice (MacIntyre 2007).

Virtues are the necessary means to attain the goods internal to communal practices, according to the interpretation of Aquinas, which means that "the end cannot be adequately characterized independently of a characterization of the means" (MacIntyre 2007, 184). It is a significant difference in the classical understanding of virtues by contrast to the functionalist interpretation of virtues presented by philosophers attracted by utilitarianism. According to the latter interpretation, virtues are external means to an end such as social welfare. It means that the aim can be achieved by other means and defined without any appeal to virtues. In the classical virtue-based approach, one cannot understand the meaning of the good physician independently of the characterization of the medical virtues. Pellegrino and Thomasma term such medical virtues as: fidelity to trust, compassion, *phronesis* (prudence), justice, fortitude, temperance, integrity (Pellegrino and Thomasma 1993; see also Gelhaus 2012a, b; Gelhaus 2013). These virtues are derivable from the ends of the profession and are linked with duties and principles; they rest upon the caring bond and the public trust (commitment to care) which constitute the meaning of the medical practice. The relationship between the physician and the patient plays a significant role in understanding the meaning and the value of the medical profession.

The principle-based moral doctrines support consumer or negotiated contract models which are usually patient-centered and in which the physician is reduced to the technical role of healthcare provider (Helén 2004); they promote autonomy as the primary value (Pellegrino and Thomasma 1993). This approach can be characterized as mainly procedural, instrumental, legalistic, economic, technical, and based on minimal trust or even on a mistrustful attitude. Pellegrino and Thomasma criticize this approach as misleading since the relationship between the physician and the patient is not equal; the vulnerability and dependence of the patient is neglected in this model, as well as the asymmetry of knowledge. They also claim that this approach corrupts the meaning of the medical practice which rests on trust. In the contractualist model, the physician is aimed solely at fulfilling the terms of the agreement which does not cover such important aspects of medical practice as beneficence, compassion, fidelity to trust, self-effacement etc. Thus, the authors argue for the end-oriented beneficence model which is based on the covenant of the fidelity to trust and on the unique experience of illness which disintegrates and alienates people who seek help from physicians. This approach emphasizes the significance of caring experience and the special, higher responsibility of the medical practitioners. The ends of the medical profession (such as: health, cure, technically correct and morally good decisions) are embedded in the practice, give medicine its distinctive character, structure the relationship, and determine the ethos of medical practice. Within this model, patients also have duties toward physicians (such as truth-telling, respecting the autonomy of the physician etc.). The physician is not an instrument to satisfy the desires and interests of the patient, but also first and foremost a moral agent too (an integral moral being), holding special responsibility attached to his profession and technical power.

By ethos of practice one may understand a complex unity of such aspects as values, habitual dispositions, characteristic traits, style of life, standards, aims and ideals etc. Ethos is based on the rational ordering of the values which constitute the axiological structure of a distinctive group of persons belonging to the same profession or social class (e.g. the chivalric ethos, the gentleman's ethos, the medical ethos). Such a professional ethos (the internal morality of a profession) cannot be inconsistent with personal values; the moral integrity of the person which is one of the virtues requires inner harmony and unity – the right ordering of the parts to the whole (including the integration of values, as well as the integration of different dimensions of human life and different social roles). What is more, it is assumed that such an ethos requires from the members of the group more than from an average person because of their special status and special responsibility built upon it. The ethos of the medical profession consists of special and general virtues and the moral obligations that are deducible from the kind of activity one is engaged in.

According to virtue ethics, moral person is aimed at her excellence and not just at obeying the minimum of moral requirements expressed in duties which are about to guarantee social cooperation and peaceful coexistence. Thus, the virtue-based approach constitutes maximalist (perfectionist) ethics in contrast to the minimalistic (contractualist – mainly formal and procedural) mainstream ethics. What is more, a virtuous person is virtuous not by respecting the principle but by being habitually disposed to respect it, and she considers her duty to be a part of her character, of who she is – she could not do otherwise.

The moral problems or dilemmas that we face in reproductive genetics do not arise from biotechnological development but rather from problems with understanding the medical profession and its aims (Pellegrino and Thomasma 1993), as well as from the problem with defining such significant human practices as reproduction. To elucidate the problem, let me invoke the distinction between phenotypic and genotypic preventive techniques in medicine emphasized by Eric T. Juengst. By phenotypic prevention one means avoiding the manifestation of a particular phenotype (a disease) and rests on the following assumptions: (1) there is a living individual who may benefit from it; (2) diseases are best defined at the level of their actual symptoms; (3) the disease is distinct from the person it burdens (Juengst 2009, 482–483). Genotypic prevention, on the other hand, is aimed at avoiding the birth of people with particular genotypes and rests on such assumptions as: (1) there are societal benefits (e.g. reducing healthcare costs) from the prevention of the birth of a person with a disease (individuals whose births are avoided cannot be beneficiaries); (2) diseases are best understood at the level of genotype; (3) diseases and their burdens are not fully distinct from their bearers (Juengst 2009, 484). The conceptual difference between these two kinds of genetic prevention in medicine should be recognized, irrespective of the assumed ontological and normative status of a fetus or an embryo. Phenotypic prevention is one of the major ends of medicine, yet genotypic prevention remains highly controversial since it requires an external value judgment about human life:

> Genes are not, like germs, external infectious agents that can be kept (or cleaned) out of a living person's body. (…) That means that to justify geno-prevention, someone (parents or society) must make the judgment that the burden of coping with cases of a disease outweighs any other value that individuals with a given genotype might bring to a family or community (Juengst 2009, 484).

Assuming that medical practice is primarily aimed at the good of the patient and reproductive counseling also includes the good of the progeny, we have to remember that this aim cannot be reduced to its negative aspect – respecting the autonomy of the patient (non-directiveness). The good of the patient encompasses the whole well-being of the patient, not only what is medically good, but also what is defined as good by the patient, and what is understood as the good for human beings as such and human beings as spiritual beings (Pellegrino and Thomasma 1993). If a medical intervention cannot benefit any particular human being but rather prevent its birth, one may question its moral justification. Yet if one understands the medical profession as being aimed at both kinds of prevention, then it would lead to different conclusions concerning genetic counseling. The main point is that the virtue approach to medical ethics, in contrast to rationalist principle-based approach, does not promise any simple, universal and general solution to the problems considered in this chapter. The thesis is that the interpretation and application of the principles of genetic counseling in the afore-mentioned cases are determined by our understanding of the aim of medicine and may differ and evolve depending on those who participate in the social practices which are at stake here.

The main objection to the virtue approach is that it is not sufficient for medical ethics since in a pluralistic modern society we are unable to define the objective good (flourishing) of the patient and thus this approach may lead to subjectivism and relativism (cf. McDougall 2007; Saenz 2010;[5] Holland 2011[6]). Principles, on the other hand, are universal and general. To refute that objection one may claim for combining both views: virtue ethics with principles and duties, claiming that they do not compete or replace each other, but they rather complement each other, constituting a more complete picture of moral life.

This is the way partly afforded by the doctrine of *prima facie* duties developed by W.D. Ross (Ross 1930; Audi 2004), which inspired the theory of four principles in medical ethics created by Beauchamp and Childress (Beauchamp and Childress 1989). Childress and Beauchamp combined virtues with principles in healthcare, and just as Ross, they emphasized the crucial role of prudence (practical wisdom) in solving conflicts of principles. Yet they neglected the problem of the nature of medicine and the intrinsic relationship between prudence, other virtues, and the nature of medicine as Pellegrino and Thomasma claimed (Pellegrino and Thomasma 1993). The difference between the principle-based approach of Childress and Beauchamp and the virtue-based approach of Pellegrino and Thomasma is that within the former principles are applied to the physician-patient relationship and required because of the doctrine. Within the latter approach, principles are derived from the relationship, and combined with the virtues; they are required because of the relationship, not because of a doctrine; they are internally related to medical practice (they are interpreted according to the aim of the medical practice and to the meaning of the relationship) as it was argued above. This approach is very promising in the context of medical practice, although not free of ambiguities and thus requires more attention and further exploration.

3.6   Concluding Remarks

The objections to Kantian and utilitarian rational principle-based moral enterprise which constitute the so-called princiNplism may be summarized in three main arguments: (1) the problem of justification (intentions vs. consequences); (2) abstract nature of principles; (3) minimalistic, simplistic and reductionist approach to the complexity of moral life. One may significantly enhance medical ethics, as well as

---

[5] Rosalind McDougall claimed in her paper that the virtue-based approach to the ethics of reproduction justifies genetic selection for deafness and other impairments if they are shared by the parents since flourishing of a child may be differently defined depending on the environment-specific characteristics. Carla Saenz, in her response to McDougall, aptly points out that this argumentation does not do justice to virtue ethics at all (McDougall's line of argumentation is based on "minimal requirement test", while virtue ethics requires excellence); Saenz concludes that virtue ethics is not sufficient to justification of moral permissibility of reproductive choices, claiming for prima facie principles instead.

[6] Stephen Holland addresses the important distinction between application of virtue ethics to the personal morality of an individual and to societal decisions of legalization certain practices or medical procedures; he discusses the main objections to the latter and ponders the rejoinders to critics which could help in enhancing the methodology of virtue ethics.

improve our understanding of moral thought, by putting principles and duties into a broader picture of moral reasoning which is rooted in the character of the moral person and determined by the physician-patient relationship based on caring and trust.

Acknowledgments The writing of this article was funded by the Polish National Science Centre (Dec-2013/10/E/HS5/00157).

References

Anscombe GEM (1981) Ethics, religion, and politics. University of Minnesota Press, Minneapolis

Aristotle (2011) Nicomachean ethics. Trans. RC Bartlett. In: SD Collins. Chicago University Press, Chicago/London

Asch A (2003) Disability, equality, and prenatal testing: contradictory or compatible? Fla State Univ Law Rev 30:315–342

Audi R (2004) The good in the right: a theory of intuition and intrinsic value. Princeton University Press, Princeton

Audi R (2016) Means, ends, and persons: the meaning and psychological dimensions of Kant's humanity formula. Oxford University Press, New York/Oxford

Beauchamp T, Childress J (1989) Principles of biomedical ethics, 3rd edn. Oxford University Press, New York

Biesecker BB (2003) Genetic counseling, practice of. In: Post SG (ed) Encyclopedia of bioethics, vol 2, 3rd edn. Thompson & Gale, New York, pp 952–955

Buchanan A, Brock DW, Daniels N, Wikler D (2007) From chance to choice: genetics and justice. Cambridge University Press, Cambridge/New York

Caplan AL (1993) Neutrality is not morality: the ethics of genetic counseling. In: Bartels DM, LeRoy BS, Caplan AL (eds) Prescribing our future: ethical challenges in genetic counseling. Aldine de Gruyter, New York, pp 149–165

Davis DS (1997a) Genetic dilemmas and the Child's right to an open future. Hastings Cent Rep 27(2):7–15

Davis DS (1997b) Genetic dilemmas and the child's right to an open future. Rutgers Law J 28:549–592

Davis DS (2010) Genetic dilemmas: reproductive technology, parental choices, and children's futures. Oxford University Press, Oxford

Diamond C (1990) How many legs. In: Gaita R (ed) Value and understanding: essays for Peter Winch. Routledge, New York/London, pp 149–178

Engelhardt TH Jr (1996) The foundations of bioethics. Oxford University Press, New York

Feinberg J (1992) Wrongful life and the counterfactual element in harming. In: Feinberg J (ed) Freedom and fulfillment. Philosophical essays. Princeton University Press, Princeton, pp 3–36

Gelhaus P (2012a) The desired moral attitude of the physician: (I) empathy. Medical Health Care and Philosophy 15(2):103–113

Gelhaus P (2012b) The desired moral attitude of the physician: (II) compassion. Medical health care and. Philosophy 15(4):397–419

Gelhaus P (2013) The desired moral attitude of the physician: (III) care. Med Health Care Philos 16(2):125–139

Hare R (2002) The abnormal child. Moral dilemmas of doctors and parents. In: Hare R (ed) Essays on bioethics. Clarendon Press, Oxford, pp 188–189

Helén I (2004) Technics over life: risk, ethics and the existential condition in high-tech antenatal care. Econ Soc 33:28–51

Holland S (2011) The virtue ethics approach to bioethics. Bioethics 25:192–201

Juengst ET (2009) Population genetic research and screening: conceptual and ethical issues. In: Steinbock B (ed) The Oxford handbook of bioethics. Oxford University Press, Oxford/New York, pp 482–485

Kamm FM (2013) Bioethical prescriptions to create, end, choose, and improve lives. Oxford University Press, Oxford/New York

Kant I (2002) Groundwork for the Metaphysics of Morals (trans Wood A W). Yale University Press, New Haven/London

Kuhse H, Singer P (2006) Bioethics. An anthology. Blackwell Publishing, Oxford

Kymlicka W (2001) Contemporary political philosophy. An introduction. Oxford University Press, Oxford

MacIntyre A (2007) After virtue, 2nd edn. Notre Dame University Press, Notre Dame

McDougall R (2007) Parental virtue: a new way of thinking about the morality of reproductive actions. Bioethics 21:181–190

Mill JS (1871) Utilitarianism. Parker, Son, and Bourne, London

Murray RF Jr (2003) Genetic counseling, ethical issues. In: Post SG (ed) Encyclopedia of bioethics, vol vol 2, 3rd edn. Thompson & Gale, New York, pp 948–952

Oakley J, Cocking D (2001) Virtue ethics and professional roles. Cambridge University Press, Cambridge

O'Neill O (2005) Autonomy and trust. Cambridge University Press, Cambridge

Ossowska M (2004) Podstawy nauki o moralności [The Foundations of the Studies of Morality]. Vol. 1, DeAgostini, Warszawa

Parens E, Asch A (2003) Disability rights critique of prenatal genetic testing: reflections and recommendations. Ment Retard Dev Disabil Res Rev 9:40–47

Parfit D (1987) Reasons and persons. Clarendon Press, Oxford

Pellegrino ED, Thomasma DC (1981) A philosophical basis of medical practice. Oxford University Press, New York/Oxford

Pellegrino ED, Thomasma DC (1993) The virtues in medical practice. Oxford University Press, New York/Oxford

Press N, Ariail K (2003) Genetic testing and screening. In: Post SG (ed) Encyclopedia of bioethics, vol 2, 3rd edn. Thompson & Gale, New York, p 1002

Ross WD (1930) The right and the good. Oxford University Press, Oxford

Saenz C (2010) Virtue ethics and the selection of children with impairments: a reply to Rosalind McDougall. Bioethics 24:299–506

Sandel M (2007) The case against perfection. Ethics in the age of genetic engineering. The Belknap Press of Harvard University Press, Cambridge MA/London

Sandel M (2010) Justice: what's the right thing to do? Farrar, Straus and Giroux, New York

Savulescu J (2001) Procreative beneficence: why we should select the best children. Bioethics 15:413–426

Savulescu J, Kahane G (2009) The moral obligation to create children with the best chance of the best life. Bioethics 23:274–290

Singer P (2011) Practical ethics. Cambridge University Press, Cambridge/New York

Soniewicka M (2015) Failures of imagination. Bioethics: disability and the ethics of selective reproduction. Bioethics 29:557–563

Toon P (2014) A flourishing practice? Royal College of General Practitioners, London

Walzer M (1985) Spheres of justice. A defense of pluralism and equality. Basic Books, Oxford

Chapter 4
A Bioethic of Communion: Beyond Care and the Four Principles with Regard to Reproduction

Thaddeus Metz

4.1 Introducing a Communal Ethic from Africa

With the rise of economic globalization and the spread of English have come greater cross-cultural engagements, including an improved awareness of the philosophical traditions of other parts of the world. It is now routine for a medical ethics textbook or anthology to include snippets of Islamic or Confucian moral approaches, for example. However, non-western bioethics are usually there for merely comparative purposes and are rarely taken seriously as alternatives to views that have lately dominated English-speaking research on morally right decision-making in a health-care context. The latter are well-known to be the four principles of James Childress and Tom Beauchamp, featuring a principle of autonomy, and the ethic of care.

In this chapter, I aim to give a fresh philosophical challenge to these influential western accounts of permissible action in a medical setting. Specifically, I draw on relational norms salient in the sub-Saharan African philosophical tradition to advance a novel principle of right action, and then apply it to several controversies concerning human procreation. According to this moral theory, an act is right just insofar as it treats people's capacity to commune with respect, where communing is a matter of identifying with others and exhibiting solidarity with them (spelled out below).

This communal ethic differs from the dominant, western approaches in several ways. Briefly, for now, it is like the ethic of care, and unlike the four principles, in that its content is relational and it is meant to be morally foundational. However, it is unlike the ethic of care, and instead is like the four principles, for taking a theoretical form and implying that welfare does not exhaust what has moral relevance about other individuals.

T. Metz (✉)
University of Johannesburg, Johannesburg, Republic of South Africa
e-mail: tmetz@uj.ac.za

© Springer International Publishing AG, part of Springer Nature 2018
M. Soniewicka (ed.), *The Ethics of Reproductive Genetics*,
Philosophy and Medicine 128, https://doi.org/10.1007/978-3-319-60684-2_4

Despite being grounded on African ideals, I contend that the communal ethic is not merely 'for Africans' and in fact has implications for reproduction that will be found intuitively attractive to a broad, global audience, including even many who currently hold a western bioethic. Although I lack the space to argue that the communal principle is more attractive than its rivals, I do aim to show that it is a prima facie attractive rival to them, meriting such weighing up elsewhere.[1]

In the following, I begin by spelling out the principle of communion, which I have in the past discussed in bioethical contexts such as the nature of health and illness, informed consent, standards of care, animal experimentation, and research ethics (Metz 2010a, b, 2012a, b, 2017). I demonstrate how it is informed by salient sub-Saharan reflections about morality, and note some of its appeal, including to those beyond Africa, as a foundational ethic. Then, I tease out its implications for the central controversy in reproductive ethics, namely, abortion, showing that it naturally justifies a moderate approach that is widely appealing. I next invoke the communal ethic to address various moral issues regarding reproductive genetics, specifically, selective abortion, surrogacy when there is no genetic link between the fetus and the prospective caregivers, enhancement, and confidentiality. I conclude by highlighting what is different and attractive about the communal ethic in relation to the four principles and the ethic of care.

4.2 A Principle of Communion

In this section, I first sketch some of the African sources of the communal ethic, then spell it out, and finally motivate it as a prima facie strong alternative to western moral perspectives common to encounter in bioethical debates. It is only in the following sections that I apply the principle to controversies about human reproduction.

4.2.1 Communion and the Capacity for It

Of the various philosophical interpretations of sub-Saharan moral thought (on which see Metz 2015), I have argued that a fundamentally relational one is most defensible and should be of particular interest to a global audience. Instead of conceiving of morally right action in terms of what honors or promotes a good intrinsic to a person, such as her welfare, autonomy, or life, my favored ethic places a certain

---

[1] My main reason for drawing on the African tradition is that there are under-explored ideas salient in it that are philosophically promising. For additional reasons to develop African ideas, see Behrens (2013).

way of relating between individuals at the ground of how to treat others.[2] The following comments from African thinkers express such an approach:

> (I)n African societies, immorality is the word or deed which undermines fellowship (Kasenene 1998, 21).
>
> Social harmony is for us (indigenous Africans – ed.) the *summum bonum* – the greatest good. Anything that subverts or undermines this sought-after good is to be avoided like the plague (Tutu 1999, 35).
>
> (O)ne should always live and behave in a way that maximises harmonious existence at present as well as in the future (Murove 2007, 181).

I do not take these comments at face value, for they have counterintuitive implications regarding human rights. As they stand, they variously suggest that certain (harmonious or communal) relationships are good for their own sake, it is always wrong to undermine them, and one should promote them as much as possible. However, if existing relationships alone were finally valuable, then a person not in relationship with an agent would seem to lack a moral status relative to her. If it were always wrong to act in ways that undermined the relevant relationship, then coercion and other forms of force would be categorically impermissible, even when directed against aggressors in order to protect innocents. And if one were supposed to maximize the relevant relationships, then it would be permissible to use any means whatsoever, including being violent towards innocents, whenever doing so would promote harmony in the long run.

To remedy these defects, while retaining a relational approach, I advance a principle according to which individuals have a dignity in virtue of their communal nature that demands respect. After spelling out what is involved both in being able to relate communally and treating that capacity with respect, I show how the ethic plausibly grounds human rights and some other intuitive moral categories.

By 'communion' or 'harmony' I mean the combination of two logically distinct relationships that are often implicit in African characterizations of how to live well. Consider these quotations:

> Every member is expected to consider him/herself an integral part of the whole and to play an appropriate role towards achieving the good of all (Gbadegesin 1991, 65).
>
> Harmony is achieved through close and sympathetic social relations within the group (Mokgoro 1998, 17).
>
> The fundamental meaning of community is the sharing of an overall way of life, inspired by the notion of the common good (Gyekye 2004, 16).
>
> (T)he purpose of our life is community-service and community-belongingness (Iroegbu 2005, 442).
>
> If you asked *ubuntu* (the Nguni catchword for African morality – ed.) advocates and philosophers: What principles inform and organise your life? What do you live for?....the answers would express commitment to the good of the community in which their identities were formed, and a need to experience their lives as bound up in that of their community (Nkondo 2007, 91).

---

[2] For a different approach to an African bioethics, which focuses more on vitality as a basic value to be promoted, see Tangwa (2010).

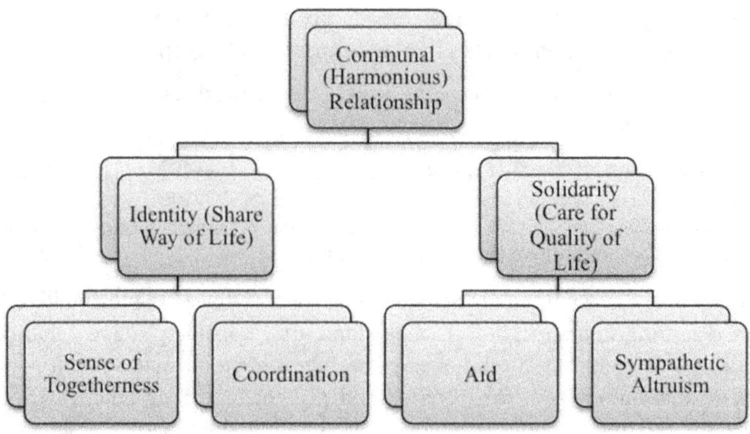

Fig. 4.1 Schematic representation of communion

In these and other sub-Saharan characterizations of how to commune or harmonize two logically distinct relationships are repeatedly mentioned, namely, considering oneself part of the whole, being close, sharing a way of life, belonging, and experiencing oneself as bound up with others, on the one hand, and then achieving the good of all, being sympathetic, acting for the common good, serving the community, and being committed to the good of one's society, on the other.

Elsewhere I have worked to distinguish and reconstruct these two facets of a communal relationship with some precision (e.g., Metz 2013). For an overview, consider Fig. 4.1.

It is revealing to understand what I call 'identifying' with others or 'sharing a way of life' with them (i.e., being close, belonging, etc.) to be the combination of exhibiting certain psychological attitudes of cohesion and cooperative behavior consequent to them. The attitudes include a tendency to think of oneself as a member of a group with the other and to refer to oneself as a 'we' (rather than an 'I'), a disposition to feel pride or shame in what the other or one's group does, and, at a higher level of intensity, an emotional appreciation of the other's nature and value. The cooperative behaviors include being transparent about the terms of interaction, allowing others to make voluntary choices, acting on the basis of trust, adopting common goals, living together, and, at the extreme end, choosing for the reason that 'this is who we are'.

What I label 'exhibiting solidarity' with or 'caring' for others (i.e., acting for others' good, etc.) is similarly aptly construed as the combination of exhibiting certain psychological attitudes and engaging in helpful behavior. Here, the attitudes are ones positively oriented towards the other's good and include an empathetic awareness of the other's condition and a sympathetic emotional reaction to this awareness. The actions are not merely those likely to be beneficial, that is, to improve the other's state, but also are ones done consequent to certain motives, say, for the sake of making the other better off or even a better person.

By the communal ethic advanced here, it is not this *relationship* that has a basic moral value, but rather an individual's natural *capacity* for it. Typical human beings, for example, have a dignity insofar as they are in principle *able* both to be communed with and to commune. The highest moral status accrues to us, beings that by nature can be both *objects* of communion, viz., able to be identified with and cared for by others, and *subjects* of it, able to identify with and care for others.

By this ethic, characteristic human beings are special relative to anything in the animal, vegetable, and mineral kingdoms. However, the communal ethic also entails that many animals have a standing as beings that can be directly wronged. In particular, those creatures with which we can commune as objects have a partial moral status, even though they lack a full moral status (a dignity) for being unable to be subjects of communion. Cats, pigs, and giraffes, for all we know, cannot identify with and exhibit solidarity towards others in the ways defined above, but, since we can with them, e.g., we can support their goals, sympathize with them, and help them, they matter for their own sake from a moral standpoint, unlike rocks and plants.

4.2.2   Respecting the Capacity for Communion

When it comes to how to treat others, the ethic prescribes respecting or honoring a being insofar as it can be party to a communal relationship. In the first instance, that means communing with others as objects, rather than ignoring them, let alone subordinating and harming them, which are the discordant or anti-social opposites of communion. In addition, one must commune with others in ways that do not degrade, and that ideally support, their capacity to commune as subjects. That means, for example, treating people as equals and not sacrificing another's capacity to commune for the sake of something worth less than it.

Supposing the capacity to commune has a dignity, sometimes honoring it will include actions that seek to promote the capacity, i.e., creating people, as well as its actualization, fostering communion. However, such pursuit of outcomes must be 'deontologically regulated', in at least two ways.

For one, actual communal relationships of which one is a part have some priority relative to merely possible relationships one could have and the actual relationships of others. To honor communion means in the first instance sustaining one's own ties, even if cutting off extant ones were foreseen to result in marginally more (sites of) communion, whether for oneself or elsewhere in the world. Such is a philosophical reconstruction of the special obligations often accorded to kin in traditional African societies (on which see Appiah 1998).

For another, to honor the capacity for communion entails that it is normally wrong to seek to realize it, even amongst one's own relations, by using a discordant means against innocents, where discord consists of relationships that are the opposites of communion, i.e., acting on an 'us versus them' attitude, subordinating, harming, and doing so consequent to hatred, cruelty, or the like. Respecting others insofar as they are capable of communion means not aiming to foster it by being discordant with those who have themselves respected communion. However, it can mean directing discordance towards those who have misused their capacity to commune, if it is necessary to get them to stop or to compensate their innocent victims.

A principle of honoring individuals insofar as they are by nature capable of communion entails that wrongdoing, in respect of innocents, is a matter of either failure to commune with them, and so being indifferent to others, or, worse, discordance. The latter means that those who have not misused their capacity to commune are treated as separate and inferior, subordinated, treated in harmful ways, and acted upon consequent to cruelty or similar negative attitudes. These anti-social ways of relating to those who have done no wrong are arguably what make it wrong to torture, kidnap, rape, and engage in other human rights violations as well as kinds of wrongdoing such as lying, breaking promises, and stealing. Such a fundamental account of the nature of wrongfulness differs from the western moral theories that it is constituted by merely degrading autonomy, failing to maximize utility, violating rules that would be reasonable for all to accept, or breaking God's commandments.

4.2.3 Contrasts with the Four Principles and the Ethic of Care

I close this section by briefly differentiating the communal principle from its most prominent bioethical rivals, the four principles and the ethic of care. Important differences will, however, also emerge in the following, applied sections of the chapter, where I strive to show that the communal principle has certain plausible implications that its rivals do not.

With regard to the ethic of care, it is normally not understood to take the form of a theory, i.e., an attempt to posit a basic, general principle that clearly entails and plausibly explains all other (or at least a wide array of) particular moral duties. Instead, adherents to a care ethic normally eschew the systematic appeal to principle of the sort I advance.

In addition, the ethic of care is not normally interpreted as including an ideal of human dignity, whereas the African moral theory does. This theory entails that what gives all people a dignity, and hence entitles them to human rights, is their robust natural capacity to relate communally with others (which, in turn, calls for communing with them, at least when innocent).

Finally, the Afro-communal ethic includes everything that care ethicists prize, precisely under the heading of 'caring' for others or exhibiting 'solidarity' with them, but it also includes a certain kind of relationship that they typically do not, or at least not explicitly. In particular, the communal principle instructs agents to prize people insofar as they can share a way of life with others, roughly a matter of enjoying a sense of togetherness and participating in cooperative projects. Elsewhere I have argued that this relationship, of identifying with others, is essential to bring into a relational morality in order to avoid concerns about paternalism and similar objections that plague the ethic of care because of its exclusive focus on well-being as a moral value (see Metz 2013).

Turning to the four principles, the communal ethic is like them for taking form of a principle. However, it purports to be 'the' single, basic moral principle, the one that grounds all other, less comprehensive ones. Childress and Beauchamp are well known for having advanced the four principles as 'mid-level' common ground among interlocutors and not invoking any allegedly basic principle, which they take to be more contested than their favored four.

Another difference between the communal ethic and the four principles is that the former is inherently relational, unlike the latter. The concepts of autonomy, beneficence, harm, and justice have usually been interpreted in individualist ways, not making any essential reference to anyone but a given person acted upon. For example, autonomy is normally conceived as an individual's ability to achieve whatever goals she elects to adopt; the goals need not be ones involving other persons. In contrast, the communal ethic instructs agents in the first instance to relate positively to other persons, by sharing a way of life with them and caring for their quality of life. Instead of grounding informed consent on a principle of autonomy, for example, the communal ethic does so on a demand for people to share a way of life with each other, which means (among other things) interacting cooperatively and not manipulatively (Metz 2010a, 6–7, 2010b, 160–161). And while some notions of beneficence and harm are implicit in the communal ethic's prescription to relate to others in a caring way, they include an essential relational dimension as well, e.g., an important way to care for others will be to protect and cultivate *their* ability to care for and share a way of life with others.

4.3 The Morality of Abortion

Before considering some of the ethical issues regarding genetics and reproduction, I first address the key issue of abortion, a woman's termination of her pregnancy that is foreseen to kill the entity in her womb. Besides being of interest in its own right, addressing abortion will help bring out and clarify some facets of the communal principle, enabling me to more easily address genetic issues in the following section. Furthermore, I submit that the communal principle grounds a moderate approach to the morality of abortion that many readers will find plausible, which is some evidence in its favor.

The striking thing about the four principles and the ethic of care is that neither perspective is grounded upon, or even routinely associated with, a theory of moral status. For whatever reasons, adherents have been reluctant to posit an account of which beings have moral standing, why they do, and to what degree (on which see, e.g., Hodges and Sulmasy 2013; Metz and Miller 2016, 8). However, it is natural in the context of abortion (and similar moral controversies that concern gametes, embryos, and animals) to make a comprehensive ethical appraisal in the light of how morally significant the beings involved are. The communal principle includes such an account.

Recall that, by the ethic I advance, an agent is to honor individuals insofar as they have the natural capacity to commune, where some beings can be both subjects and objects of communion and so have a full moral status, and others can be only objects of it, with a partial moral status. Here is what follows from the application of this principle to abortion.

First off, normal, adult human beings can be subjects of a communal relationship; they by their nature can think of themselves as a 'we', cooperate with others, sympathize with them, and help them for their sake. In addition, they can also be objects of a communal relationship, meaning that they are essentially the kinds of beings that can be thought as a member of a relationship or group, cooperated with, commiserated with, and aided. So, a typical pregnant human being has a full moral status, a dignity.

In contrast, a fetus instead has at best a partial moral status, depending on how far along it is in its development. Certainly by the sixth month (and probably a bit earlier), fetuses can feel pain and more generally have interests that enable us to have emotional reactions toward their flourishing (e.g., sympathy), to help them, and to do so for their sake. They might also have aims of the sort that animals do that we could either retard or promote. Since we can commune with them, but they cannot commune with us, they have a partial moral status at that stage.

However, zygotes, embryos, and very early fetuses probably lack a moral status, by the present principle.[3] Besides being obviously unable to exhibit identity and solidarity themselves, we also cannot do so with them; they lack the capacity to be communed with, even if they have the potential for it in the way a tree does not. They have no aims that we could support, and they have no quality of life for which we could care (even though they have life as such). It might be that characteristic human beings can enjoy a sense of togetherness, a sense of 'we', with an embryo or young fetus. However, if this capacity to relate does ground some kind of moral status, it is significantly less than that had by a third-term fetus, with which our capacity to relate is qualitatively greater.

Moral standing is one thing, and right action is another. Moving from considerations of how important an embryo or fetus is to how to treat it, the communal principle instructs an agent to honor a being insofar as it is capable of being party to relationships of sharing and caring. As the woman has a dignity, her urgent interests come before those of a fetus, so that if there were an unavoidable choice to be made between her life and that of even her third-term fetus, it would be permissible for her and medical professionals to make the tragic decision to abort.

Furthermore, abortion is permissible for less than urgent interests on the part of the woman, when there is merely an embryo at stake or when the fetus is very young. As the latter beings lack a moral status altogether, or at most have one that is very

---

[3] A view that starkly differs from some other interpretations of African moral thought, e.g., Tangwa (2007).

low, abortion would not degrade them, at least when done for the sake of the woman's ability to relate in other, more substantial ways, say, with her extant children.

I submit that this account of abortion fits the secular ethical judgment of many enquirers particularly well. As noted, the four principles and the ethic of care do not invoke an account of moral status, but such seems essential when having to choose between the urgent interests, such as the life, of the pregnant woman and her late-term fetus.

Furthermore, note that prominent moral theories that do invoke an account of moral status have unwelcome implications, at least for most interlocutors. Kantianism, which accords a moral status only to beings with the capacity for autonomy, counterintuitively entails that late-term fetuses (and new-borns) lack a moral status altogether. A utilitarianism ascribing full moral status to any being capable of feeling pleasure and pain counterintuitively entails that a late-term fetus has a moral status equal to the woman's, as does a principle of respect for human life. Although there are of course ways that these views can be adjusted to try to avoid these implications, the communal principle stands out for naturally being able to make sense of a gradational approach to abortion: a fetus lacks a moral status early on, since it is incapable of being communed with as an object, but acquires one later upon acquiring that capacity, albeit one less than the dignity had by the woman insofar as she can herself commune as a subject.

I cannot argue in this chapter that the communal principle makes the best sense of the abortion debate. I submit, however, that it offers a novel and reasonable account that should not be neglected and that demonstrates some of its explanatory power.

4.4   Ethical Issues in Reproductive Genetics

In this section, I address four major moral controversies in reproductive genetics, highlighting respects in which the communal principle either entails conclusions that differ from commonly held ones or grounds rationales for them that differ. Specifically, I consider some ethical issues regarding selective abortion, surrogacy when there is no genetic link between the fetus and its intended caregivers, enhancement, and confidentiality.

4.4.1   Screening and Selective Abortion

In the previous section I argued that abortion in the early stages is normally permissible, since the embryo or very young fetus lacks a moral status; it cannot be party to a communal relationship, neither as subject nor as object. Or at most it has a very low moral status, as a being with which we can (merely) share some emotional attachment, a status that is not enough to outweigh the moderate interests of the mother, given her dignity. Does it follow from this that terminating a human pregnancy because the fetus lacks the desired sex is invariably permissible?

The mere fact that a being lacks a moral status, or has a low one, does not necessarily mean that one may treat it however one likes, as the treatment could wrong other beings that have robust standings. There are forms of selective abortion that I argue are probably wrong for this reason.

It appears possible to determine the sex of a fetus reliably at 7 weeks, by analyzing the fetal DNA that is present in the mother's blood. Note that, upon reflection, it would not *always* be immoral on balance to terminate a pregnancy in the light of this knowledge. Hypothetically, suppose there were a serious shortage of girls throughout the world, risking the demise of the human race, or imagine that a dictator required parents to have only girls for a 2 year period, on pain of death to them and the doctors if they had boys instead. These 'fantastic' thought experiments show that selective abortion is not categorically wrong, and that it depends on the purpose for which it is undertaken. If it were done to continue the human species or to prevent murders of beings with a dignity, it would be morally justified, on the whole.

However, other purposes are more suspect. If one were to abort a fetus because one thought that girls are worth less than boys – perhaps one believed they are less intelligent – then the action would be discriminatory. Although it would not wrongfully discriminate against the fetus, for it lacks a moral status, it plausibly would do so against females generally, a large majority of whom have a dignity.[4] It would treat females as 'less than', would convey a sexist judgment, which is wrongful even if, as in the present case, doing so would neither impair anyone's autonomy, nor harm or be uncaring with regard to anyone (remember that the fetus, although alive, does not yet have a quality of life).[5]

It does not follow that selective abortion should be criminalized. Not all moral wrongs should count as legal wrongs. After all, it would be unjust to put people in jail or fine them for seeking to adopt a child of a particular sex, even if they did so because they thought that members of that sex were smartest. For one, non-punitive reactions, such as education about biology and its influence on intelligence, would likely be sufficient to prevent selective abortion. For another, the scarce resources available to the criminal justice system should be directed against more serious wrongs, ones in which the victims are not females in the abstract, but rather are specific women, e.g., as in the case of rape.[6]

One might wonder whether the communal principle would prescribe (rather than proscribe) selective abortion since one sex is more disposed to commune as subjects

---

[4] This explanation differs from a more common, traditional one in the African context, which Segun Gbadegesin has articulated: 'The use of genetic knowledge for choice of sex is not looked upon favorably because it is considered tampering with the work of God' (1998, 193).

[5] One might suspect that the fourth principle, of justice, could account for the discrimination, but this principle is usually interpreted as a macro-level account of how to allocate medical benefits.

[6] I am open to the idea that part of what makes rape wrong is the more 'general' or 'group' consideration of sexism, of targeting women because they are women, but am suggesting that a necessary condition for prioritization with respect to criminal justice is the presence of a more specific, immediate victim.

than another. Suppose it turned out that females are on average better able than males to commune in virtue of their nature, not nurture. And then imagine that parents were inclined to select in favor of girls on that basis. Would the communal principle morally forbid, permit, or perhaps even require this form of selective abortion?

I believe it would forbid it. The principle instructs one to treat beings with respect in accordance with their capacity to be party to a communal relationship with us; it does not say to promote communion as much as possible à la consequentialism. Even if it were true that females were somewhat better able to commune than males, it would not be to such a degree as to make a moral difference. Just as degrees of moral status do not track degrees of rational capacity for the Kantian, so degrees of moral status do not track degrees of communal capacity, by my African-inspired ethic. Instead, everyone has the same dignity upon reaching some threshold.[7] So long as females and males were in the same ballpark with regard to the capacity to exhibit communion, as they indeed are, they both have a dignity, and it would be disrespectful, because discriminatory, to select against males, even supposing one did so in order to maximize the amount of communion in society.

4.4.2   Surrogacy without a Genetic Link[8]

In this section I set aside many controversies about surrogacy, such as whether there is something morally objectionable about using in-vitro fertilization or paying for a surrogate beyond compensation for her discomfort, loss of time, and the like. In contrast, I address the rarer question of whether it is permissible to arrange a surrogate to give birth to a child who will not be genetically related to one of the 'parents' who intends to rear it.

Imagine a couple that has tried for over a decade to get pregnant, but has been unable to do so. Suppose that he is sterile and she is now in her late 40s, without viable eggs and unable to gestate a fetus. And yet both still long to be involved with a child from the start, perhaps one that is likely to have some features similar to theirs. May they rightly use a surrogate mother who would carry an embryo fertilized by donor gametes that the couple has selected from a bank?

Considerations of autonomy straightforwardly ground an affirmative answer. In addition, so long as the child reared in such circumstances would have a life well worth living,[9] there need not be any maleficence or failure to care. The communal principle also entails the conclusion that it would be permissible to create a child

---

[7] With degrees of moral status below dignity being based on large differences of ability to be communed with, e.g., between a human baby, a cow, and a fish (on which see Metz 2012a, 399–400).

[8] Much of this section borrows ideas and phrasings from Metz (2014).

[9] As recent evidence suggests, cited in Metz (2014, 35).

without a genetic link to one of its caregivers, but for a different reason, roughly that there would be no disrespect of familial relationships in doing so.[10]

Other, more conservative thinkers such as Leon Kass have argued that such an arrangement would in fact undermine or otherwise degrade the family. Speaking of creating a child who would not have a genetic link to those who rear it, Kass says that techniques such as IVF and embryo transfer would serve:

> not to ensure and preserve lineage, but rather to confound and complicate it...Properly understood, the largely universal taboo against incest, and also the prohibitions against adultery, defend the integrity of marriage, kinship, and especially the lines of origin and descent. These time-honored restraints implicitly teach...clarity about who your parents are, clarity in the lines of generation, clarity about who is whose...This means, concretely, no encouragement of embryo adoption or especially of surrogate pregnancy (2002, 55–57).

Some of Kass's rationale is consequentialist, to the effect that when origins and parentage become opaque, the prospects of what he calls 'civilized community' (2002, 57) decline. However, another part of it is non-consequentialist and, specifically, a matter of the 'respect owed to our humanity on account of the bonds of lineage, kinship, and descent....The navel, no less than speech and the upright posture, is a mark of our being. It is for these sorts of reasons that we find the Brave New World's Hatcheries dehumanizing' (Kass 2002, 52–53). When people are foreseeably created without a genetic link to those who will take care of them, then for Kass they are objectionably treated as inhuman, or perhaps human nature in the abstract is degraded.

I believe the claim that such surrogacy would degrade human nature has been adequately criticized already (Buchanan 2011, 52–74; Metz 2014, 36–37). In any event, here I want to show that there are other conceptions of what is valuable about human nature and familial relationships that are plausible and ground a conclusion different from Kass'.

Although it is true that often traditional African peoples valued kinship, the present philosophical reconstruction of their partialism does not accord genetic lineage any ethical significance in itself. Social ties matter morally, while biological ties do not. More specifically, according to the communal principle, normal adults have a dignity in virtue of their capacity to be party to relationships of identity and solidarity. Treating them with respect means (in part) enabling them to actualize this capacity and taking care not to interfere with their actualizations of it, i.e., with the relationships themselves. And there is nothing genetically essential to these relationships, ones of enjoying a sense of togetherness with others, participating with them on a cooperative basis, and helping them out of sympathy and for their sake. Arranging

---

[10] This is another place where my reconstruction of African ethics might have implications that differ from the intuitions of traditional sub-Saharan peoples. Segun Gbadegesin remarks that they 'will not entertain any counsel against natural reproduction because of the belief that one should bear one's children' (1998, 188). However, later in the same text he suggests, 'Surrogate parenthood is not all that strange' for Africans since it is present in a way in polygamous marriages, in which African women have often helped to rear one another's biological children (1998, 193).

a surrogate to give birth to a child whom genetically unrelated caregivers would love does not appear disrespectful of such relationships.

In fact, just as the marriage of two people who are genetically unrelated to each other warrants respect (as a way to respect those party to it), so their creation and parenting of a child genetically unrelated to them merits the same kind of respect. Communion, not blood, is arguably what counts about family.

4.4.3 Genetic Enhancement

Is it morally permissible to modify individuals genetically in order to give them abilities beyond the norm? For example, if one could alter the DNA of an embryo so that the eventual child would be more intelligent, creative, or empathetic, would it be right to do?

Like selective abortion, there is no question that enhancements would be justified for certain reasons. As Stephen Hawking has pointed out, the sun will not last forever and in order to survive as a species we will probably need to modify our genes in order to be able to travel in space for long times in search of a new planetary home.

However, the harder question is whether enhancements are justified for less urgent purposes. A principle of respect for autonomy seems readily to entail an affirmative answer, as neither the parent's nor the child's autonomy would necessarily be impaired, or otherwise degraded, by the former authorizing genetic modification of the embryo from which the latter grew. In addition, so long as enhancements were safe, in the sense of not threatening the child's health or other facets of her well-being, the care ethic would appear to permit them.

In contrast, the communal principle provides some reason to doubt the moral permissibility of enhancing, or at least enhancing in a particular way. It grounds a criticism of genetic enhancement different from the usual suspects of contending that it would: degrade human nature qua biological configuration à la Kass (see the previous sub-section on surrogacy); consist of playing God; foster injustice in the form of greater inequalities between the rich and the poor; and evince base desires for perfection and total control. Instead, a principle requiring one to prize communion entails that enhancing would be morally objectionable roughly insofar as it put distance between people.

More carefully, one may reasonably suspect that enhancements would threaten to make it harder for people to share a way of life. Recall that one part of what is involved in this way of relating is enjoying a sense of togetherness, i.e., thinking of oneself as a 'we', cultivating a sense of 'who we are', feeling good about being with others, and appreciating and even taking pride in others' accomplishments. To be avoided are more divisive attitudes, of thinking of oneself as an 'I' as opposed to others, developing senses of identity that are defined in opposition to each other, disliking other people's presence, and being envious of others.

A second respect in which one can share a way of life, for many in the sub-Saharan tradition, is undergoing the same fate as one's fellows, at least in certain ways. As the most influential sub-Saharan political philosopher of the past 20 years remarks, African communitarianism 'advocates a life lived in harmony and cooperation with others, a life of mutual consideration and aid and of interdependence, a life in which one shares in the fate of the other' (Gyekye 1997, 76). The most salient examples in the African philosophical tradition are living with a family and participating in the rituals of one's society. However, there are plausibly additional ways to ride in the same boat, as it were. For example, if there were a choice of distributing two units of a burden on ten people or ten units on one person, the communal principle would prescribe the former, despite the fact that there would be double the aggregate amount of harm than the latter. Similarly, it would likely entail that it would be more apt to distribute one unit of a benefit on ten people rather twenty units on only one.

Now, enhancements would probably interfere with people's ability to share a way of life in these various ways, at least if some were substantially enhanced and others were not. First, large differences in abilities could be expected to impede psychological identification, as great inequalities of wealth tend to do. It is well known that it is harder to form friendly and loving relationships across class divides. In addition, studies show that large gaps between rich and poor tend to foster indifference to others' suffering on the part of the former as well as feelings of envy on the part of the latter. The present point is not the common one to make, that by allowing enhancements the rich would get biologically richer and become all the more superior, but rather that biological hierarchy would likely have negative influences on people's ability to share a sense of self with one another, like economic stratification does.

Second, if some were to enhance while leaving others behind, then those authorizing the enhancing would probably be failing to honor the value of undergoing the same fate. Giving a handful of people qualitatively superior intelligence, for example, would be akin to distributing a large amount of wealth on a minority, instead of opting for a more egalitarian and inclusive allocation.

I do not mean to suggest that honoring the communal value of sharing a way of life requires 'levelling down' or demands that everyone live identical lives. Rather, the point is that enjoying a sense of togetherness and undergoing the same fate can be important instantiations of this value, and that they would be much harder to realize if enhancements were distributed such that only some people received them and their abilities were noticeably greater.

Of course, there is nothing intrinsic to enhancements that would require that sort of distribution. Perhaps they could be allocated more equally. Or perhaps if they were allocated unequally, they would not facilitate *great* divergence in abilities. Or maybe if there were only some who were genetically gifted, the value of sharing a way of life could still be honored if they were to use their gifts to lift up others in society. Although there are these ways to avoid the concern about divisiveness, the point is that divisiveness is a relevant moral concern on the face of it, and that it has not been salient in discussion about the ethics of enhancement. The communal principle highlights an under-theorized objection to genetic enhancement, an objection that, if it does not forbid it altogether, would require that it be undertaken in certain ways rather than others.

4.4.4 Confidentiality Regarding Genetic Tests

Upon having given informed consent to undergo a genetic test, under which circumstances may the tester disclose the obtained information to someone other than the one tested? When may the tester share the genetic data with a third party supposing the one tested has not given consent for him to do so?

There are a variety of cases discussed in the literature, but I focus on two. First, suppose that a mother is concerned about passing a serious disease onto her offspring, and a genetic test reveals that she indeed has one (which may or may not be the initial one in which she was interested). Suppose, too, that it is a disease that her siblings likely also have, unbeknownst to them. Under which conditions may the tester tell her siblings?

The second case concerns paternity. Imagine that a genetic test reveals that a woman's male partner incorrectly believes that he is the biological father. Under which conditions may the tester tell him that he is mistaken?

There are two dominant answers to these questions in the western literature. On the one hand, some maintain that the privacy of the individual is paramount, such that genetic information may not be disclosed to third-parties unless doing so were necessary to prevent an imminent serious harm to them. Respect for autonomy is of course the main rationale for this restrictive approach, but sometimes a concern 'not to disrupt families' is also given as a reason not to disclose. On the other hand, some maintain that, while individual privacy matters, public goods often matter more and more frequently permit infringing an obligation to keep the information secret.

The communal principle entails an approach that is distinct from both of these. According to it, the one tested has strong obligations to aid those with whom she has intensely communed, roughly, family members. Aid can come in the form of promoting their health and other needs, but it can also be a matter of fostering their self-realization as communal-moral beings. That is, a particularly important way to help one's family members is by enabling them to exhibit identity and solidarity with others. And identifying with others, recall, is neither the mere absence of force and fraud, nor even the mere presence of peaceful interaction. In addition, genuinely *sharing* a way of life with others means relating to them on the basis of knowledge of fundamental facts about their history, and so can require revealing truths known to one party and not yet to another.

Furthermore, because the individual's duties to aid family, including what westerners call 'extended' family, are weighty (Gyekye 1997, 61–75), it is reasonable to deem family members to have a stake in becoming aware of her illness and playing a role in discussing how she ought to treat it. Such is the view of several other bioethicists working in the African tradition (Gbadegesin 1998, 194; Kasenene 2000, 349–356; Dube 2009, 170–171; Murove 2009, 192–199; cf. Metz 2010b, 160–161).

Consider, now, some of the implications of the communal principle for the first case, regarding a serious disease revealed by genetic testing that the mother's siblings probably also have. Central to consideration of whether the tester may disclose this fact to the siblings is the point that the mother has an obligation to do so. If she did not do so, she would (probably) be dishonoring the communal relationship. Just as slaughtering a goat and failing to offer some of the meat to one's family would be considered theft by many traditional African peoples (Metz and Gaie 2010, 278), so failing to offer one's brother or sister weighty information one has acquired about their health is wrongful.

In contrast to the 'pro privacy' approach mentioned above, the communal principle entails that ignorance on the part of the siblings would itself partially constitute a 'disrupted' familial relation. Disruption is not merely a matter of bad feelings by the communal ethic, but is also a function of the absence of beneficent treatment. And in contrast to the more permissive approach, the communal principle does not reduce the moral considerations that the tester must weigh to a duty not to harm the patient and a duty to help her siblings, à la the four principles; in addition, the fact that the former would be wronging the latter in not telling them is a relevant ethical factor.

It does not *obviously* follow that a tester may disclose the information. However, the communal principle entails that there is more moral reason to do so than normally highlighted by the dominant approaches in the literature. The tester would have reason to urge the one tested to share the information with her siblings. In addition, the tester would have reason to urge the siblings to speak with the one tested about the results of the test. It could also be, in the final analysis, that the tester should tell them about the results, for the sake of their relationship with the one tested, if it were unlikely to become one of aid on its own.

Now consider the paternity case. Like the restrictive, 'pro privacy' approach mentioned above, the communal principle entails that familial relationships matter morally, and, indeed, for their own sake. That means that a broken family would be a serious ethical cost. However, unlike that approach, the communal principle also entails that there is a moral cost to a relationship based on the withholding of important truths such as paternity. The mother has a duty, of some weight, to the one who thinks (because of her) that he is the biological father to inform him that he is not. It does not immediately follow that the genetic tester may tell the partner if the mother failed to do so. However, the tester would probably not be wronging *the mother* if he were to disclose to the partner, supposing she were refusing to, with the more salient moral concern being the likelihood of the prospective child not having enough support.

Such an approach differs from the more permissive one mentioned above, in that the latter would normally weigh the mother's interests against those of the father. However, by the former, communal ethic, for the tester to disclose to the mother's partner would not be to disrespect her capacity to commune, i.e., her interests carry less weight since she is not respecting her partner's capacity to commune.

4.5 Conclusion: The Promise of the Communal Bioethic

I close this chapter by briefly highlighting what has made the communal ethic, grounded on African relational ideals, distinct from and appealing in relation to the four principles and the ethic of care, the dominant western approaches to right action in bioethical contexts. According to the former principle, one is to honor individuals in virtue of their natural capacity to commune, i.e., to share a way of life and to care for others' quality of life. A being may be able to exhibit communion, in which case she has a full moral status or a dignity that demands respect, or a being may be able merely to be communed with by us, in which case it has a partial moral status that also deserves some (albeit lesser) consideration.

Neither the four principles nor the ethic of care typically invoke the concept of dignity (or even moral status), while the relationality of care is merely welfarist, lacking the dimension of sharing a way of life. For every bioethical matter discussed in this chapter, the key concepts of dignity and relationality were invoked and, I submit, made prima facie good sense of how to understand each of them. Recall that among the key claims were that: a moral status of a fetus becomes greater with its development but remains short of dignity; aborting individuals of a certain sex would, while not wrongly them, disrespect other, dignified beings of that sex; arranging a surrogate to give birth to a child genetically unrelated to its caregivers would not degrade people's capacity to relate in a familial way, i.e., communally; genetic enhancement is morally suspect insofar as it would make it more difficult for people to share a way of life; and when deciding whether to infringe a duty to maintain confidentiality, the fact that the patient has failed to respect her communal relationships in not sharing information is a morally relevant factor. Supposing the reader finds these claims and those like them to be worthy of consideration, this chapter has shown the communal ethic to be a promising new approach to thinking about the ethics of human reproduction.

References

Appiah A (1998) Ethical systems, African. In: Craig E (ed) Routledge encyclopedia of philosophy. Routledge, London

Behrens K (2013) Towards an indigenous African bioethics. S Afr J Bioeth Law 6(1):32–35

Buchanan A (2011) Better than human: the promise and perils of enhancing ourselves. Oxford University Press, New York

Dube M (2009) 'I am because we are': giving primacy to African indigenous values in HIV&AIDS prevention. In: Murove MF (ed) African ethics: an anthology of comparative and applied ethics. University of KwaZulu-Natal Press, Pietermaritzburg, pp 188–217

Gbadegesin S (1991) African philosophy: traditional Yoruba philosophy and contemporary African realities. Peter Lang, New York

Gbadegesin S (1998) Genetic screening, sickle cell anemia, and the African and African-American perspectives. In: Kegley JA (ed) Genetic knowledge: values and responsibility. ICUS, Lexington, pp 183–195

Gyekye K (1997) Tradition and modernity: philosophical reflections on the African experience. Oxford University Press, New York

Gyekye K (2004) Beyond cultures: perceiving a common humanity. Council for Research in Values and Philosophy, Washington, DC

Hodges KE, Sulmasy DP (2013) Moral status, justice, and the common morality: challenges for the principalist account of moral change. Kennedy Inst Ethics J 23(3):275–296

Iroegbu P (2005) Beginning, purpose and end of life. In: Iroegbu P, Echekwube A (eds) Kpim of morality ethics: general, special & professional. Heinemann Educational Books, Ibadan, pp 440–445

Kasenene P (1998) Religious ethics in Africa. Fountain Publishers, Kampala

Kasenene P (2000) African ethical theory and the four principles. In: Veatch RM (ed) Cross-cultural perspectives in medical ethics. Jones and Bartlett, Sudbury, pp 347–357

Kass L (2002) The meaning of life – in the laboratory. Public Interest 146(Winter):38–73

Metz T (2010a) African and western moral theories in a bioethical context. Dev World Bioeth 10(1):49–58

Metz T (2010b) An African theory of bioethics: reply to Macpherson and Macklin. Dev World Bioeth 10(3):158–163

Metz T (2012a) An African theory of moral status: a relational alternative to individualism and holism. Ethic Theory Moral Prac 15(3):387–402

Metz T (2012b) 'Giving the world a more human face': human suffering in African thought and philosophy. In: Malpas J, Lickiss N (eds) Perspectives on human suffering. Springer, Dordrecht, pp 49–62

Metz T (2013) The western ethic of care or an afro-communitarian ethic? Finding the right relational morality. J Glob Ethics 9(1):77–92

Metz T (2014) Questioning South Africa's 'genetic link' requirement for surrogacy. S Afr J Bioeth Law 7(1):34–39

Metz T (2015) African ethics. In: LaFollette H (ed) The international encyclopedia of ethics. Blackwell Publishing Ltd, Malden, pp 1–9

Metz T (2017) Ancillary care obligations in light of an African bioethic: from entrustment to communion. Theor Med Bioeth 38(2):111–126

Metz T, Gaie J (2010) The African ethic of *ubuntu/botho*: implications for research on morality. J Moral Educ 39(3):273–290

Metz T, Miller SC (2016) Relational ethics. In: LaFollette H (ed) The international encyclopedia of ethics. Blackwell Publishing Ltd, Malden, pp 1–10

Mokgoro Y (1998) *Ubuntu* and the law in South Africa. Potchefstroom Electron Law J 1(1):15–26

Murove MF (2007) The Shona ethic of *ukama* with reference to the immortality of values. Mank Q 48(2):179–189

Murove MF (2009) African bioethics: an exploratory discourse. In: Murove MF (ed) African ethics: an anthology of comparative and applied ethics. University of KwaZulu-Natal Press, Pietermaritzburg, pp 157–177

Nkondo GM (2007) *Ubuntu* as a public policy in South Africa: a conceptual framework. Int J Afr Renaiss Stud 2(1):88–100

Tangwa GB (2007) Moral status of embryonic stem cells: perspective of an African villager. Bioethics 21(8):449–457

Tangwa GB (2010) Elements of African bioethics in a western frame. Langaa RPCIG, Bamenda

Tutu D (1999) No future without forgiveness. Random House, New York

Chapter 5
Parents, Special Obligations and Reproductive Genetics

Wojciech Lewandowski

5.1 When Does Parental Love Prevail?

One of basic bioconservative objections to the possibility of parental decisions about the genetic makeup of their children is the fact that this kind of action would be contrary to the principle of unconditional parental love. The core of the disagreement in this debate is not merely based on different theoretical approaches on what parental love is and whether reasons of parental love should have priority over others. It seems that difficulties in formulating universal moral principles concerning parental decisions are based on the complexity of points of view which prospective parents could adopt. First of them is perspective of an individual pursuing to her own ends. A child's existence and its traits are referred to the prospective parents' project on how their life should look like and what is important to consider their life as happy and fulfilled. The second perspective is a point of view of an individual who has special obligations to some existing people. Prospective parents should take into account their special obligations to their actually existing children, other family members and other people which can be affected by the existence of the new child. The third perspective is the impartial point of view of a rational individual who is aware of the equal moral status of all other people. From this perspective, the possible value of the existence of the new child and his rights is impartially balanced against the value of possible alternate options such as adoption or refraining from having a new child. The fourth perspective is the point of view of someone who is forming a new relationship and at the same time creates special obligations based on reasons which aren't reasons for other rational agents.

The plurality of these perspectives requires the character of prospective parents' reasons to be specified and for the formulation of criteria of comparison between them to make rational decisions. But there is even more difficult problem as the

W. Lewandowski (✉)
John Paul II Catholic University in Lublin, Lublin, Poland
e-mail: adalberto@kul.pl

© Springer International Publishing AG, part of Springer Nature 2018
M. Soniewicka (ed.), *The Ethics of Reproductive Genetics*,
Philosophy and Medicine 128, https://doi.org/10.1007/978-3-319-60684-2_5

existence of a child can radically change the hierarchy of reasons, leaving parents with diametrically different reasons for actions than they have had before the child was brought into existence. The clearest example of this situation is when a child's traits don't meet the parents' original expectations. In bioethical debates, the idea of unconditional parental love is generally accepted. According to proponents of deciding about child's genetic makeup it is wide enough to be easily compatible with the idea of the widest parental autonomy. An example of such argumentation is Ronald M. Green's view and who, in his book "Babies by Design. The Ethics of Genetic Choice". formulates the following principle:

> Parental Love Almost Always Prevails (PLAAP) – most parents end up loving whatever child they receive, no matter how much he or she conforms to – or disappoints – their pre-birth expectations (Green 2007: 114).

According to Green this principle allows to avoid the bioconservative objection based on the assumption that the failure of biomedical procedures aimed at the birth of a healthy child having certain desired traits will result in his rejection and the lack of parental love. The psychological and common sense nature of this principle may be used as a supplement to the idea of unconditional parental love. The application of PLAAP to discussions about gamete donation, embryo selection, abortion or human enhancement is intended to demonstrate that the use of genetics to design desired traits isn't wrong because even if the child doesn't possess a given trait, parents will still love him. Combining PLAAP with a consequential obligation not to harm one's child genetically (Green 1997) would allow the justification of all of the mentioned interventions.

It seems that the main challenge to argumentation based on PLAAP in its primary, common sense form is that it is a double-edged sword. Advocates of the sanctity of life view may use this principle against genetic selection and abortion as unnecessary interventions, because even if the child will be born with serious genetic defects or other undesired traits, the parents still will end up loving him. Combining PLAAP with the assumption about the full moral status of human embryos and fetuses and the deontological prohibition of killing them would justify the wrongness of these interventions. The strength of both types of argumentation based on PLAAP depends on answering the question of whether it is possible to predict a change of one's original preferences as a result of becoming a parent. Furthermore it is necessary to ask if there are some normative reasons justifying special obligations for maintaining or changing these preferences in situations when the child's health or traits differ from predicted ones.

Green discusses four bioconservative arguments supporting the claim that deciding about a child's genetic makeup is incompatible with unconditional parental love. The first of them refers to the risk of the situation in which parental love, based on affirmation of the child, is replaced by a critical evaluation of its traits. Formulating expectations towards child's physical and intellectual capacities at the stage of embryonic development may result in prejudice and negligence if it turns out that it doesn't meet these expectations (President's Council on Bioethics 2003: 54–55).

The second argument concerns the danger of an unfair assessment of a future child's achievements when most of its successes will be recognized as the effect of genetic interventions and most of its failures will be interpreted as the child's own responsibility. The third argument is based on the child's right to an open future and says that genetic interventions may constrain a child's development and possibilities of deciding about its own future (Feinberg 1980). This argument has the greatest strength in relation to "selection to disability" situations when parents decide to use genetic selection in order to have a child with the same disability as them. It is also present in the debate about human enhancement and states that future decisions of enhanced child might be constrained due to programmed innate predispositions and the family's expectations. The fourth argument concerns the risk of altering the moral aspects of the relationship between parents and children. If some expected trait is a reason for bringing a child into the world then it can be interpreted as giving priority to instrumental values over the child's dignity.

It is worth noting that all of the above arguments can be interpreted as referring to the risk of possible conflicts between parents', other people's and child's expectations which will negatively affect the child's life. Debates in this area would require determining the likelihood and frequency of the attitudes leading to these outcomes. Their mere possibility, illustrated by the examples of withdrawing from the surrogacy contracts or the voluntary termination of parental rights to the children who are born handicapped, is insufficient to justify the contradiction between genetic interventions and unconditional parental love. The main argument in this context should recall the risk of the dissemination of these kinds of attitudes. The basic aim of PLAAP would show that this risk is overestimated as there are evolutionary grounded psychological mechanisms regulating the care of offspring regardless of its state of health, appearance, sex or capacities. According to this argument, the number of undesired attitudes wouldn't be increased significantly as compared with the situation in which genetic procedures weren't available. In addition, some of these procedures, such as moral enhancement, could allow the reduction in the frequency of undesirable behavior.

There is another interpretation of bioconservative arguments. According to Green, they are based on the supposed analogy between the attitude of a customer who acquires some goods incompatible with his original preferences and the attitude of parents whose child is born with traits other than originally preferred. The exact description of this analogy would require reference to the customer satisfaction theory. The first theories aimed at explaining the process in which a customer formulates a judgment about his own satisfaction were based on Festinger's dissonance theory (Festinger 1957) and claimed that if the product doesn't meet original expectations, the consumer experiences a cognitive dissonance which produces pressures for its reduction by either changing our expectations or perceptions of the product (Cardozo 1965; Engel et al. 1968; Howard and Sheth 1969). The widest accepted theory is that using the Expected-Disconfirmation model (Oliver 1977). It implies that the level of expectations toward a product before its acquisition is seen as a standard to which the experience of using it after the purchase is compared.[1] One of the basic assumptions in the satisfaction theories is that concerning the dynamic nature of expectations and the possibility of changing them. Purchased vacations may differ from primary expectations but the new, unpredicted experiences gained during trip may give the reason for the high evaluation of one's satisfaction (Botterill 1987). Using the theory of satisfaction as a support for PLAAP would lead to the thesis that even if the 'parent-customer' analogy is right, new valuable experiences of raising a child with previously unpredicted traits can affect feelings of satisfaction and support parental love. The problem for this kind of argumentation would be that the frequency of changing expectations under the influence of new experiences may be too low to justify PLAAP. Furthermore, in many cases this kind of changing expectations may be mere rationalization. The main challenge to the parent-customer analogy therefore is to justify the role and value of new experiences in the process of changing primary expectations.

Green rejects this analogy as a support for PLAAP. According to him, most of the exceptions to it are not related to disappointment and unfulfilled expectations about child's traits (Green 2007, 114–115). The failure to establish a relationship with one's child has its roots in other factors such as immaturity, problems in marriage, psychological problems or burdens related to child's serious health or developmental problems (Pound 1982). Green's argument is supported by another psychological theory, called attachment theory. It states that parent-child attachment is one of the fundamental evolutionary shaped behavioral systems which increases the probability of passing genes on to the next generation (Bowlby 1958). In its early years of life, a child develops a strong bond with his primary caregiver. The compatible behavioral system, the caregiving system, is developed by parents (Solomon and George 1999). The model of caregiving includes a representation of the self as a caregiver, the perception of a child as helpless and deserving help and a representation of caregiver-child relationship. Behaviors such as the seeking of closeness are activated by the signals associated with potential danger or the child's suffering. Most research on the development of the caregiving system concerns the changes that occur in the mother's perception of the future and the actual child. Ideas about child's appearance and traits are strongest in about the seventh month of pregnancy, then they start to become less clear, leaving room for the perception of child's actual traits (Fava Vizziello et al. 1993; Stern 1995). Important to the development of the caregiving system are the parent's previous experiences, marital relationship, support network, economic situation, parturition and its subjective interpretation, contact with the infant and how it is perceived (Lewis 2008). Exceptions from PLAAP can be explained by deficits in any of these areas. The discrepancy between the actual and predicted experiences of relationship can be quite easily overcome under favorable conditions. The above studies show that

---

[1] Theories alternate between those based on comparing expectations and the perception of the product, taking into account among other: customer's values (Westbrook and Reilly 1983), judgments about causes of dissatisfaction (Pearce and Moscardo 1984) or judgments concerning the fairness of the transaction (Oliver and Swan 1989).

becoming a parent involves changing one's perspective, including the change of one's priorities, and value hierarchy. The problem is to what extent should prospective parents take into account the possibility of changing our perspective in their decisions about using genetic interventions.

5.2 Becoming a Parent as a Transformative Experience

In her book "Transformative experience" Laurie Paul starts with the simple observation that people recognize their new experiences as valuable because they enrich their cognitive abilities, understanding and knowledge. According to her, the value of these experiences is subjective which means it is possible to grasp from a first-personal perspective what it is like to have that experience (Paul 2014, 13–14). Gaining new experience, for example that of a new taste, allows us to predict how the future experiences of the same kind would look like and whether they would be preferable for us. Among these new experiences there may occur those which are *personally transformative*, i.e. radically changing our perspective, such as traumatic experiences, overwhelming success, religious conversion or becoming a parent. The problem in making decisions which may lead to transformative experiences is that we cannot estimate the subjective value of these experiences before they occur. In decisions concerning becoming a parent we should assume that it will change our perspective and determine our future evaluations. According to Paul, we cannot rationally choose actions leading to transformative experiences merely on the basis of our predictions on how it would be like to have such experience. The criterion of assessing one's personal decisions as rational is their compliance with the rules of normative decision theories but not in the case of decisions like becoming a parent. For example, if one of the basic rules is the maximization of expected value then this value depends on actual dispositions, beliefs and desires and they can change as a result of transformative experience. It means that decisions about actions which may result in such experiences are always made under conditions of uncertainty and we cannot apply to them any standard normative decision-making models as we don't know what we will value after the change and how high will be our risk-aversion (Paul 2013, 31–32).

According to Paul, the experience of becoming a parent is personally transformative as psychological changes associated with development of a system of caregiving experienced from the first-person perspective cannot be predicted. I will call her standpoint the unpredictability of parental love.

> Unpredictability of Parental Love – if you've never had a child, it is impossible to make an informed, rational decision by imagining the outcomes based on what it would be like to have your child, assigning subjective values to these outcomes, and then modeling your preferences on this basis (Paul 2013, 18)

One of the essential elements of the ignorance about what our parental preferences will look like is the uniqueness of the new child. Ignorance about the traits of the future child means that we cannot know in advance what it is like to be a parent of that and not another child. It means that this skeptical position may be quite easily extended into situations of parents who already have children and make decisions about the genetic makeup of the new one. Can we use the psychological observations described in previous paragraph to reject Paul's skepticism? They imply that when we are parents of a child with traits other than originally expected, in favorable circumstances the objective probability that our parental love will prevail is much higher than the probability of negative experiences. Our subjective estimation of probability may be different but relying merely on subjective estimation without taking into account statistics may be seen as an obvious bias. According to Paul, objective probability is of little help when it comes to the subjective value of an experience. The inability to assess whether winning the lottery will be important to us is different from the inability to estimate the probability of winning this lottery. The evidence derived from scientific research and testimonies from family or friends may only show that having a child will change us in the way that our new preferences will be satisfied by taking care of him. It doesn't, however, constitute a reason for actions aimed to satisfy preferences which we had before the child's existence. According to Paul, the main reason to undergo such a kind of transformative experience is to discover who we will become.

The skeptical account poses a problem as it, at least in one aspect, eliminates PLAAP and doesn't resolve moral problems which PLAAP was designed to resolve. Questions about parental autonomy including the right to decide about child's genetic makeup or a child's right to an open future remain unanswered. Having very high expectations towards a future child seems as unjustified as rejecting any expectations. The reason based on the discovery on who we will become seems to be of little help as this discovery may be unsatisfactory. We may find ourselves as horrible parents and our child as not deserving our attention. It seems that we need PLAAP as our hope that we will find ourselves in favorable circumstances but we also need some normative principles which will guide our actions and could be applied even when our preferences will change.

One way to avoid skepticism in deciding about a child's genetic makeup is to abandon the first-person perspective and to justify our choices merely with agent-neutral reasons. Even if we can't know what the child will be like and what parents we will be like, what we *can* know is that our agent-neutral reasons, as the reasons for all rational agents, will remain the same. It means that no matter whether we will be satisfied or disappointed about child's traits, all of our actions should be performed in accordance with agent-neutral reasons. Argumentative strength of PLAAP as a merely descriptive, common sense and psychologically supported principle would be diminished as agent-neutral reasons can be used to justify normative version of it.

> PLSBN (Parental Love Should Be Neutral) – it doesn't matter what your attitude towards your child will be. You should always choose the universally best outcome in a given situation

PLSBN implies that that no matter whether parents will be satisfied or disappointed about a child's traits, all of their actions should be performed in accordance with agent-neutral reasons. Furthermore, their expectations should be subordinated to the result of calculating which option is impartially best to everyone. Changing the nature of the principle from describing psychological regularity into ethical obligation entails significant consequences. First, it no longer applies to the statistical majority of parents but to the whole set of rational agents. Second, it is objective and may serve as a guide for parents who don't know what to do in the situation of dissatisfied expectations. And finally, the consequent application of this principle requires that not only after child is born but also at the stage of deciding about child's genetic makeup, parental expectations should have been formed in accordance with agent-neutral reasons.

One of the objections against PLSBN is that it seems paradoxical or even self-contradictory as it assumes that parental love can be reduced to agent-neutral reasons. This objection can be weakened if we differentiate between two meanings of "prevailing parental love" as (1) valuing a child's actual traits over originally preferred or (2) as acceptance of the child even if he possesses traits seen as less valuable. The former corresponds with the idea of transformative experience but is possible to understand only from first-person perspective and thus, it faces the challenge of skepticism. The latter can be compatible with the idea of neutrality, assuming that in many cases it would be impartially better if the parents accept the child.

Another way to guarantee that parental perspective will resist the transformative experience is a normative principle, which categorically requires the acceptance of one's child.

> PLSAP (Parental Love Should Always Prevail) – all parents should love their child no matter if he satisfies or dissatisfies expectations which they had before he was born.

Unlike PLSBN, this principle assumes a first-person perspective. At the same time it could be the basis for moral obligations even if skepticism based on the fact of transformative experiences is a true position. It is categorical, first, because it implies that parents should act contrary to their desires if the child's traits don't meet their expectations.[2] Second, it doesn't take into account any exceptions. Regardless of child's condition the reasons based on the fact of being his parent outweighs any other reasons. The "almost always" clause is absent here as internal and external factors described in previous paragraph can be only explanation why people sometimes don't obey this principle but they can't revoke its universal character. The idea standing behind this principle is that we have special obligations which can be handled only from first-person perspective. The categorical character of PLSAP is a major challenge, because it requires the justification of how the fact of being a parent can generate such strong obligations and justifies partiality towards one's child at the cost of resignation from achieving an optimal outcome. The main way to include the "almost always" clause at the same time as maintaining its resistance to skepticism is to treat this principle as a *prima facie* moral requirement

---

[2] Principle contrary to this one would be an imperative of loyalty to one's original desires but if agent has undergone transformative experience reason based on previous desires seems to be weak.

which can be outweighed by some stronger reasons generating other kinds of obligations. The character and validity of both principles described above depends on the how the complex web of parental reasons can be balanced before and after child begins to exist.

5.3 Continuity of Parental Reasons

There are serious difficulties in searching for reasons which might overcome skepticism based on the fact of transformative experiences. The first of them is that prospective parents make their decisions with a mixture of different kinds of reasons and it is often hard to find out which one is decisive. It is hard to expect that we find among them one conclusive reason which can be used as a moral justification for all procreative decisions in all contexts of reproductive genetics. However, it seems possible to establish some criteria of comparing these reasons and to justify that at least some combination of these reasons provides us with stronger justification of our decisions than any other combinations and that this justification will resist transformative experience. The second difficulty is that in the context of reproductive genetics we cannot refer directly to the most imposing reasons related to parental love, because the object of this love doesn't exist yet. According to Harry Frankfurt, the fact that parents love their children even before their birth, regardless of their possible traits can be explained by the fact that love doesn't require experience of value of whom it is addressed (2004, 38–39). Even if this is true, the question remains whether it is possible to love someone who doesn't exist.[3] The third difficulty is the assumption of parental autonomy. As long as none of parental reason leads to objective harm, the parents have right to pursue to have a child with any reasons they have. But such a negative criterion is not enough to justify parental love. Fourth, it is difficult to classify prospective parents' reasons. In contemporary ethics there is a debate about the status of reasons concerning special obligations to people who share with special relations an agent. Some views claim that the justification of these obligations can be reduced to some agent-neutral reasons such as consequential principle to maximize happiness (Railton 1988), principle that all agents should obey obligations which they acquired by voluntarily entering into some relationship or that agents who brought other persons into existence are responsible for them (Prusak 2013). Other views search for the foundations of these obligations in agent-relative reasons based on an agent's ground projects (Stroud 2010), intrinsic value of a relationship (Scheffler 1997) or the value of individuals with which this relationship is shared (Keller 2013). Some threads from this discussion may be useful in trying to handle the phenomenon of becoming a parent.

Diane Jeske formulated an important distinction between reasons for entering into some relationships and reasons for staying in the relationship once it is formed

---

[3] It seems appropriate to distinguish between love, where its object exists, and readiness to love, where this condition is not fulfilled.

(Jeske 2008, 45–46). The usefulness of this distinction to the problem of prospective parents' reasons is that it doesn't presuppose the existence and identity of a child and, what follows it could help to explain different perspectives before and after becoming a parent grounding special obligations in reasons for staying in the relationship. According to Jeske, parental reasons can be seen as reasons of intimacy which means that they have three features. First, they are agent-relative (as opposed to agent-neutral), i.e. they are not necessarily reasons for any other agent which could have been in the same causal position to do or promote that action. Furthermore, they are fundamental (and not derivative), as they are not constituted in the whole or in part by some other reason, and finally, they are objective – which means they are not grounded by, and only by, the fact that an agent takes some interest in (or cares about) some action or some causal consequences of that action (Jeske 2008, 10–11). Jeske claims that the reasons for entering into relationship may be either subjective agent-relative, i.e. based on agent's desires, or agent-neutral, but reasons to stay are always objective agent-relative. Forming a friendship with some person can be justified with agent-neutral reason that friendship is valuable to anyone and subjective agent-relative reason that friendship with this particular person will be easy and pleasant to maintain (Jeske 2008, 45–46). Once a relationship is formed there occur objective agent-relative reasons of intimacy based on seven features of a relationship: positive attitude, mutuality, special concern, desire to share time, considered effort to achieve some level of knowledge about each other, spending time in each other's company or causal interaction in some other relevantly similar manner and parties' history with one another as the evidence of concern (Jeske 2008, 47–61). Special obligations which arise from relationship are based on the voluntarist assumption that the only way to acquire these obligations is through some voluntary actions, such as entering into a relationship and developing it (Jeske 2008, 127).

There are some good arguments for grounding parental reasons to enter into a parent-child relationship with long-term desires. The decision about becoming a parent is an important part of our life and always has its beginning in our desire and belief that having a child will make us happy. On the other hand these reasons may not resist the transformative experience and so seem to be the weakest reasons to stay in a relationship after this experience occurs and child's traits aren't satisfactory. Any change on the level of desires is followed by the change on the level of reasons. The desire to end some relationship causes the elimination of reasons for caring for a person within this relationship. The existence and strength of this reason depends merely on how much it costs to prolong this relationship. There are relationships which could undergo these kinds of calculations, such as business relationships or some non-instrumental relationships like the one with co-workers and partners but in stronger relations, such as family, the view which bases reasons only on desires would leave the agent without any reasons for caring about the other person. There are attempts to avoid this objection by replacing long-term desire-based reasons with the category of ground projects which are not necessarily self-interested but make a life meaningful. A ground project is, however, still insufficient as a reason for special obligations. We can have altruistic projects which are merely supererogatory and hence still be free to abandon them. It seems that we need some other reasons, like other people's expectations, to be added to our ground projects to justify the existence of special obligations but then those other reasons seem to be the basis for these obligations (Scanlon 1998).

Arguments for agent-neutral reasons as reasons to enter into a parent-child relationship concentrate on the objective value of this relationship as one of the most important ways of distributing happiness. As such, these reasons would be compatible with PLSBN and allow it to be justified that in most cases accepting one's child unconditionally is an optimal solution. There are, however, serious objections against them as reasons to stay in relationship with child. First, such justification would require weighing the value of our own actual relationship against the potential value of possible relationships that we could form with other persons (Jeske 2008, 38ł Kolodny 2010, 41). It could be used to justify, for example, killing or allowing the newborn child to die in order to bring another child into existence. Priority of neutrality implies that if a child's traits do not meet expectations then there are no moral reasons strong enough to maintain relations with him. If there is the possibility to create another child which will have desirable traits and with which we could form another valuable relation, the agent-neutral reasons would imply that this is our primary reason. Second, it requires that we weigh the value of our own current, actual relationship against the potential value of possible relationships other person could form with our child. If other people could be better parents for our children than us, then agent-neutral reasons would support ending such a relationship. The third objection is Bernard Williams' famous 'one thought too many' argument, according to which justifying a preference which we have toward our loved ones with agent-neutral reasons is unnecessary, unconvincing and doesn't do justice to our real practical reasoning and motivations (Williams 1981). All of these arguments show that when some kind of parental reasons are only externally connected with the future child's anticipated reasons and interests it doesn't support special obligations to the child once he comes into existence.

Studies about the reasons why people decide to have a baby show not only reasons related to parents' desires or to an estimation of what would be optimal for everyone but also some kinds of reasons which cannot be reduced to any of them, such as those based on expectations, interests or claims of other existing people, relationship's value or possibility to satisfy child's needs (Langdridge et al. 2005). These reasons can be qualified in the manner shown in Table 5.1.[4]

Most of these reasons may be applied to decisions of having biological child as well as to adoption or surrogate motherhood. I decided to qualify the relationship-centered reasons to agent relative ones because as parents we value a relation with our child due to the fact that *we* are the part of this relation and it will be *our* child we will share it with. Furthermore, there is the possibility to qualify reasons like "I want to share what I have and what I know with a child or "I would give a child a good home" to distinguish a class of "child-centered" reasons as they seem to concentrate on the child's interests and needs, however in the moment of making

---

[4] I decided not to include religious reasons as they require more profound discussion as regards where to qualify them.

Table 5.1 Reasons to enter into a parent-child relationship

| Subjective agent-relative | Objective agent-relative | | Objective agent-neutral |
	Other-existing people-centered	Relationship-centered	
It would be fun to have a child around the house	My family would be pleased if I had a child	I feel it would make us a family	I want to invest in the future
Raising a child would be fulfilling	My partner would be pleased if I had a child	I want the special bond that develops between a parent and child	I want to help shape the next generation
So that in my old age I would have someone to care about me	To carry on our family name and traditions	It would give me something to strive for	
I want to receive love and affection from a child	It would be something that is a part of both of us	I want to share what I have and what I know with a child	
Biological drive	Good for my relationship with my partner	I would give a child a good home	

decision they aren't fundamental because the child and his value, interests and needs, which can be basis for these reasons don't exist yet.[5] These reasons seem to be derivative from relationship-centered reasons and become fundamental once the child begins to exist.

The presence of other-existing-people-centered and relationship-centered reasons as reasons to enter into a parent-child relationship shows that objective agent-relative reasons can be considered as the basis for moral guidance in situations when a child's traits don't meet expectations. The main argument for including these reasons is that they guarantee the continuity of moral perspective, can resist transformative experience just like agent-neutral reasons but at the same time they don't abstract from first-person perspective. There is a significant difference between the strength of these kinds of reasons. Other-existing-people-centered reasons seem to be weaker than relationship-centered or even subjective ones. Even if they are often taken into account, they rather support some other kinds of reasons. If they collide with parental autonomy, are too demanding or not supported by agent-neutral reasons, they are easily outweighed. The main argument against them as the main basis for parental moral principles is analogous to that against the priority of agent-neutral reasons. Because they are only externally connected with the anticipated child's reasons, interests and needs, there is a danger that in conflict situations special obligations to other people will prevail over special obligations to the child. On the other hand, relationship-centered reasons can be easily combined with any other kind of reasons. Furthermore, from this perspective the child's future traits do not have the same importance as from any other perspective. All relationship-centered

---

[5] These kinds of reasons, however, can be seen as fundamental in cases of adoption.

reasons preserve all their strength even if the child's traits are different than originally expected.

One of the most difficult problems with views based on the plurality of perspectives is whether there is a possibility to balance these reasons. If relationship-centered reasons have absolute priority, they will support categorical character of PLSAP. If not, this principle will be a mere *prima facie* moral requirement. Another possibility is that these reasons cannot be compared at all. Agent-neutral reasons would be the strongest from an impartial perspective and agent-relative from agent's point of view. This possibility, based on the Sidgwickian dualism of practical reason, cannot, however, resist skeptical challenge. It is impossible to formulate the exact set of rules of comparing these reasons, although there are some points which could be made.

When Green introduces PLAAP he recalls two approaches: of William Ruddick and of Adrienne Asch. According to Ruddick, the parents' role is the role of guardians and gardeners. The role of the guardian is to protect the child's future interest. The role of the gardener is to shape the child according to one's preference. To maintain the balance between these two roles, Ruddick suggests the rule according to which parents should, as long as resources and prediction allow, foster (1) life prospects, any one of which is eventually realized by a child would be acceptable to both parent and child; (2) life-prospects which jointly encompass the range of likely futures the child may encounter in adult life; and (3) characteristics in a child and in her- or himself that make for the child's eventual independence of parents (Ruddick 2012, 172).[6] Among these conditions, only two and three can remain current, even after transformative experience. The first one can be included once the child begins to exist. This would support the view that even if future people don't have the right to an open future, there are reasons not to decide about his traits and so in the moment of making decision about becoming a parent, subjective agent-relative reasons should be weighted lower than relationship-centered ones.

According to Adrienne Asch, genetic selection or eugenic abortion is morally wrong because it assumes that some trait or quality outweighs all the other facts about the child that could be discovered if he was born. Abandoning the readiness to give birth to a child due to the undesirable result of preimplantation or prenatal diagnosis is would negatively affect the idea of the parent-child relationship as it can be seen as an exclusive club (Asch 2000, 239). The main argument against Asch's position is that it confuses reasons to stay in a relationship, which should imply taking care about other person independently from the traits which he possesses, and reasons to enter into relationship which commonly include some kind of selection (Kamm 2005). But Asch's view may not be based on confusion. In the context of reproductive genetics, the trait criterion is not the only criterion for forming a relationship but also the criterion of existence which is different from other

---

[6] These conditions are different from those originally formulated by Ruddick which Green refers to. In the original version parents should foster life prospects which (1) jointly encompass the futures that the parents and those they respect deem likely and (2) individually, if realized, would be acceptable to both parent and child (Ruddick 1979).

situations in which the relationship is formed. Furthermore, the parent-child relation is irreversible, which means that it is supposed to be life-long and ending such a kind of relationship would involve serious harm. Finally, this relationship is based on the phenomenon of inheritance whose main sense is to transmit every value which the parents achieved during their lifetime to their children.

Despite the difficulties in combining parental perspectives, it is possible to formulate the necessary conditions for all prospective parental decisions which say that, as long as child doesn't exist, parental reasons should include agent-relative relationship-centered reasons. Furthermore, the coherence with relationship-centered reasons makes any other reasons stronger. In other words, when making a decision about having a child or about using some genetic interventions, an option is more preferable than others if it is more coherent with the relationship-centered agent-relative reasons for having children, i.e. aimed at building life-long, irreversible interpersonal relationship related to the parent's cognitive, emotional and physical abilities and based on the idea that parents transmit to children all of the valuable means (including knowledge, ideals etc.) which can be freely used by them in new, creative ways.

Acknowledgments   The writing of this article was funded by the Polish National Science Centre (Dec-2013/10/E/HS5/00157).

References

Asch A (2000) Why I haven't changed my mind about prenatal diagnosis: reflections and refinements. In: Parens E, Asch A (eds) Prenatal testing and disability rights. Georgetown University Press, Washington, DC, pp 234–260

Botterill TD (1987) Dissatisfaction with a construction of satisfaction. Ann Tour Res 14:139–141

Bowlby J (1958) The nature of the child's tie to his mother. Int J Psychoanal 39:1–23

Cardozo RN (1965) An experimental study of consumer effort, expectations and satisfaction. J Mark Res 2(3):244–249

Engel JF, Kollat TD, Blackwell RD (1968) Consumer behavior. Holt, Rineheart, Winston, New York

Fava Vizziello G, Antonioli ME, Cocci V, Invernizzi R (1993) From pregnancy to motherhood: the structure of representative and narrative change. Infant Mental Health J 14(1):4–16

Feinberg J (1980) The child's right to open future. In: Aiken W, LaFollette H (Edss) Whose child? Children's rights, parental authority and state power Rowman and Littlefield, Totowa, p 124–153

Festinger L (1957) A theory of cognitive dissonance. Stanford University Press, Stanford

Frankfurt HG (2004) The reasons of love. Princeton University Press, Princeton

Green RM (1997) Parental autonomy and the obligation not to harm one's child genetically. J Law Med Ethics 25(1):5–15. doi:10.1111/j.1748-720X.1997.tb01389.x

Green RM (2007) Babies by design. The ethics of genetic choice. Yale University Press, New Haven/London

Howard JA, Sheth NJ (1969) The theory of buyer behavior. Wiley, New York

Jeske D (2008) Rationality and moral theory: how intimacy generates reasons. Routledge, New York

Kamm FM (2005) Is there a problem with enhancement? Am J Bioeth 5(3):5–14

Keller S (2013) Partiality. Princeton University Press, Princeton

Kolodny N (2010) Which relationships justify partiality. Philos Public Aff 38(1):37–75

Langdridge D, Sheeran P, Connolly K (2005) Understanding the reasons for parenthood. J Reprod Infant Psychol 23(2):121–133

Lewis M (2008) Interactional model of maternal-fetal attachment: an empirical analysis. J Prenat Perinat Psychol Health 23(1):49–65

Oliver RL (1977) Effect of expectation and disconfirmation on postexposure product evaluations: an alternative interpretation. J Appl Psychol 62(4):480–486

Oliver RL, Swan EJ (1989) Consumer perceptions of interpersonal equity and satisfaction in transactions: a field survey approach. J Mark 53:21–35

Paul LA (2014) Transformative experience. Oxford University Press, Oxford

Pearce LP, Moscardo MG (1984) Making sense of tourists' complaints. Tour Manag 5(1):20–23

Pound A (1982) Attachment and maternal depression. In: Parkes CM, Stevenson-Hinde J (eds) The place of attachment in human behavior. Tavistock, London, pp 118–130

President's Council on Bioethics (2003) Beyond therapy: biotechnology and the pursuit of happiness. The President's Council on Bioethics, Washington, DC

Prusak B (2013) Parental obligation and bioethics: the duties of a creator. Routledge/Taylor & Francis Group, New York/London

Railton P (1988) Alienation, consequentialism and the demands of morality. In: Scheffler S (ed) Consequentialism and its critics. Oxford University Press, Oxford, pp 134–171

Ruddick W (1979) Parents and life prospects. In: O'Neill O, Ruddick W (eds) Having children: philosophical and legal reflections on parenthood. Oxford University Press, New York, pp 124–137

Ruddick W (2012) Children's rights and parents' virtues. In: Cafagna AC, Peterson RT, Staudenbaur CA (eds) Philosophy, children and the family. Plenum Press, New York, pp 165–173

Scanlon T (1998) What we owe to each other. Belknap Press, Cambridge, MA

Scheffler S (1997) Relationships and responsibilities. Philos Public Aff 26:189–209

Solomon J, George C (1999) The development of caregiving: an attachment theory approach. In: Cassidy J, Shaver PR (eds) Handbook of attachment: theory, research and clinical application. Guilford Press, New York, pp 649–670

Stern DN (1995) The motherhood constellation. BasicBooks, New York

Stroud S (2010) Permissible partiality, projects and plural agency. In: Feltham B, Cottingham J (eds) Partiality and impartiality: morality, special relationships and the wider world. Oxford University Press, Oxford, pp 131–149

Westbrook RA, Reilly MD (1983) Value-precept disparity: an alternative to the disconfirmation of expectations theory of customer satisfaction. In: Bogozzi PR, Tybouts A (eds) Advances in consumer research, vol vol 10. Association for Consumer Research, Ann Abor, pp 256–261

Williams B (1981) Persons, character and morality. In: Moral luck: philosophical papers 1973-1980. Cambridge University Press, Cambridge/New York, pp 1–19

Chapter 6
Moral Virtue and the Principles of Practical Reason

Adriana Warmbier

6.1 Introduction

The recent growth of knowledge concerning reproductive genetics has provided prospective parents with the possibility to decide whether they wish to have a child with a specific genetic makeup. For many, this improvement of reproductive choices is viewed as an opportunity offered to prospective parents, to afford their future children with the best chance of the happiest life. In response to the possibilities of employing reproductive technologies like IFV and Prenatal Screening for selection, Julian Savulescu has formulated the so-called Principle of Procreative Beneficence, according to which he has argued that parents have a moral obligation to use selective technology in order to bring to birth the best child that they can possibly have. As he puts this:

> couples (or single reproducers) should select the child, of the possible children they could have, who is expected to have the best life, or at least as good a life as the others, based on the relevant, available information. (Savulescu 2001)

This claim, which has given rise to an ethical controversy, has been extensively discussed by both its proponents and the adversaries (see e.g. Parfit 1984; Harris 2001; Glover 1992; Bennett 2009; Hotke 2014; Stoller 2008). Those who oppose the moral obligation, where choice is possible, to bring the "best" child into the world have put forward a number of counter-arguments among which the pivotal are: (1). the proposed principle discriminates against the disabled and thus it leads to eugenics (Sparrow 2007, 2011; Bennett 2009; Sandel 2007; Shakespeare 1998); (2). such an obligation will undermine reproductive autonomy (Bennett 2009; Sandel 2007; Habermas 2003); (3). it will trigger off the impulse to mastery and control over the potential child and thus disfigure the relation between parents and

A. Warmbier (✉)
Jagiellonian University, Kraków, Poland
e-mail: adrianajoanna@gmail.com

© Springer International Publishing AG, part of Springer Nature 2018
M. Soniewicka (ed.), *The Ethics of Reproductive Genetics*,
Philosophy and Medicine 128, https://doi.org/10.1007/978-3-319-60684-2_6

children (Sandel 2007; Habermas 2003; Kass 2003). The criticism over the Principle of Procreative Beneficence focuses not only on the consequences which would follow from the establishment of such an obligation, but also, however, rarely, it appeals to the philosophical assumptions on which this principle is based (Bennett 2009; Stoller 2008; Parker 2007; Soniewicka 2015). The theoretical foundations of the claim that potential parents are morally obliged to choose to give birth to a child who will live a worthwhile life stem from the utilitarian outcome-orientation and from its approach to happiness. But since the utilitarian reasoning rests too much on reductive assumptions about what a worthwhile life is and disregards an agent-centered conception of person, which is constitutive for a normative character of human action, the alleged moral obligation formulated by Savulescu seemed to be groundless. This is not to deny that prospective parents are morally required to act in the best interests of their children. But we may differ about what sense of the "best interest" this principle is taking into account. If moral requirements which pertain to reproductive decisions are based on what particular societies value the most they will confirm their preferences regardless whether such preferences lead to discrimination against those who do not fit with them. As Sandel notes "there is something wrong with the ambition (…) to determine the genetic characteristics of our progeny by deliberate design" (Sandel 2007). It is because in doing this we presume that the worth of a person is contingent upon specific factors that we regard to be of high value. Thus some lives are believed to be of a higher quality than others. The proponents of the Principle of Procreative Beneficence put the lower merit on those who are disabled, however, they often explicitly deny doing so. Some, like Peter Singer, goes even further, and claim that "killing a disabled infant is not morally equivalent to killing a person. Very often it is not wrong at all" (Singer 2011). Frances Kamm in turn argues that selective reproduction does not discriminate against people who are disabled but it is aimed at disability itself taken as a property which is less worthy than others (Kamm 2013). The problem with this argument is that its conclusion is false. Properties seem to be always embedded in their holder, in which they emerge as having one value or another. In making a reproductive decision we do not refer to the disability itself, taken as a property which is detached from its holder, but we refer to the person who is defined by its properties. If this is so, then the faulty premise is the one that claims that properties may operate in isolation from their carriers.

According to Savulescu's account, the scope of application of the Principle of Procreative Beneficence means that there is no harm in attempting to avoid giving birth to a disabled child, thus, as he presumes, no theoretical or practical conflict arises here. The problem occurs if prospective parents choose to do otherwise when they have a chance to decide whether they want to have a child "with the best opportunity of the best life" or not. "The choice to have a disabled child is wrong for Savulescu, not because it would harm the resulting child, but because it is to bring about a worse life than could have been the case" (Parker 2007). And to allow it to happen, as he claims, is against a rational attitude towards reproductive decision-making. This standpoint rests primarily upon two underlying assumptions. The first one repudiates the view that a person is of absolute value and thus her worth cannot be compared to the worth of the other. Such a concept of personhood, which appeals to the fundamental value of human beings, is held by both Christian philosophers (i.a. Boethius, Augustine, or Thomas Aquinas) and Kant. The second assumption pertains to the notion of happiness that, according to Bentham's account, has been reduced to a specific state, in particular, to a state of feeling good about something. Thus the notion of "happy life", as the utilitarians hold (see: Singer 2011; Hare 2002; Savulescu 2001), rests in cost and benefit analysis which assumes that one may calculate the costs and benefits of a reproductive choice in accordance with the principle of utility. There are sound reasons to believe that since the utilitarian principle focuses on the outcomes rather than on agency, and therefore treats people instrumentally, it can hardly provide a strong basis for an approach to biomedical research and medical practice, nor it can serve as an useful guidance in reproductive decisions (see: O'Neill 2007; Kołakowski 1971). The utilitarian claim for the maximization of utility is not an appropriate principle when considering individuals' choices about reproduction for it does not afford a justification for the theoretical foundations of this claim. Furthermore, the very category of the "best life" which is constitutive for the utilitarian line of reasoning is vague and thus inadequate when referring to a particular person (for a critical discussion on this issue see: Parker 2007).

The problem with the concept of the "best life" consists in the fact that it is practically unworkable unless it is applied with reference to other complex ethical concepts such as those of human flourishing, individual fulfilment of life, and of what it is to live a life worth living. I endorse Parker's claim that this is not to say that the significance of these concepts would need to be established before that of the "best life" could be understood and used as the basis for interpretation, but rather to highlight that any coherent use of the Principle of Procreative Beneficence in ranking possible lives would unavoidably involve ranking the characteristics of embryos in relation to a cluster of complex, rich and interdependent moral concepts (see: Parker 2007). Hence, since the category of the "best life" to which Savulescu's principle appeals is always contextual and can be applied only within moral framework – a set of concepts and practices – which refers to an ample account of agency, such concepts as best life, self-development and a sense of life satisfaction cannot be reduced to the "simple elements or constituent parts which might be identified through the testing of embryos" (Parker 2007; see also: Bennett 2009). The best possible life relies on different factors and, what is the key issue here, these factors are closely interrelated with a particular person, while the best or better life mentioned in the Principle of Procreative Beneficence is said to be best or better *for* anybody; "and if it is not best or better for anybody, but simply best or better *period* (whatever that may mean), then it is difficult to see how we can be obliged to bring it about" (see: Herissone-Kelly 2006). Moreover, it is hard to specify and predict which conditions may or may not affect one's life in a way that it would be unlikely or likely to go well.

In contrast to the advocates of the Principle of Procreative Beneficence, I argue that the notion of the "best prospects for a happy life" ought to be rather captured in terms of an agent-relative reasons than in the context of impersonal reasons. Our understanding of ourselves as being persons is primarily based on an agent-centered account of personhood thus a proper concern for the happiness of future children needs to appeal to the key concept of agency which stands for the capacity to initiate the choice and for the capacity to attribute responsible authorship for one's choices to oneself. To elucidate this standpoint, let me turn to virtue ethics with its fundamental concepts of practical wisdom and *eudaimonia*, which, as I argue, may provide us with a deeper understanding of the notion of "good life" and with an explanation of what truly determines the quality of individual and one's self-fulfillment. With this framework at hand, I attempt to address the requirement of a "reasonable chance of a happy life". I discuss the possibility of applying agent-based ethics to the debate upon medical and bioethical issues concerning reproductive genetics, demonstrating that virtue ethics may be regarded as an attractive ethical approach to guide biomedical decisions.

6.2 The Category of the "Best Life": An Utilitarian and Aristotelian Approach

Within the criticism of the moral obligation of procreative beneficence many forceful, well-formed, and plausible arguments have been already presented. It seems that the most adequate response to the obligation to create only the best children is the one which explores the theoretical foundation of this claim. The Principle of Procreative Beneficence is based on a notion of impersonal harm which, as Bennett rightly argues, "turns out to be not only intuitively very appealing, but also an abstract concept that is difficult to pin down, analyse and thus criticize" (Bennett 2009; see also: Parker 2007). The classical formulation of the concept of impersonal harm is offered by Derek Parfit (1984). Parfit constructs several thought experiments which are aimed at answering the question of whether we may reasonably state that something is harmful if no definite person is identifiable. Parfit argues that harm can only be identified for currently existing persons and since future persons do not yet exist, therefore they do not have definite identities. He concludes that if there is no "definite person" to whom harm can be assigned we cannot assess the moral quality of the effect of our actions on persons who do not yet exist. And since those for whom our actions may be considered as right or wrong do not exist, we cannot explain the wrongness or rightness of our actions referred to particular future people in terms of person-affecting harm. Thus Parfit claims that in consideration of the effects of our actions today on future generations we should apply the non-person-affecting principle. Those who uphold the Principle of Procreative Beneficence support their standpoint by referring to the non-person-affecting principle. Furthermore, the obligation to choose to give birth to worthwhile but impaired lives rests on an idea of "making the world a better place than it could otherwise have been, not in terms of any individual person's welfare, but in terms of creating the greatest total score for what is regarded as the goods of life" (Bennett 2009). Following this line of reasoning, Harris, Savulescu, and Bostrom go one step further, claiming that we may maximize the welfare of future people using a wide range of biomedical technologies along with the most controversial one, namely genetic engineering. According to their demand, if what used to be implemented to maintain or restore health may now allow us to expand human capacities above normal levels we would act unreasonably if we refrain from using new opportunities to create healthier, longer lived and altogether "better" individuals (see: Savulescu 2001; Savulescu and Kahane 2009; Harris 2007).

The advanced version of postulates concerning enhancing human capacities was offered by Nick Bostrom who states that the very idea of human enhancement is a "way of thinking about the future" that is based on the assumption that the human species is neither final, ultimate nor perfect in its form, but represents merely a phase of a wider evolutionary process that will soon give rise to cognitive systems surpassing man in intelligence and other cognitive capacities (Bostrom 2003). The obligation of procreative beneficence thus stands for the idea that parents have a moral responsibility to provide the best life possible for their children and this might imply enhancing their physical or cognitive capacities through biomedical means. Bostrom and other supporters of this stance argue that an extension our intellectual and physical capacities, an increase of human health-span, and of control over our own mental states and moods will provide future children with access to basic goods and thus will widen their potential life plans (see: Schaefer et al. 2014). The proponents of biotechnological intervention claim that in enhancing our human condition prospective parents may increase both the child's dignity and capacity for autonomous choice (see: Bostrom 2005; Schaefer et al. 2014). However, such claims have serious underlying weaknesses. Bostrom and others assume that human dignity is contingent upon external factors and thus may be considered in terms of increase or decrease. Yet the concept of dignity, as it is understood in a long philosophical tradition, does not have conditional worth but it is of absolute value. This overriding of moral worth of persons, which is the same regardless of their quality of life, seems to be one of the most controversial points concerning the idea of bioenhancing human condition, but I shall not pursue it further in this paper.

Biomedical interventions, which are believed to enhance our mental and physical capacities, do not seem as attractive for some as the idea of using biotechnological means in order to improve our moral conduct.

> Future may depend on making ourselves wiser and less aggressive. If safe moral enhancements are ever developed, there are strong reasons to believe that their use should be obligatory, like education or fluoride in the water, since those who should take them are least likely to be inclined to use them. That is, safe, effective moral enhancement would be compulsory. (Persson and Savulescu 2008)

Savulescu, Persson, and Douglas regard the idea of moral enhancement as very promising for, as they claim, biomedical interventions may result in having "morally better motives". They assume that enhancing the "biological" factor that plays a part in the process of making moral choices, whether through biomedical or genetic interventions, will increase the probability of having "morally better future motives" (Douglas 2008). Thus we need to explore the possibility of using a "science of morality" which would enable one to acquire dispositions that make it more likely that one will arrive at the correct judgment of what is the right thing to do, and also more likely to act on that judgment (Savulescu and Persson 2012). Likewise, Mark Walker postulates the Genetic Virtue Project which proposes to discover and enhance morality using biotechnology genetic correlates of virtuous behavior. His arguments rely on the assumption that virtues have biological correlates.

> The companions in innocence point applies to the idea of promoting virtue: much of our (pre-theoretic) ethical practice assumes that virtues are important. An enormous amount of energy is spent attempting to socialize people into being virtuous, as in teaching children to be truthful, just, and caring. If the "Genetic Virtue Project" is wrong in attempting to promote virtue as a means of making people morally better, then much current socialization and education is mistaken as well (Walker 2009).

All those claims are the preferences about the sorts of world we would like to live in and the sort of children we would like to have (see: Bennett 2009). They go with a general approach to human behavior, which, as it stands, is contingent upon many internal and external factors. Much of the inner tensions that we experience as well as the inconsistency of our behavior, are generated as a result of the conditioned character of our motivational process. The proponents of the Principle of Procreative Beneficence implicitly believe that once we manage to reduce, that conditioned character of our nature through the use of applied science and biotechnological means, we will increase the probability of becoming better persons and improving one's merit (for a critical discussion see: Habermas 2003; McKibben 2004; Kass 2003; Warmbier 2015). Those who oppose biomedical interventions viewed as an alternative method for bringing about a better person do not deny the conditioned aspect of our nature. On the contrary, they take this aspect seriously and thus, holding it as their fundamental context, they attempt to explain the very idea of the "best life", the agent's own good, and self-development.

We all agree that human flourishing is a kind of moral obligation. The supporters of the Principle of Procreative Beneficence attempt to justify their demands for affording the future children the best chance of the happiest life by appealing to the category of the "best life". For some, it might be viewed as a counterpart of the ancient conception of a "good life". But there are reasons for not drawing such a simple analogy too soon. First and foremost, the category of the "best life" which underlies the claims in favor of the utilitarian Principle aimed at enhancing human capacities, stands for a mere possible state of affairs, disregarding the perspective of the particular person one who is involved in this state. The problem that arises here goes to the heart of transhumanists' adoption of impersonal reasons for applying the category of the "best life". The agent's perspective is a central issue within the ancient idea of the "good life". Aristotle notes that "we intuitively believe that the good is something of our own and hard to take from us" (Aristotle 1999, 1095b 26–27). Plato in turn makes this point when he says that "the soul is the source both of bodily health and bodily disease for the whole man (...) So it is necessary first and foremost to cure the soul if the parts of the head and of the rest of the body are to be healthy" (Plato 1973, 156E-157A). For the thinkers of Antiquity, the concept of human flourishing was inextricably related to time. They assumed that self-development and self-fulfillment have little to do with a state which one may reach instantly as a result of one's decision to alter oneself, rather, as they account for, it is a long-term process which requires both a specific attitude of distance toward one's intentions, inclinations, and beliefs, i.e. the ability to call them into question, and an awareness of the grounding of one's choices (see: Warmbier 2017). This is not to suggest that the Greeks were not aware of the fact that some of our physical and intellectual qualities are hereditary and thus are not the outcome of our own efforts and formation, indeed they were, but rather to emphasize that any child, regardless of whether he or she will possess desirable physical and intellectual traits or not, still has a reasonable chance of a worthwhile life.

6.3 Eudaimonic Ethics and the Principles of Reproductive Decisions

The rational approach to the reproductive decision-making, as Harris and Savulescu claim, means that if we do not prevent the birth of an impaired or disabled child, we choose to make the world a worse place than it could have been. We tend to think that impairment precludes a worthwhile life and that disabled people live lives which are overwhelmingly dominated by suffering. Thus, prospective parents should bear in mind that an impaired child who will start a life does not have much of an opportunity to make the best of it. However, I say that disability may not make impossible to have a life worth living. And it is not to suggest that disability is a good thing, but rather to underline the fact that one cannot assess the quality of the future child's life by referring to any particular feature of an embryo. In the previous sections I argued what are the weaknesses of the concept of the "opportunity of the best possible life" in relation to reproductive choice. Now let me turn to the ancient concepts of practical wisdom and *eudaimonia* and elucidate why eudaimonic ethics is to be treated today as a serious alternative to a utilitarian account of the principle that, as some claim, ought to guide reproductive decisions.

I share with ancient philosophers the hope and belief that virtue concepts such as those of *phronesis* and *eudaimonia* have an especially fruitful explanatory power which may provide us with a deep insight into the conditions of one's happiness and a sense of life satisfaction. Aristotle believed that "we may take it as agreed (…) that each person has just as much happiness as he has virtue, practical wisdom, and the action that expresses them (…) Good luck and happiness are necessarily different. For chance or luck produces goods external to the soul, but no one is just or temperate as a result of luck or because of luck" (Aristotle 1998, 1323b 21-29). And this in turn should lead us to conviction that "as far as we can, we ought to (…) go to all lengths to live a life in accord with our supreme element; for however much this element may lack in bulk, by much more it surpasses everything in power and value" (Aristotle 1999, 1177b 34-35 – 1178a 1-2). Virtue ethics has had a profound effect on reflection upon the decision-making process in various fields of applied ethics, affording fresh and distinctively virtue ethical approaches to bioethics and related areas. Some claim that:

> although virtue ethics has sometimes been criticized as offering no practical guidance, its strong entrance into the field of applied ethics suggests that it may be in the area of action guidance that virtue ethics shines brightest (…) Virtue ethicists have paved their way into practical ethics by defending a new conception of what it is for a theory to be 'action guiding', pointing out the naivety of expecting an action-guiding theory to produce solutions to ethical problems by itself, and thinking of action guidance instead as modeling those features of agents that make deliberation and choice ethically responsible and, in a word, done well. On this view, theories guide action not so much by telling agents what to do as by directing their development into ethically mature agents who are better able to tell for themselves what to do. (Russell 2009; see also: Hursthouse 1999; Louden 1990; Broadie 2012)

Plato believed that people reach perfection when they try to reach the good. The "good life" was a central category for ancient ethics. Aristotle believed that realizing virtuous ends is what induces one to the good life in general. Virtue makes one's end the right one, and *phronesis*, which is the excellence of practical reasoning, allows one to recognize or specify the contents of one's end correctly (Aristotle 1999, 1142b 28–33, 1144a 7–10; Aristotle 1915 I.18). The final end is the flourishing of one's life. The good life depends on what one chooses to do and what sort of person one becomes. As Aristotle holds, living a life worth living strongly involves the agency. The point of this dependency is to emphasize the autonomous and individual character of the very process striving for self-fulfillment. When Aristotle talks about the fundamental goal of living well, he introduces the notion of *eudaimonia* (Aristotle 1999 I, 5; I, 7; Aristotle 1973 I, 3). The final end of *eudaimonia* is presented as a reflective framework insofar as one can define the content of that end for oneself and can specify the reasons for pursuing it. As Julia Annas notes:

> The ethics of virtue, as we have seen, takes shape within the framework of a search for an adequate specification of my final end. In ancient ethics this is the entry point to serious ethical reflection; it is taken to be a deep fact about us that we do have such a final end and that when we start reflecting on our lives we do not rest until we have brought the whole of our lives into reflective focus. Further, it is taken for granted that this final end is happiness, though happiness is understood weakly and in an unspecific way. (Annas 1993)

If we bear this in mind, we can see that there is a vast gap here between virtues ethics and utilitarian approaches. The classical ethics which begins from the agent's own good (agent's own concerns and projects) along with its key concepts of practical wisdom and *eudaimonia*, may provide us with a deeper understanding of human behavior and the process of self-development and self-fulfillment. The Aristotelian concept of *phronesis*, which is the virtue of practical reasoning, is a complex disposition and not mere practical intelligence, for *phronesis* is never morally neutral, but it is always bound up with one's final end (with *eudaimonia*), the end which determines one's attitude and normative beliefs. Both of these points are stressed in Aristotle's account of the way in which one acquires the disposition to shape one's patterns of action. The crucial point here is that without one's moral engagement there is little prospect of success in enhancing the processes of cognition, motivation, and action, which would include correcting the patterns of one's emotional reactions, breaking a bad habit and establishing a good one, and, what is most important, understanding the very need for this change. Putting it differently, an effective improvement of one's disposition to arrive at morally right judgements and to act on those judgements might be more likely achieved if we viewed it from the agent standpoint. With this framework at hand, which involves a cognitive-affective conception of virtue along with its core notions of practical reason and eudaimonistic flourishing, agent-based virtue ethics may provide true practical guidance in our moral growth (see: Warmbier 2017). And this approach, despite its weaknesses (see: Louden 1984, 1990), can most definitely be an adequate response to the current debate over the moral obligation to provide the best life possible for the future children.

6.4 Concluding Remarks

The Socratic' question of "How should one live?" or "What kind of person should I be?", which appeals to the issue of self-development and self-fulfillment, cannot be cut off from a plausible conception of the "best life". What may offered new directions forward in the debate on the principles of reproductive decisions is an agent-based virtue ethics that results from attending to the idea of practical reason and related conception of *eudaimonia* and personality. Our sense of self-fulfillment depends on our relationship to our choices and actions. Thus, it does matter whether we are discussing the idea of enhancing one's moral dispositions in terms of a possible paternalistic intervention, or in the context of an agent-centered conception of person. The forms of flourishing ought to be adjusted to our grasp of ourselves – to us as rational agents who view themselves as authors of their decisions and actions. The Principle of Procreative Beneficence may determine the kind of child the prospective parents will have, but not the kind of person their child will be. The question of whether this principle should guide decisions about which of a range of possible future children ought to be brought into the world remains unanswered since it is impossible to make a realistic assessment of the quality of their life by referring to any particular feature of an embryo.

References

Annas J (1993) The morality of happiness. Oxford University Press, Oxford
Aristotle (1915) Works (trans:Smith JA, Ross WD). Clarendon Press, Oxford
Aristotle (1973) Ethics (trans:Ackrill JL). Faber & Faber, London
Aristotle (1998) Politics (trans: Reeve CDC). Hackett Publishing Company Inc., Indianapolis/Cambridge
Aristotle (1999) Nicomachean ethics (trans: Irwin T). Hackett Publishing Company Inc., Indianapolis/Cambridge
Bennett R (2009) The fallacy of the principle of procreative beneficence. Bioethics 23(5):265–273

Bostrom N (2003) The transhumanist FAQ – a general introduction (Version 2.1.). http://www. nickbostrom.com/. Accessed 10 Jun 2016

Bostrom N (2005) In defence of posthuman dignity. Bioethics 19:202–214

Broadie S (2012) Aristotle and beyond. Essays on metaphysics and ethics. Cambridge University Press, Cambridge

Douglas T (2008) Moral enhancement. J Appl Philos 25(3):228–245

Glover J (1992) Future people, disability, and screening. In: Laslett P, Fishkin J (eds) Justice between age groups and generations. Yale University Press, New Haven, pp 127–143

Habermas J (2003) The future of human nature. Polity Press, Cambridge

Hare R (2002) The abnormal child. Moral dilemmas of doctors and parents. In: idem essays on bioethics. Clarendon Press, Oxford

Harris J (2001) One principle and three fallacies of disability studies. J Med Ethics 27:383–387

Harris J (2007) Enhancing evolution. The ethical case for making better people. Princeton University Press, Princeton

Herissone-Kelly J (2006) Procreative beneficence and the prospective parents. J Med Ethics 32(3):166–169

Hotke A (2014) The principle of procreative beneficence: old arguments and a new challenge. Bioethics 28(5):255–262

Hursthouse R (1999) On virtue ethics. Oxford University Press, Oxford

Kamm FM (2013) Bioethical prescriptions to create, end, choose, and improve lives. Oxford University Press, Oxford/New York

Kass L (2003) Beyond therapy: biotechnology and the pursuit of happiness. Prepared for the president's council on bioethics. Regan Books, New York

Kołakowski L (1971) Killing handicapped babies as the fundamental problem of philosophy. In: Choices on our conscience. An international symposium on human rights, retardation and research, Washington, DC

Louden RB (1984) On some vices of virtue ethics. Am Philos Q 21:227–236

Louden RB (1990) Virtue ethics and anti-theory. Philosophia 20:93–114

McKibben B (2004) Enough: staying human in an engineered age. St. Martin's Press, New York

O'Neill O (2007) Autonomy and trust in bioethics. Cambridge University Press, Cambridge

Parfit D (1984) Reasons and persons. Clarendon Press, Oxford

Parker M (2007) The best possible child. J Med Ethics 33(5):279–283

Persson I, Savulescu J (2008) The perils of cognitive enhancement and the urgent imperative to enhance the moral character of humanity. J Appl Philos 25(3):162–177

Plato (1973) Laches and Charmides (trans: Sprague RK). The Bobbs-Merrill Company Inc., Indianapolis/New York

Russell DC (2009) Practical intelligence and the virtues. Oxford University Press, Oxford

Sandel M (2007) The case against perfection. Ethics in the age of genetic engineering. Harvard University Press, Cambridge, MA/London

Savulescu J (2001) Procreative beneficence: why we should select the best children. Bioethics 15(5/6):413–426

Savulescu J, Kahane G (2009) The moral obligation to create children with the best chance of the best life. Bioethics 23(5):274–290

Savulescu J, Persson I (2012) Unfit for the future: the need for moral enhancement. Oxford University Press, Oxford

Schaefer GO, Kahane G, Savulescu J (2014) Autonomy and enhancement. Neuroethics 7(2):123–136

Shakespeare T (1998) Choice and rights: eugenics, genetics and disability and equality. Disability & Society 13(5):665–681

Singer P (2011) Practical ethics. Cambridge University Press, Cambridge/New York

Soniewicka M (2015) Failures of imagination: disability and the ethics of selective reproduction. Bioethics 29(8):557–563

Sparrow R (2007) Procreative beneficence, obligation, and eugenics. Genomics, Society and Policy 3:43–59

Sparrow R (2011) A not-so-new eugenics. Hastings Cent Rep 41:32–42

Stoller SE (2008) Why we are not morally required to select the best children: a response to Savulescu. Bioethics 22(7):364–369

Walker M (2009) Enhancing genetic virtue. A Project for twenty-first century humanity? Politics Life Sci 2(28):27–47

Warmbier A (2015) Moral perfection and the demand for human enhancement. Ethics in Progress 6(1):23–37

Warmbier A (2017) Moral enhancement. Enhancing motivational process and agent-based ethics. In: Warmbier A (ed) The idea of excellence and human enhancement. Reconsidering the debate on transhumanism in light of moral philosophy and science. Peter Lang GmbH, Frankfurt am Main. (in print)

Chapter 7
Context Counts – Bioethics in the Age of Globalization

Aeddan Shaw

7.1 Introduction

Environmental ethics has long been the poor relation to its more glamourous counterparts in bioethics, namely medical and animal ethics. However, as this article will argue, a number of issues in medical bioethics, including reproductive genetics, require a broader environmental context in order to fully understand them, a context which can be afforded by the practice-consequentialism of Attfield. The other different dominant perspectives in medical bioethics, from utilitarianism to virtue ethics, offer competing and equally persuasive accounts and arguments for and against issues ranging from stem cell research to human augmentation. Yet their increasing fragmentation and often attention to a single issue in isolation, however, means that the ramifications and consequences of a given choice of action are rarely considered, often with considerable environmental consequences. Reproductive genetics is almost always considered from an anthropocentric perspective whilst broader bioethical deliberations such as Singer's, which attempt to shed light on them from a pathocentric one, are often considered to be misguided or deliberately provocative. A genuine biocentric perspective accords all forms of life, together with the biosphere as a whole, the right to develop and flourish and therefore be considered in deliberations which may affect them. Since this biocentric perspective is accorded by Attfield's practice consequentialism, a brief outline of exactly why such a perspective is necessary is required.

A. Shaw (✉)
Ignatianum University of Education and Philosophy, Kraków, Poland
e-mail: aeddan.shaw@ignatianum.edu.pl

© Springer International Publishing AG, part of Springer Nature 2018

M. Soniewicka (ed.), *The Ethics of Reproductive Genetics*,
Philosophy and Medicine 128, https://doi.org/10.1007/978-3-319-60684-2_7

7.2   The Need for a Biocentric Perspective

Whilst previous accounts and theories have tended to focus on the here and now, with the long term consequences of deontic acts rarely considered – the greatest good for the greatest number is not for all time but for a particular time and place. In other words, as Attfield has noted, "when the classical texts of ethics from Plato to Kant were written, the impacts of human action were seen as affecting almost exclusively the contemporaries of the agent, and any long term outcomes could be disregarded as serendipitous and unpredictable side effects" (Attfield 2009, 225). Our ethical decisions have far reaching consequences and it is increasingly clear that our lack of consideration for the ecological and generational ramifications of our actions are having a dramatic impact on the world around us. Whilst it may seem odd to charge a perspective in bioethics as being anthropocentric – after all, ethical considerations concern human conduct and actions – on the other, it seems equally strange to deliberate over bioethics without considering the broader context of the *bios* in which they are located, especially when the consequences of the decisions taken may have an impact upon both the environment and future generations of humans. An example may permit this connection to be made more clearly and convincingly. The invention of the female contraceptive pill in the 1950s and its subsequent diffusion from the 1960s has revolutionized birth control and empowered millions of women. It is often touted not only as a tool for liberation and freedom but also as a potentially effective weapon in the fight against global warming, since decelerating the growth of the human population is one clear way of reducing the burden upon the ecosystem. Contraception and birth control campaigns in the developing world have tended to focus on the promotion of female contraception as a way of reducing the typically higher birth rates to be found there. However, a number of studies (e.g. Schwindt et al. 2014) have shown that an artificial estrogen found in the pill, ethynylestradiol or EE2, has been finding its way into the water supply, with devastating impacts on fish populations. Male fish exposed to high concentrations of the hormone typically display intersex features, with considerable reproductive consequences for the species. Greater use, especially in the developing world, could have considerable knock on ramifications, not only for the environment and the economy but also potentially for reproductive health. This raises a potentially troublesome question: should the female contraceptive pill therefore be championed or should other measures be pursued? We will return to this question later on in the paper.

Furthermore, the various competing anthropocentric accounts give the lie that many bioethical conflicts are indissoluble yet, as this paper will argue, the version of practice-consequentialism rooted in a biocentric perspective as championed by Attfield, perhaps allows for new solutions to be proffered. Attfield's practice-consequentialism adds two crucial aspects to be considered when making any deontic decisions: the impact upon the environment and upon future generations, with both united by the key concept of flourishing. Equally, however, such an approach raises in turn some awkward questions which will need to be addressed in the future. Having sketched out the argument for a biocentric perspective, a brief outline of the differences between practice-consequentialism and other varieties of consequentialism is required before we turn to some case studies.

7.3 Practice Consequentialism

Traditional consequentialist theories such as utilitarianism tended to hold that the rightness or wrongness of a particular act may be judged according to one moral value such as the greatest good for the greatest number. However, the main issue with such approaches lies in the fact that they are, as Peterson has said (2013), essentially one-dimensional – they fail to account for other aspects which may have equal weight to the criteria chosen. As a result, a number of competing and contrasting versions of consequentialism have arisen at present which can be broadly divided into multi-dimensional and practice-consequentialist versions and which aim at addressing this problematic aspect. The former, such as those proposed by Peterson or Carter, generally incorporate a number of different perspectives and assign them given values. Carter's multi-dimensional consequentialism, for example, adopts three values as contributing to the overall value of the world: the total number of worthwhile lives, equality and the average level of utility. Peterson offers a subtly different notion of multi-dimensional consequentialism, claiming that Carter's idea is in fact one-dimensional, despite its best intentions (Peterson 2014). His account essentially concludes that multiple aspects, such as moral well-being and equality, need to be considered when judging the deontic value of a given act but that this ultimately means that they are irresolvable. He draws the conclusion therefore that such concepts are essentially non-binary – they are somewhat right and somewhat wrong since any increase in one aspect may lead to a reduction in the value of the other (Peterson 2012). For example, we may choose a course of action which promotes equality but perhaps restricts our well-being. Whilst both Carter and Peterson offer fascinating accounts, the practice-consequentialism of Attfield seems to perhaps offer more as it allows more traditional, "binary" conclusions such as right and wrong to be ventured by affording two crucial dimensions missing from the work of both Carter and Peterson. Firstly, their version of consequentialism does not necessarily consider the impact of moral actions on the wider world – although arguably one could easily add it to their considerations. Secondly, Attfield affords an equally important and connected factor to bear in mind – the impact of our choices upon future generations. As a result, this paper will examine Attfield's practice-consequentialism rather than other types and, having briefly sited Attfield's work in the broader context of environmental ethics, it will proceed to consider some of its applications.

7.4 The Concept of Flourishing

One last clarification remains before putting Attfield's philosophy and biocentrism in the spotlight. It is important to stress what it is not and what makes it distinct, especially as a number of contemporary theories of environmental ethics are underpinned by two particularly problematic assumptions. The first concerns their definitions and understanding of the environment whilst the second are often to be found in their views of the role of mankind. In the case of the former, much of environmentalist thought is indelibly marked by traces of Rousseau and especially the concept of the Noble Savage and the role of nature in mankind's development. The first instance concerns a fundamental presupposition which underpins the perhaps the founders of environmentalism, the Ecosophy of Arne Naess and the Animism of Freya Matthews. Both seem to posit that the activity of man has changed the balance in the natural order and needs to be redressed, returning the Earth to a former, Eden-like idyllic state. By encompassing the Earth as a part of the Self, Naess envisages a greater respect and degree of care for the environment. Whilst such aims are laudable, it is unclear as to exactly how one can reconcile the desires and material needs of 6 billion people with an environment that has already undergone so much change. Whilst the Animism of Matthews does envisage a role for synergy, the meshing of new organisms with existing ecosystems rather than simply returning them to a pristine state, it is unclear as to what can drive this process. In reality, much of the environment is anthropogenic - made and shaped by man – and we have little idea of what an original state of nature would look like. Something like 2/3 of the world's edible crops are manmade, the result of millennia of cross breeding and selection. Many of the existing mammal species are also domesticated variants – there are a startling 1000 types of sheep, all of which have been developed from one 'natural' type. As any biologist will tell you, diversity is the key to survival and man has undoubtedly had a role in ensuring the extraordinary diversity of the world today and a role that Attfield sees as coming with a responsibility to ensure that it flourishes.

The second issue concerns environmental egalitarianism, the idea that all elements of the environment are of equal value, and this is where Attfield's idea of flourishing comes into play. Attfield argues that all species have the right to flourish, by which he means:

> One of the elements present in the flourishing of members of a species, or so I will maintain, consists in the development of those potentials in the absence of which from most of its members a species would not be recognizable as the species which it actually is in our world (Attfield 1995)

Thus, if something helps to develop the potentials which belong to a species, it can be considered to be advantageous. Anything which extends or exceeds these potentials, however, would not be regarded as helping it to flourish since, even if it were to improve them in a way which is advantageous – the ability for pigs to fly in Attfield's example – it would not be helping the individual to flourish, to improve within the limitations of its own capacities. Thus Attfield's idea of flourishing means that we have a responsibility akin to that of a gardener: encouraging the harmonious development of an ecosystem and the individual species within it so that it can support both the gardener and the inhabitants of the garden. It is a poor gardener that takes more than they need, that takes out more from the soil that they return to it – as any farmer will tell you, this ultimately spells problems for future generations.

In environmental ethics, the practice-consequentialist approach is persuasive and this is largely due to its unique contribution in the form of the concept of flourishing. Let us consider some concrete examples from biomedical ethics concerning reproductive genetics and examine to what extent it may be of use in this field, particularly the notion of flourishing. Decisions over whether or not to undertake biomedical research generally focus on utilitarian considerations such as wellbeing or those derived from virtue ethics. The notion of flourishing, however, affords a broader and richer backdrop since rather than considering will this be in the interests of a particular group or individual, it invites the consideration of to what extent this will benefit the biosphere as a whole. Furthermore, flourishing helps the practice-consequentialist position avoid falling into a number of *reductio ad absurdum* type problems. For example, one could claim that in the interest of the environment, it would be better to limit or even cease human reproduction – after all, we cause the greatest damage to the earth's current ecosystem and our removal would undoubtedly be beneficial for many other species. However, thanks to the concept of flourishing, this charge is unfounded as humanity is also accorded an equal right to flourish and develop – our removal would deny us this and not necessarily encourage other species to flourish. This is largely because we are currently living in what has been termed the Anthropocene – we have shaped the Earth to such an extent that our removal would have dramatic consequences for the wildlife and ecosystem as a whole. Just as the loss of a species from the lower rungs of the food chain can have dramatic consequences, so can the removal of those higher up, especially when they play such a crucial role in the shaping of the biosphere as we do. For example, our last remaining 'wilderness', the Amazon, actually appears to be manmade: over centuries the native peoples have planted, cultivated and husbanded the forest to supply a dazzling array of fruits and nuts. Its startling biodiversity is not a matter of 'nature' but rather of nurture, man as *Homo Hortualis,* a constant gardener who has 'improved' his environment rather than destroyed it. No other lifeform can do this, despite the attempts of some[1] to posit that the world possesses an organizing self, a Mother Nature which is capable of shaping and developing a harmonious ecosystem. History is full of examples to the contrary – from the population explosion of the American bison to rabbits in Australia, 'nature' regulates itself about as well as Wall Street does, falling victim to an endless cycle of boom and bust. Mankind, therefore, plays a crucial role in ensuring the flourishing of other species – and should do more rather than less to control its environment.

What then is the lesson here for us? That the role of humans is akin to that of the gardener: encouraging the harmonious development of an ecosystem that can support both the gardener and the inhabitants of the garden. The remainder of this paper will consist of posing a number of questions to the practice-consequentialist camp

---

[1] Such as Freya Matthews.

and exploring a number of case studies, considering especially the impact of certain deontic decisions in the field of reproductive genetics upon the environment and future generations.

7.5   Practice-Consequentialism Case Studies

Let us begin with a first and natural application which may be seen in terms of refining and considering implementations of the precautionary principle in biomedical ethics and particularly in reproductive genetics. The application of the principle in science in general often meets with opposition since, as Stirling has indicated, "some scientists fear that irrational anxieties over particular issues mean that public engagement will lead to indiscriminately technophobic or anti-science results" (Stirling 2012). The precautionary principle seen in light of flourishing would argue that no actions or research should be taken which would have an undue or irreversible impact upon the environment, particularly when it affects the ability of other species to flourish. Let us consider a few concrete examples in reproductive genetics and try to determine whether or not the precautionary principle thus understood should be applied.

7.5.1   In Vitro Fertilization

IVF or In Vitro Fertilization enables couples who would otherwise be unable to have children to reproduce and has been utilized in the last 30 years to bring over 5 million children into the world (Knapton 2016). Disregarding the somewhat facile argument that more people have an adverse effect on an already over strained environment, are there any other reasons for caution to be applied in the utilization of IVF? There have been some concerns that IVF could negatively impact fertility or that so-called test tube babies may have heightened risks of congenital illnesses or even shorter life expectancy. A recent longitudinal study in Sweden revealed the following:

> Outcomes of pregnancies after IVF were studied in Sweden over a period of 25 years and revealed a decrease of multiple pregnancies, a decrease of preeclampsia and premature rupture of membranes, and an increased risk for cerebral palsy, possibly for attention-deficit and hyperactivity disorder, for impaired visual acuity and for childhood cancer, although stressing that these outcomes were generally rare, even after IVF. (Harper et al. 2013)

Thus despite the slightly heightened rates of childhood cancer and decreased visual acuity, IVF does not seem to have any other broader consequences either for the environment or human reproduction per se. The 5 million otherwise (literally) unconceivable IVF babies and their right to flourish, therefore, seems to clearly outweigh the potentially heightened risks involved. The other controversial aspect of IVF regarding discarded embryos is also not particularly problematic for practice-consequentialism since, although it means some embryos are unable to be implanted and come to term (and thus denied the chance to flourish), to prohibit IVF would also do so, together with those who are chosen to be brought to term.

7.5.2 Pre-implantation Genetic Diagnosis

Peter Sykora, in this volume (2016), cites the example from gene therapy that lay people may advocate banning gene therapy outright based on a misunderstanding of whether it affects the germline or not. Rightly or wrongly, practitioners and researchers often seek to reserve judgement over the direction of future research but their own considerations are often limited to the immediate impact of the research, overlooking or deliberately ignoring some of the potential consequences. Practice-consequentialism, on the other hand, is primarily concerned with those very consequences. In this volume, others have written about Preimplantation Genetic Diagnosis (PGD), where embryos are tested prior to implantation. The debate has tended to focus around either utilitarian arguments or those from virtue ethics, depending on where the person stands, resulting in a stalemate of sorts. Practice-consequentialism potentially affords the opportunity to "zoom out" and consider the debate in the broader context of the biosphere and future generations. PGD would at first seem to be relatively unproblematic for practice-consequentialism since it allows the embryo with the greatest potential to flourish to be chosen. Yet does this not lead to a slippery slope type situation, wherein presumably the practice-consequentialist would advocate wide ranging screening and testing of embryos in order to ensure only those with the greatest capacity to flourish are chosen? In other words, does practice-consequentialism lead us to eugenic conclusions?

This, I believe, would be to misjudge and misunderstand the practice-consequentialist position and especially the role of flourishing. It should be recalled that this is with regard to its capacities – there is nothing with regard the ideal capacities of the species as a whole, merely the capacities of the individual. If, therefore, a decision is to be made in regards to the embryo to be implanted with the others discarded, the practice-consequentialist would surely advocate the selection of the embryo most likely to flourish the most. If, on the other hand, all embryos are to be retained and implanted, there is nothing within practice-consequentialism that requires embryos which are less likely to flourish to the full capacities of the germline species to be discarded.

7.5.3 The Female Contraceptive Pill

In closing this section, let us return to our example from earlier in the paper concerning the female contraceptive pill. Do the unquestionable benefits of the pill outweigh the potentially negative side effects for practice-consequentialism? Here the issue seems to be less clear cut. The pill undoubtedly allows women to have greater reproductive freedom, to avoid unwanted pregnancies and thus affords them a greater capacity to flourish in their personal lives. On the other hand, there is the link with the negative impact upon other species and, perhaps less clearly proven to date, with human reproductive capacity. It seems to me that practice-consequentialism would still advocate for the championing of the pill but in conjunction with improved filtration and water treatment systems, particularly in conurbations where the pill is more widely used. Thus practice-consequentialism would not argue for the discontinuation or banning of the pill but rather a heightened awareness of the environmental impact that it may have and an accompanying plan to mitigate its adverse effects.

7.6 Some Broader Considerations

Finally, let us consider another even more problematic example, that of biomedical research. What would practice-consequentialism advocate in a situation when a potential cure has been found for a serious condition but the active ingredient is only to be found in a plant which grows in a relatively limited geographical area? Let us take the EBC-46 compound as our example, derived from the blushwood bush found in the Atherton Tablelands of Northern Queensland in Australia (BEC Crew 2014). Expense, environmental restrictions and costs prohibit making the drug synthetically or in controlled conditions yet the drug has the potential to effectively combat many inoperable tumors. Does this potential application justify turning this incredibly diverse region into a monoculture devoted to the production of one plant? This example would seem to present practice-consequentialism with a conundrum as it would presumably argue in favor of retaining the diversity of the region at the expense of the potential cure. Too many species are denied the opportunity to flourish in order to aid a number of representatives of one species and thus the practice-consequentialist would not advocate the large-scale production of the plant, at the expense of providing a cure for those suffering from certain cancers. This may seem conclusive until we factor in the aspect of future generations: if this treatment has the ability to effectively rid mankind of cancerous skin tumors, such a course of action would be justified.

At this stage, one needs to be careful in utilizing the argument for future generations and one should stress the crucial aspect of flourishing. One could equally argue that turning this particular area into a golf course would provide satisfaction and pleasure to generations of golfers as yet unborn and thus the increase in utility would also merit the decision to destroy this particular ecosystem. However, whilst golf undoubtedly provides pleasure to some, it does not constitute something which contributes to the flourishing of individuals or helps us to reach our full potentials whilst the avoidance of lengthy medical treatment or even the fatal consequences of some skin tumors certainly does.

Another important area concerns the scope for determining future biomedical research, especially with regards its focus. At present, biomedical research can be divided into two main directions and motivations. The first seeks to extend life by combatting certain typically late-onset illnesses such as heart disease, cancer or diabetes whilst the second is devoted to exploring potential cures for conditions which may impact at any stage in an individual's life, anything from Crohn's to glaucoma. Given the broader context outlined above, practice-consequentialism would presumably argue for a shift in focus in biomedical research from seeking to extend life to enhancing the quality of a "natural" lifespan. Let us examine the reasons for this in a little more detail.

Biomedical research can undoubtedly help to ensure that the individual can flourish to the maximum possible level, enhancing the quality of life for the individual in question. As we grow older, the uncomfortable truth is that our bodies gradually begin to fail us – to wilt and die to use the gardening metaphor which seems to naturally accompany the notion of flourishing. Extending our lifespan does not seem to lead to greater flourishing – indeed, the opposite is almost true, with more conditions and issues arising as we grow older from Parkinson's to dementia. We are increasingly able to keep our bodies together with operations, drugs and treatment but despite our best efforts, our minds are often unable to keep up. I would not argue that practice-consequentialism advocates euthanasia – again, this would be to grossly misunderstand the position – but rather it would seek to draw a line under what exactly are our biomedical priorities. Does the consideration of future generations impact upon this as well? Unlike the ecosystem example above, it would seem that it does not as future generations would essentially face the same conundrum as our own: to prolong life or to improve the quality of that which we have. Whilst it may seem like a jagged little pill to swallow, practice-consequentialism would argue for the latter.

7.7 Conclusions

This paper has sought to outline some potential responses that a practice-consequentialist position might afford to some enduring debates in bioethics and especially the crucial concept of flourishing. Whilst at first they may seem to be strange bedfellows, augmenting bioethical considerations with an environmental dimension seems to make the previously intractable much more achievable. Although they remain "non-binary" in Peterson's memorable term, they are arguably much more persuasive than those which lack the biocentric perspective.

References

Attfield R (1995) Value, obligation and meta-ethics. Rodopi BV Editions, Amsterdam
Attfield R (2009) Mediated responsibilities, global warming, and the scope of ethics. J Soc Philos 40(2):225–236

BEC Crew (2014) Super-effective cancer-fighting berry discovered growing in Australia. Science Alert 7 Oct. http://www.sciencealert.com/news/20140710-26300.html. Accessed 23 June 2016

Harper JC, Geraedts J, Borry P et al (2013) Current issues in medically assisted reproduction and genetics in Europe: research, clinical practice, ethics, legal issues and policy: European Society of Human Genetics and European Society of Human Reproduction and Embryology. Eur J Hum Genet 21(Suppl 2):S1–S21. doi:10.1038/ejhg.2013.219

Knapton S (2016) Test tube babies could die sooner. The telegraph 16 Feb. http://www.telegraph.co.uk/news/science/science-news/12158072/Test-tube-babies-could-die-sooner.html. Accessed 6 Sep 2016

Peterson M (2012) Multi-dimensional consequentialism. Ratio 25:177–194

Peterson M (2013) The dimensions of consequentialism. Cambridge University Press, Cambridge

Peterson M (2014) Consequentialism and environmental ethics. Review of: Hiller A, Ilea R, Kahn L (eds.) Consequentialism and environmental ethics, Routledge. Notre Dame Philosophical Reviews 04(33). http://ndpr.nd.edu/news/48009-consequentialism-and-environmental-ethics-2/. Accessed 22 June 2016

Schwindt AR, Winkelman DL, Keteles K, Murphy M, Vajda AM (2014) An environmental oestrogen disrupts fish population dynamics through direct and transgenerational effects on survival and fecundity. J Appl Ecol 51:582–591. doi:10.1111/1365-2664.12237

Stirling A (2012) Opening up the politics of knowledge and power in bioscience. PLoS Biol 10(1):1–5

Sykora P (2016) Germline gene therapy in the era of precise genome editing: how far can we go? In: Soniewicka M (ed) The ethics of reproductive genetics – between utility, principles, and virtues. Springer, New York

Chapter 8
Conscientious Objection of Health Care Workers in the Context of Genetic Testing

Jakub Pawlikowski

8.1 Introduction

Human dignity, based on conscience and reason, is the foundation of any system of positive law and the source of inalienable human rights and freedoms. Freedom of conscience occupies a special place among the fundamental human rights and freedoms, and is a prerequisite for the development of every human person and the basis for the existence of a responsible democratic society that creates a state, based on justice and moral integrity of citizens (Sulmasy 2008). Freedom of conscience is not only the right to create and possess one's own beliefs (*forum internum*) but also the right to manifest his or her beliefs (*forum externum*). Consequently, the right to conscientious objection is therefore a manifestation of personal beliefs and originates directly from freedom of conscience (Evans 2003). It should be emphasized that conscientious objection as a fundamental human right does not result from the established laws, but directly from constitutional provisions and various international agreements on freedom of conscience (e.g. Art. 18 of the International Covenant on Civil and Political Rights, art. 10 of the Charter of Fundamental Rights of the European Union, art. 18 of the Universal Declaration of Human Rights of the United Nation). The right to conscientious objection can be limited only by an enactment of law (e.g. included in conscience clauses), if it is necessary for the protection of the fundamental rights and freedoms of others and ensuring national security, public order, health and morality (art. 18 of the ICCPR). The imposed restrictions cannot simultaneously violate the principle of equality in the face of law, non-discrimination and ideological impartiality of the public authorities.

J. Pawlikowski (✉)
Medical University of Lublin, Staszica 4/6, Lublin, 20-059, Poland

Harvard T.H. Chan School of Public Health, Harvard University, Cambridge, MA, USA
e-mail: jpawlikowski@wp.pl

© Springer International Publishing AG, part of Springer Nature 2018
M. Soniewicka (ed.), *The Ethics of Reproductive Genetics*,
Philosophy and Medicine 128, https://doi.org/10.1007/978-3-319-60684-2_8

This is particularly essential in the context of determining the obligations of healthcare professionals in relation to patient's rights.

The concept of conscientious objection may have various content ranges. In its broadest sense it is an opposition to the existing legal regulations, and even social and political ones (e.g. in the form of civil disobedience) independent of the existence of legal codification of objection (an extralegal dimension). In the legal doctrine, it is possible to distinguish two understandings of conscientious objection: broad and narrow. In the first case it will refer to both action and omission of an action, and in the second only to the refusal of performing certain actions based on religious, ethical and philosophical premises (Pawlikowski 2014; Evans 2003).

Conscientious objection should not be confused with the conscientious clause. Conscientious objection should thus be understood as a moral stance (this problem is timeless, e.g. "Antigone" by Sophocles), whereas a conscience clause is a legal structure regulating the expression of conscientious objection. The conscience clause contains provisions and regulations that both confirm and limit conscientious objection due to other important values such as national security, or life and health of the patient. The conscience clause usually creates a legal possibility of refusing legally imposed obligations because of their conflict with religious beliefs and moral values, and can also indicate other, substitutive forms of fulfilling of this obligation. Initially this found application in relation to the military service by providing an opportunity of fulfilling this service in a different form. Along with the process of transforming the armed forces into a professional army, the clause lost its importance in this area (however in some countries it is currently important, e.g. Bayatyan v. Armenia). It must be noted that the broad contemporary discussion on conscientious objection concerns also other professions, e.g. judges, lawyers, officials, teachers.

8.2   Conscientious Objection in Medical Practice

The problem of the conscientious objection of healthcare workers has recently become the subject of serious interest from both the scientific and socio-political perspective. This in turn resulted in a dynamic growth of scientific literature on this subject, the increase of stances of various expert groups, international resolutions (e.g. Resolution 1763 of Council of Europe, The right to conscientious objection in lawful medical care, 7 October 2010), legislative initiatives, and courts' and tribunals' rulings. In the Pubmed-Medline database there are over 250 articles on the conscience clause, over half of which are from the past 10 years. In the medical literature, the subject of conscientious objection began to appear in 1953, however the oldest articles do not apply to healthcare workers, but to the general requirement of the vaccination of children and the parent's rights to evade this obligation (e.g. Knack 1953). In the 1970s and 1980s, the discussion focused on the physicians' and other healthcare workers' refusal to perform certain procedures. It coincided in time with the decriminalization and legalization of abortion and other procedures that could harm the principle of respect to human life.[1]

In a global sense, conscientious objection in medical practice can occur for a number of procedures related to both the initial period of human life (abortion, prenatal diagnosis, postcoital contraception, intrauterine devices and certain methods of assisted reproduction- particularly those that lead to the creation of supernumerary embryos and preimplantation selection) and its end (euthanasia, medically assisted suicide, participating in death penalty), but it also applies to situations such as brain death criteria, transplantation or blood transfusion (e.g. Jehovah's Witnesses), refusal to discontinue persistent therapy (some orthodox Jewish movements) and the use of drugs and vaccines, whose production is based on embryonic cells and fetal tissue (Shanawani 2016; Garcia Calvente and Lomas-Hernandez 2016; Zampas and Andion-Ibanez 2012; Orr 2013; Kato 2013; Boehnlein 2013; Magelssen 2012; Wicclair 2011; Pope 2010; Harris et al. 2011; Lawrence and Curlin 2009; Lawrence et al. 2010; Pawlikowski et al. 2010; Curlin et al. 2008; Wicclair 2008). Conscientious objection can be raised by both non-religious and religious people originating from different traditions such as Christian, Jewish, Muslim, Buddhist and others (Schenker 2005; Sachedina 2005; Keown 2005; Wenger and Camel 2004).

Apart from physicians, conscientious objection also concerns nurses, pharmacists (refusing to sell contraceptives, anti-implantation and abortive drugs) and laboratory diagnostics (e.g. diagnosis for the in-vitro procedure) (Davis et al. 2012; Duffy 2010; Brock 2008; Spreng 2008).[2] In literature, this issue is also discussed in the context of support staff (e.g. a paramedics transporting a patient to an abortion clinic), administrative staff (preparing and issuing documents being the basis for termination of a pregnancy) and even medical students (participating in an abortion, teaching physical examination to Muslim students in the context of the ban on touching the opposite sex established in the law of Islam) (Card 2012; Strickland 2012). Also observed are patients' (or their parents'/guardians') objections towards some medical interventions (vaccination, newborn screening tests) (Beard et al. 2016; Drabiak-Syed 2010; Bailey et al. 2009).

It should be emphasized that the healthcare workers' refusal to perform certain procedures is not very common and there must be special conditions and premises for it to occur. For instance, the Polish Ministry of Health in 2013 send a letter to 406 hospitals, in order to collect data relating to the number of doctors referring to the conscientious clause in case of: a) refusing to perform abortion, b) refusing to issue a referral for prenatal testing. Replies were received from 375 entities, of which only three hospitals indicated that a doctor relied on the "conscience

---

[1] In the countries of the Eastern Bloc, these discussions did not reverberate, because in the totalitarian regimes freedom of conscience was not treated as a natural right of every human being, but as a civil right, the scope of which has been authoritatively determined by the state.

[2] In the Republic of Poland, the right to conscientious objection is legally guaranteed to physicians, nurses and midwifes, however the demand for including the conscience clause in other professional acts is also reported among other medical professionals, especially pharmacists and laboratory diagnosticians.

clause". These data may indicate that the conscience clause is rarely referred to or that there is a different way of dealing with these situations not necessarily according to the scheme described in the conscience clause (Report of Government Council of Republic of Poland 2015).

The motivation behind refusals is multifaceted, and can be the cause of both ordinary difficulties in doctor-patient (or his family) communication (e.g. about the treatment method), or may result from other premises such as lack of medical justification for the procedure (e.g. use of antibiotics in viral infections, ineffective and persistence treatment in the end of life), organizational and legal restrictions (e.g. exceeding limits of benefits contracted with the National Health Fund) and moral or religious beliefs. In the context of the conscience clause, the most important issue are those resulting from moral (which may be independent from religious beliefs) or religious premises (e.g. doctrinal teaching of religious community of which the person is a member) (Pawlikowski 2014). Apart from these moral or religious aspects, there may appear substantive points arising from different professional experiences and assumptions concerning the functioning of various organs and structures (e.g. reactivity of the nervous system), which in turn may lead to objections towards some standards of conduct (e.g. diagnostic criteria of brain death or indications for blood transfusion) (Kato 2013).

In the context of genetic testing, the most important issue connected with the conscientious objection of health care workers are preimplantation and prenatal diagnosis. However, it should be noted that from the patient's perspective the genetic testing screenings of newborns are also raised as discussed question. The newborn screening is frequently done as a preventative health measure for the clear benefit of the child because currently early treatment for some of the conditions is available (e.g. phenylketonuria and congenital hypo-thyroidism). It is particularly difficult to establish whether genetic testing should be made in non-curable and non-preventive diseases or whether the state can obligate the practitioner to perform additional research. History has witnessed complaints issued against the state health department for not obtaining parents' consent for the storage and use of their child's samples, indicating this action as a violation of their child's privacy and property rights. In *Bearder v. Minnesota*, the lawsuit was dismissed and the state health department continues to collect, retain, and use samples unless parents refuse. In *Beleno v. Tex. Dept. of State Health Services* an agreement was reached whereby the state health department was obliged to post research results of samples, inform the plaintiff of how their child's samples were used and to destroy over four million NBS it had obtained without parental consent (Drabiak-Syed 2010; Bailey et al. 2009; Bailey et al. 2008).

8.3 Conscientious Objection in the Context of Preimplanation Diagnosis

Preimplantation Genetic Diagnosis (PGD) is a combination of techniques in genetic testing performed on embryos during in vitro fertilization (IVF). PGD was developed for testing the genetic mutation/malfunction thus relieving parents of the decision on abortion. The classic IVF method is composed of several stages: controlled ovarian hyperstimulation, egg retrieval; collection of semen; insemination; fertilization and embryo culture; assessment of embryo's development and (usually) preimplantation selection on the second or third day after insemination, the transfer of 1–2 embryos (in previous years, more than two embryos were transferred) into the female's uterus (there may be several transfer attempts).

Although most in the medical community accept in vitro fertilization, it is not uniform in its attitude (about 15–17% physicians raise objection) (Pawlikowski et al. 2010). One of the reasons for the opposition of IVF is the preimplantation diagnosis, which is enumerated as one of the major problems associated with IVF (Draper and Chadwick 1999). Other most frequently raised problems relate to the additional embryos, children's rights (including the right to know their genetic identity), parental rights (in case of gametes donation and a "surrogate mother"), market trade in gametes, the possibility of having children by homosexual couples, single women, women beyond the conception age; *post mortem* fertilization, the creation of embryos for instrumental reasons (e.g. research), harmful environment for development of the embryos, because which may disturb the epigenetic processes in the first hours after fertilization leading to a higher incidence of certain diseases (Esteves and Bento 2013; Diasi and Maher 2013; Goold 2005; Devolder 2005; Schenker 2005). Although discussions have weakened slightly in recent years, however IVF is still seen as a procedure, which cannot be forced upon the medical conscience (Brakman and Fozard 2008; Schotsmans 1998).

Preimplantational diagnosis may be a subject of conscientious objection sometimes independent of the acceptance of the IVF procedure itself. The most important reason for the objection to PGD is the association of this procedure with embryo destruction. Those who object to this procedure raise the risk of potentially killing a healthy human being and state that the selection of human beings is contrary to the basic principles of medical ethics (Midro 2013). It causes moral dilemmas among those who consider an embryo a human being and not separate the moment of conception from the moment of hominization. Doubts are also raised by PGD predictive value (risk of false positive results). Some reports indicate that its application does not always increase the number of pregnancies, what is more, the embryos in their preimplantation phase frequently have a mosaic karyotype (some cells are normal, some are abnormal) and during further stages of the development the organism may eliminate the abnormal cells, leaving the initial diagnostic test a false positive (Mertzanidou et al. 2013). Conscientious objection to PGD may, however, be reinforced by religious reasons. This applies particularly to the Catholic Church, which has a negative opinion on this issue (*Congregation for the Doctrine of Faith* 1988; 2008). Other faiths and religious traditions do not occupy such a restrictive stance. However, members of these religions may turn to conscientious objection in the case of the destruction of embryos (Sivaraman and Noor 2016).

Conscience objection in the context of PGD is usually a logical consequence of the principle of respect for human life from the moment of conception. It can be assumed that PGD would not cause controversy, if it was carried out on gametes (e.g. oocytes) or for the benefit of embryos e.g. early treatment (currently only in hypothesis). Objections towards preimplantation diagnosis are not raised frequently, as in many countries IVF procedures are not a part of standard medical care, and thus are beyond the scope of physicians' obligations (they are usually performed by specialized infertility treatment centers where the personnel accepts it). If IVF in conjunction with PGD was a part of the services provided in gynecological and/or obstetric wards, or as an obligatory part of the pre- or postgraduate course, it would probably more frequently be the subject of conscience objection.

8.4 Conscientious Objection in the Context of Prenatal Genetic Testing

The problem of conscientious objection frequently appears in the context of prenatal genetic testing – a technique which is performed on a fetus. Prenatal diagnosis and associated genetic counseling, in principle, determines the risk of the offspring having genetic or congenital defects. It is also aimed at the detection of a defective gene, predicting the effects of the disease, providing information on the treatment and care possibilities for children with disabilities in order to improve their and their family's quality of life. This is undoubtedly the main purpose of prenatal diagnosis. Prenatal diagnosis in its essence is a medical diagnostic method and, as such, raises little controversy. However, the moral assessment of prenatal testing can be different and it depends on the aim of the diagnosis, the risk to the fetus and the validity of the testing (Fraczek et al. 2013).

Prenatal diagnosis is based on several techniques that can be divided into the invasive (amniocentesis, fetoscopy, cordocentesis, trophoblast biopsy) and non-invasive (ultrasound, biochemical testing of the markers and fetal cells in the maternal serum) and is a multistage process. The first stage is based on imaging (typically ultrasound), during which an initial diagnosis can be established and invasive diagnosis continued (e.g. amniocentesis). The diagnostic process involves both gynecology-obstetrics specialists (mainly imaging non-invasive diagnostic) and specialists in clinical genetics (invasive diagnosis, interpretation of results), and sometimes other specialists (e.g. cardiologists). It is also necessary to mention laboratory diagnostics, whose role in obtaining biological material, conducting research and obtaining results is not to be undervalued.

The positive aspect of prenatal testing is the possibility of the development of perinatology and prenatal surgery. Currently, drug therapy is possible in the direct treatment of the fetus (e.g. hyperthyroidism and hypothyroidism), in the stimulation of lung maturation, and surgery of the fetus in the case of e.g. obstructive uropathy, diaphragmatic hernia, lung lobe sequestration, spina bifida, hydrocephalus, twin reversed arterial perfusion sequence or some congenital heart defects (Fraczek et al. 2013). Prenatal diagnosis is often for the benefit of the child's life, not only through therapeutic possibilities but also through the fact that the information about a healthy child removes the parent's anxiety and leads to the continuation of the pregnancy. The above goals of prenatal diagnosis do not cause ethical controversies; they are widely accepted by the medical community and the society. A positive consequence of prenatal testing, particularly imaging, is the greater empowerment of the dignity of human being before birth.

With time, prenatal diagnosis began to be also seen as a tool for eliminating sick fetuses and an important element of "procreative autonomy of women", which allows them to decide on continuing or terminating a pregnancy. This in turn causes serious moral dilemmas and conscientious objection among some medical professionals. This objection occurs most frequently in relation to issuing referrals for abortion based on the genetic test results, less frequently in relation to the referral for more, in-depth invasive testing and least frequently in relation to the idea of prenatal diagnosis.

Conscientious objection to prenatal diagnosis is mainly caused by the possibility of abortion in the case of the detection of a severe or fatal condition. This relationship was already noticed years ago by J. Watson, winner of the Nations Nobel Prize in medicine, who noted that genetic testing may lead to abortion in cases of a fetal defect. This is confirmed by reports of the Polish Government, which indicate that almost all cases of legal abortion (95%) are the outcome of discovering prenatal birth defects, and what is more this trend is escalating (Report of Government Council of Republic of Poland 2015). The association between prenatal testing and abortion is also visible in the public opinion (Morton 2016). In addition, prenatal diagnosis is sometimes associated with a form of eugenic ideology. For these reasons, some doctors, including specialists in clinical genetics, express their objection (Midro 2014). The conscientious objection of healthcare workers in the context of prenatal genetic tests springs from the risk of violating a human being and lack of axiological consistency between prenatal selection and postnatal duty of solidarity with regard to children burdened with a disease. In recent years, as an alternative for abortion in case of detecting severe and incurable disease, is the emergence of perinatal hospices, which encompasses medical, psychological and spiritual care to parents of children with lethal diseases in prenatal and perinatal period (Dangel 2009; Leong Marc-Aurele and Nelesen 2013; Balaguer et al. 2012).

Among other reasons of conscientious objection is the possibility of diagnostic errors caused by false-positive genetic test results (e.g. in genetic mosaicism cases). The probabilistic nature of diagnostic methods in medicine is not a problem in itself, as further diagnosis, treatment or observation allows for the verification and refinement of the diagnosis. However, in the case of prenatal genetic tests, the possible consequence in the form of abortion is an irreversible event. Another problem is the questionable credibility of the prognosis of the child's development, because it overestimates the influence of the genetic factors and underestimates the role of environmental ones. There is no absolute genetic determination and mental retardation may be to some extend a consequence of social rejection (e.g. in Turner syndrome) (Midro 2014). Parents deciding on the continuation or termination of the pregnancy are usually unaware of the probabilistic nature of the genetic diagnosis.

Among the reasons against prenatal diagnosis is also the risk to the health and life of the child. There are also arguments that with such poor therapeutic possibilities for children in the prenatal stage, diagnosis in this aspect may be a sign of overdiagnosis and may cause unnecessary psychological burden to both parents and doctors (Midro 2014). The doubt arises also on the grounds of the interpretation of the category of the severity of the abnormality (Green 1993). In some countries prenatal genetic testing is conducted to aid sex selection (e.g. Korea, India, China), which is illegal in Europe (Wolf 1996).

Recommendations for doctors performing prenatal tests indicate that on the basis of test results, they are to provide "non-directive advice" i.e. enumerating possibilities of further proceedings, but not indicating any specific actions. The scope of this information is not controversial in regard to treatment options or further care; however objections are caused with the information on the possibility of terminating a pregnancy and the assumption that a genetic advice can be neutral and non-directive is considered to be unrealistic, because in practice it is difficult to maintain an attitude of axiological neutrality (Pellegrino 2002).

Religiously engaged healthcare workers may also shape their attitude towards prenatal testing based on the teachings of their religious community. Not every religious tradition has an official stance on prenatal testing. Most opinions in this regard can be found in the teaching of the Catholic Church. For instance, the Catechism of the Catholic Church (Art. 2274) states that prenatal diagnosis is morally licit, "if it respects the life and integrity of the embryo and the human fetus and is directed toward its safe guarding or healing as an individual. (…) It is gravely opposed to the moral law when this is done with the thought of possibly inducing an abortion, depending upon the results: a diagnosis must not be the equivalent of a death sentence". Article 2275 in turn states that "one must hold as licit procedures carried out on the human embryo which respect the life and integrity of the embryo and do not involve disproportionate risks for it, but are directed toward its healing the improvement of its condition of health, or its individual survival". John Paul II in *Evangelium vitae* said that prenatal diagnostic techniques are morally permissible "when they do not involve disproportionate risks for the child and the mother, and are meant to make possible early therapy or even to favor a serene and informed acceptance of the child not yet born" (n. 63). Christian ethics is therefore in favor of carrying out prenatal testing for therapeutic purposes, but does not accept them if they are to be a prerequisite for a decision on terminating a pregnancy.

In Catholic moral theology it is emphasized that from an objective point of view, referral for prenatal genetic testing and their very conduct is not directly related to a possible abortion, which in order to be carried out requires a positive test result and an additional separate medical referral. In this case, when there is uncertainty about the woman's final decision, conscientious objection is not a moral duty. On the other hand, when a doctor is certain that the result of the prenatal testing may contribute to the abortion (e.g. abortion is a direct consequence) the doctor should object to this procedure (*Congregation for the Doctrine of Faith* 1988, n. I 2).

Dilemmas of the medical conscience in the field of prenatal testing deepen through specific legal regulations which treat prenatal testing as patients' rights and physicians' obligation, regardless of the purpose of the test. The regulations usually do not directly associate prenatal diagnosis with neither therapeutic nor eugenic aims, but the context indicates the possibility of associating these regulations with the possibility of conducting abortion on the basis of finding fetal defects.

Women's' rights to perform prenatal tests have resulted in the emergence of lawsuits for damages relating to the conception and birth of a child with disabilities (wrongful birth, wrongful life, and wrongful conception). In the case of conscientious objection to prenatal testing, the doctor may be exposed to a complaint for a wrongful birth. The physician may be accused of breaching the obligation of performing prenatal tests or misinterpretation of their results. Lack of a causal relationship between the physician's action or its omission and the resulting effect in the form of damages does not cause a liability for pecuniary damage, but it can be the basis for compensation for the pain associated with the violation of a patient's right to health care benefits and the right to health information.[3] On the other hand, it is worth mentioning that article 3 of the European Charter of Fundamental Rights contains the prohibition of eugenic practices aimed at the selection of persons, although the discussion is whether or not this rule applies to unborn children.

In literature, there is a dispute as to whether prenatal tests can be the subject of conscientious objection. On the one side, under the statutory prerequisites of the conscience clause, it can be legal (Gałązka 2014), but on the other hand it could be a violation of the patient's right to information and to receive diagnostic services. It is currently discussed, whether the relationship between the activity performing of which is refused by an entity (prenatal diagnosis) and other activity which it opposes (abortion) may justify refusal. In the case of *Jean Bouessel du Bourg v. France* (n. 20747/92) the European Tribunal of Human Rights pointed out that a taxpayer's refusal to pay tax due to the fact that his money may be spent by the state to finance practices which are not accepted by the taxpayer in question (e.g. abortion) cannot be construed as a realization of the right to conscientious objection, since the link between paying tax and termination of a pregnancy is too far reaching. However, in case of prenatal testing and abortion, conscientious objection can be justified, because it is impossible to abstract from easy to reconstruct cause-effect relationships, physician's life experience, as well as common sense. Thus, if a doctor determines that an activity determines an action that in his opinion is unacceptable and considers the implementation of its fulfillment as possible, than his referring to the conscience clause should also be considered possible (it confirmed The Polish Highest Court Tribunal in its judgment from June 12, 2008 (III CSK 16/08) and Polish Constitutional Tribunal in its judgment from October 7, 2015 (K 12/14), abolishing the duty to perform procedures that are not done for medical purposes and the necessity of indicating a different entity). Probably most imaging prenatal diagnostic tests in phase 1 will continue to be done in most outpatient gynecological-obstetric centers, while there may be a significant number of refusals towards directing patients to invasive genetic examinations in phase 2, which should lead to contracting these examinations only in selected entities.

---

[3] For the first time, a compensation claim appeared in 1967 before the Court of Appeal of New Jersey in Gleitman v. Cosgrovecase. A very famous case in European Tribunal of Human Rights was R.R. v. Poland (n. 27,617/04).

Summarizing, the moral evaluation of prenatal diagnosis which manifests itself in the conscience objection to a great extent depends on the aim of diagnosis, safety and consequences. After a period of doubt as to its admissibility, caused by the fear of harmful effects associated with the examination technique, these objections began to decrease due to the improvement of their safety. Currently, most important in the assessment is the matter of the purpose of conducting such examinations. On one hand, parents are entitled to full information about their child's development, which is also associated with information on his/her diseases and treatment options in the prenatal period (and this is not the subject of contestation). On the other hand, the doctor may feel responsible not only for the parent's rights, but also the children's rights including his basic right to live, which in the conditions allowing abortion due to genetic defects may cause conscientious objection. The limitation of obligatory prenatal testing for only medical purposes (leading to treatment and care for parents and children) would contribute to solving the problem of conscientious objection in prenatal genetic testing.

8.5   Conclusions

Conscientious objection in medical practice results mainly from situations which violate the principle of respect for human life and the principle of non-maleficence (*primum non nocere*). The objection to the destruction of human life, mostly relate to the prenatal and terminal phase, which is visible in case of abortion (especially for non-medical reasons), euthanasia, medically assisted suicide, or participating in death penalty, but also lies at the root of some of the doubts raised against prenatal diagnosis (if it leads to abortion) or destruction of embryos in preimplantation diagnosis. The principle of non-maleficence is one of the motivation elements of conscientious objection in case of invasive prenatal testing and preimplantation diagnosis.

Many controversies in the moral assessment of preimplantation and prenatal genetic testing derive from the various criteria of hominization. For many healthcare workers, ontologically essential qualitative change from a biological perspective occurs in the process of fertilization, when the two haploid parental cells form a new diploid biological organism. This criterion is identified sometimes as the genetic one because at that moment specific genetic information is created, basic biological features are determined, and the organism development that has started in this moment is a continuous one. However, there are proposed other criteria of being a human person (e.g. the criterion of implantation, sufficient body organization criteria, brain development criteria, self-awareness etc.), which lead to various assessments of undertaken actions during the prenatal or the preimplantation period of the human development (Cooper et al. 2007; Warnock 1988). Legislators and courts often avoid speaking about the beginning of human life, however sometimes they waive this rule, e.g. the ruling of the Court of Justice of the European Union from October 18, 2011 in the *Oliver Brüstle v Greenpeace* (C-34/10) where the court decided that every human embryo, independent of the method it was conceived in (e.g. cloning) is endowed with human dignity.

The principle of respecting human life and the principle of non-maleficence play a special role in everyday work of a clinician and in shaping the medical ethos, but they are also universal principles, present in the majority of moral systems both on a religious and non-religious level. The conscientious objection of health care workers cannot therefore be reduced only to their individual beliefs or moral teachings of the religious community to which they belong; it is based on principles which constitute the basis of professional ethics and are part of a universal morality (Meaney et al. 2012). Refusing to conduct certain procedures for strictly religious reasons (e.g. blood transfusion) is a marginal issue and does not cause real organizational problems for the healthcare system. It must therefore be emphasized that the conscientious objection of healthcare workers is mostly based on the Hippocratic ethical tradition and on rational premises which do not require religious faith (although they may be reinforced by reasons of religious nature). Only in rare cases is it necessary to acknowledge religious beliefs to understand these motives (e.g. the status of blood among Jehovah's Witnesses). It should be underlined that the conflict between medicine and religion is uncommon, and that the majority of religious traditions highlight the value of efforts made towards the sick, and underline the confluence of medical ethics with moral principles drawn from religion.

In light of the above considerations, the question arises of whether it is justified to impose on the medical society the obligation to perform certain procedures that are the subject of conscientious objection, especially if they are done for non-therapeutic reasons and are contrary to the medical ethos. Medicine must be based on moral principles, such as the desire to help, and obedience to the law must have a deep moral justification. Otherwise, we will again be witnessing the drama of Antigone or Thomas More and other martyrs of conscience will appear successively. From the patients' perspective, acting according to conscience generates trust. The loyalty to the professional ethos and the belief that the doctor always, independent of administrative, legal and political circumstances, will stand on the patient's side and protect him, making the physician grow in respect and trust.

References

Bailey DB, Skinner D, Davis A et al (2008) Ethical, legal, and social concerns about expanded newborn screening: fragile X syndrome as a prototype for emerging issues. Pediatrics 121(3):693–704

Bailey DB, Skinner D, Roche M et al (2009) Emerging dilemmas in newborn screening. Virtual Mentor 11(9):709–713

Balaguer A, Martín-Ancel A, Ortigoza-Escobar D et al (2012) The model of palliative care in the perinatal setting: a review of the literature. Biomed Central Pediatrics 12:25

Beard FH, Hull BP, Leask J et al (2016) Trends and patterns in vaccination objection, Australia, 2002-2013. Med J Aust 204(7):275

Boehnlein JK (2013) Should physicians participate in state-ordered executions? Virtual Mentor 15:240–243

Brakman SV, Fozard F (2008) The ethics of embryo adoption and the Catholic tradition: moral arguments, economic reality and social analysis. Springer, New York

Brock DW (2008) Conscientious refusal by physicians and pharmacists: who is obligated to do what and why? Theor Med Bioeth 29:187–200

Card RF (2012) Is there no alternative? Conscientious objection by medical students. J Med Ethics 38(10):602–604

Congregation for the Doctrine of Faith (1988) Donum vitae. Acta Apostolicae Sedis 80:70–102

Congregation for the Doctrine of Faith (2008) Dignitas personae. Acta Apostolicae Sedis 100:858–887

Cooper RJ, Bissell P, Wingfield J (2007) Dilemmas in dispensing, problems in practice? Ethical issues and law in UK Community pharmacy. Clin Ethics 2:106–107

Curlin FA, Nwodim C, Vance JL (2008) To die, to sleep: US physicians, religious and other objections to physician-assisted suicide, terminal sedation, and withdrawal of life support. Am J Hosp Pall Care 25(2):112–120

Dangel T (2009) Lethal defects in foetuses and neonates: palliative care as an alternative to eugenic abortion, eugenic infanticide, and therapeutic obstinacy. In: Kuczkowski M, Drobnik L (eds) International textbook of obstetrics anaesthesia and perinatal medicine. MED-MEDIA, Warszawa, pp 137–144

Davis S, Schrader V, Belcheir MJ (2012) Influencers of ethical beliefs and the impact on moral distress and conscientious objection. Nurs Ethics 9(6):738–749

Devolder K (2005) Creating and sacrificing embryos for stem cells. J Med Ethics 31(6):366–370

Diasi RP, Maher ER (2013) Genes, assisted reproductive technology and transillumination. Epigenomics 5(3):331–340

Drabiak-Syed K (2010) Newborn blood spot banking: approaches to consent. PredictER Law and Policy Update. Indiana University Center for Bioethics

Draper H, Chadwick R (1999) Beware! Preimplantation genetic diagnosis may solve some old problems but it also raises new ones. J Med Ethics 25(2):14–20

Duffy ME (2010) Good medicine: why pharmacists should be prescribed a right of conscience. Valparaiso University Law Review 44:509–564

Esteves SC, Bento SC (2013) Implementation of air quality control in reproductive laboratories in full compliance with the Brazilian cells and Germinative tissue directive. Reprod Biomed Online 26(1):9–21

Evans C (2003) Freedom of religion under the European convention on human rights. Oxford University Press, New York

Fraczek P, Jabłońska M, Pawlikowski J (2013) Medical, ethical, legal and social aspects of prenatal diagnosis in Poland Medycyna Ogólna i Nauki o Zdrowiu 19(2):103–109

Gałązka M (2014) Polish penal law on the human being in the prenatal stage. In: Stępkowski A (ed) Protection of human life in its early stage. PeterLang, Frankfurt am Main, pp 173–174

Garcia Calvente M, Lomas-Hernandez V (2016) Emergency contraception and conscientious objection: an ongoing debate. Gac Sanit 30:91–93

Goold I (2005) Should older and postmenopausal women have access to assisted reproductive technology? Monash Bioeth Rev 24(1):27–46

Green J (1993) Ethics and late TOP. Lancet 342:1179

Harris LH, Cooper A, Rasinski A (2011) Obstetrician-gynecologists' objections to and willingness to help patients obtain an abortion. Obstet Gynecol 118(4):905–912

Kato Y (2013) Conscience in health care and the definitions of death. Croat Med J 54(1):75–77

Keown D (2005) End of life. The Buddhist view. Lancet 366:952

Knack AV (1953) Compulsory vaccination with or without conscientious objection clause. Hippokrates 24(18):565–567

Lawrence E, Curlin FA (2009) Physicians' beliefs about conscience in medicine: a National Survey. Acad Med 84(9):1276–1282

Lawrence RE, Rasinki KA, Yoon JD (2010) Obstetrician-gynecologists' beliefs about assisted reproductive technologies. Obstet Gynecol 116(1):127–135

Leong Marc-Aurele K, Nelesen R (2013) A five-year review of referrals for perinatal palliative care. J Pall Med 16(10):1232–1236

Magelssen M (2012) When should conscientious objection be accepted? J Med Ethics 38:19

Meaney J, Casini M, Spagnolo AG (2012) Objective reasons for conscientious objection in health care. Nat Catholic Bioethic Q 12(4):615

Mertzanidou A, Wilton L, Cheng J et al (2013) Microarray analysis reveals abnormal chromosomal complements in over 70% of 14 normally developing human embryos. Hum Reprod 28:256–264

Midro A (2014) Conscience clause from the perspectives of clinicians genetics. In: Stanisz P, Pawlikowski J, Ordon M (eds) Conscientious objection in medical practice – ethical and legal aspects. WN KUL, Lublin, pp 175–182

Morton J (2016) Women have the right to prenatal genetic testing and to choose abortion. The Guardian, 26 Mar 2015. https://www.theguardian.com/commentisfree/2015/mar/26/women-right-to-prenatal-genetic-testing-abortion. Accessesed 15 Mar 2016

Orr RD (2013) Autonomy, conscience and professional obligation. Virtual mentor. Am Med Ass J Ethics 15(3):245

Pawlikowski J (2014) Concerns about conscience clause from the perspectives of aims of medicine and medical ethics. In: Stanisz P, Pawlikowski J, Ordon M (eds) Conscientious objection in medical practice – ethical and legal aspects. WN KUL, Lublin, pp 145–170

Pawlikowski J, Sak J, Marczewski K (2010) Physicians religiosity and controversive medical procedures. Wiad Lek 5:5–16

Pellegrino E (2002) The physician's conscience, conscience clauses, and religious belief: a Catholic perspective. Fordham Urban Law J 30(1):240

Pope TM (2010) Legal briefing: conscience clauses and conscientious refusal. J Clin Ethics 21:163–176

Report of Government Council of Republic of Poland (2015) [Sprawozdanie Rady Ministrów z wykonywania oraz o skutkach stosowania w roku 2013 ustawy z dnia 7 stycznia 1993 roku o planowaniu rodziny, ochronie płodu ludzkiego i warunkach dopuszczalności przerywania ciąży oraz o skutkach jej stosowania, Dz. U. Nr. 17, poz. 78 z późn. zm. http://orka.sejm.gov.pl/Druki7ka.nsf/0/FC3CC580EF5568B0C1257E890026E7DB/%24File/3686.pdf. Accessesed 15 Mar 2016

Sachedina A (2005) End-of-life. The Islamic view. Lancet 366:778

Schenker JG (2005) Assisted reproductive practice: religious perspectives. Reprod Biomed Online 10(3):310–319

Schotsmans PT (1998) In vitro fertilisation: the ethics of illicitness? A personalist Catholic approach. Eur J Obst Gyn Rep Biol 81(2):235–241

Shanawani H (2016) The challenges of conscientious objection in health care. J Rel Health 55:384–393

Sivaraman MA, Noor SN (2016) Human embryonic stem cell research: ethical views of Buddhist, Hindu and Catholics leaders in Malaysia. Sci Eng Ethics 22(2):467–485

Spreng JE (2008) Pharmacists and the "Duty" to dispense emergency contraceptives. Issues Law Med 23:215–277

Strickland SLM (2012) Conscientious objection in medical students: a questionnaire survey. J Med Ethics 38:22

Sulmasy DP (2008) What is conscience and why is respect for it so important? Theor Med Bioethics 29:135–149

Warnock M (1988) Report of the committee of inquiry into human fertilisation and embryology, London. http://www.hfea.gov.uk/docs/Warnock_Report_of_the_Committee_of_Inquiry_into_Human_Fertilisation_and_Embryology_1984.pdf Accessesed 15 Mar 2016

Wenger NS, Carmel S (2004) Physicians' religiosity and end-of-life care attitudes and behaviors. Mt Sinai J Med 71(5):335–343

Wicclair M (2008) Is conscientious objection incompatible with a physician's professional obligations? Theor Med Bioethics 29:171–185

Wicclair MR (2011) Conscientious objection in health care: an ethical analysis. Cambridge University Press, Cambridge

Wolf SM (1996) Feminism and Bioethics: beyond reproduction. Oxford University Press, Oxford

Zampas C, Andion-Ibanez X (2012) Conscientious objection to sexual and reproductive health services: international human rights standards and European law and practice. Eur J Health Law 19:231–256

Chapter 9
Are There Unresolvable Dilemmas in Bioethics?

Barbara Chyrowicz

9.1 Moral Dilemmas

According to the most often quoted definition, a moral dilemma is a situation in which agent *S* ought to do *a* and ought to do *b* at the same time, however, doing both "hic et nunc" is impossible (Gowans 1987, 3). The obligations may also refer, respectively, to commission and omission; in one aspect agent *S* ought to do *a* and in another ought to refrain from doing *a*. We say then that a dilemma is a situation in which agent *S* both ought to do *a* and ~*a* at the same time, which is obviously impossible (Williams 1985, 171). In the situation of a dilemma, one may not put off fulfilling one of the obligations until later; one may not also explicitly point to one of the competing obligations as an overriding option. If it was like this we should rather state that we are dealing here with a difficult choice, and not with a dilemma. Calling a dilemma a moral one means that both the obligations have a moral dimension. Although the term "obligation" is used in nonmoral meaning, nonmoral obligations like a prudential obligation, or a legal obligation do not immediately generate moral dilemmas. Moral dilemmas also do not appear on the ground of ethical theories that assume the conditional character of a moral obligation to act. Before we pass on to the title question about unsolvable dilemmas in bioethics, let us stop for a moment and consider the problem of the unresolvability of moral dilemmas as such. The next two parts of the article will be devoted, respectively, to examples of bioethical dilemmas that – not necessarily accurately – are defined as unsolvable, that is such, in which choosing one of the obligations does not eliminate the other, and to theoretical strategies of solving such dilemmas. There are a lot of situations defined as dilemmas in bioethical discourse and I will confine myself to a few examples connected with reproductive medicine, broadly understood.

B. Chyrowicz (✉)
John Paul II Catholic University of Lublin, Lublin, Poland
e-mail: barbarac@kul.lublin.pl

© Springer International Publishing AG, part of Springer Nature 2018
M. Soniewicka (ed.), *The Ethics of Reproductive Genetics*,
Philosophy and Medicine 128, https://doi.org/10.1007/978-3-319-60684-2_9

9.2   The Problem of Unsolvable Dilemmas

We may talk about the unsolvability of a dilemma (the concepts of dilemma and conflict are used interchangeably) – independent of what it actually concerns – only when both the obligations (in the definition called, respectively, *a* and *b* have an unconditional character; if they both have a conditional character it seems sensible to look for an argument that is external to and allows indicating the superiority of, one of them. When there are two obligations involved, one of them being conditional and the other one unconditional, for obvious reasons priority ought to be given to the latter one. The conditional or unconditional character of an obligation is closely connected to the ethical theory within which the moral norms are formulated. Taking into consideration the general typology of moral theories which divides them into consequentialist and non-consequentialist, the unconditional moral obligation that gives moral norms an equally unconditional character and hence generates unsolvable moral dilemmas should be looked for in non-consequentialist (deontological) theories, that is in the ones in which recognizing a definite action as a morally required one is not exclusively dependent on the consequences that follow it. This does not mean that followers of consequentialist theories, that is of various kinds of utilitarianism, in practice will not have to struggle with making difficult choices. John Stuart Mill, one of the proponents of classical utilitarianism, states that there is not a moral system in which different obligations do not clash in a distinct conflict. Mill's advice is to refer the conflicting obligations ultimately to the principle of utility, that is to the principle saying that the option is right that will probably give more good than evil as compared to the alternative one. Maybe this is not the best criterion for resolving such conflicts – Mill admits – but it is better to use this one than none. Also, in Mill's opinion, the appearance of conflicts is not the fault of theories, but of the complicated character of human affairs, in which I think the philosopher is right. In practice – Mill continues – conflicts are resolved more or less successfully depending on the intellectual and moral values of a given person (Mill 2001, 26–27). Mill's remark about the reasons why moral dilemmas arise is important in the context of the criticism of practicing ethics within a theory that is offered today. One of the arguments put forward against the theory is "the argument from the existence of moral dilemmas". The antitheorists who quote it maintain that the fact of the existence of moral dilemmas proves the weakness of the theory. They are of the opinion that depending on the situation different principles should be referred to; sometimes utilitarian criteria will be more useful for resolving a conflict, and in another case Kantian ones; moral reflection cannot be restricted to one paradigm (Williams 2006 16–17). And if we are mentioning Kant's position, the philosopher from Königsberg is of the opinion that when one's obligations clash with one another, one of them partially or completely overrules the other (Kant 1995, 220–224). Hence both the classical utilitarianism and Kant do not recognize the existence of dilemmas – however, they do it for different reasons. Utilitarians (consequentialists) do not deny that situations involving a conflict may appear in action, however, the principle of utility that is basic for utilitarianism gives at least a theoretically simple way of resolving them, whereas Kant does not accept that there could exist a conflict of equivalent moral obligations. Do then unsolvable moral dilemmas (conflicts) really exist, or are they only difficult choices, or mistakes in recognizing the rank of moral obligations? In contemporary ethics the existence of this type of dilemmas is a matter in dispute. Opponents of the existence of unsolvable moral dilemmas most often quote the condition of consistency demanded from an ethical theory, while its followers – the simple experience of situations, in which every possible option of acting seems equally morally required. However, one may defend, I think, the existence of practically unsolvable dilemmas without rejecting the condition of consistency demanded from a theory.

Returning to Mill's remark – it is not a theory that generates moral dilemmas! It is true that lack of them in utilitarianism results also (or perhaps first of all) from a conditional character of moral obligation, but even then, if we refer to non-consequentialist theories: Kantianism, Ross's theory of *prima facie* duties, or Thomism, one cannot but admit that these theories meet the condition of consistency and that moral obligations (duties) mentioned in them do not come into conflict with each other. Should it then be admitted that moral dilemmas do not appear in theory but in practice?

An affirmative answer to the above question could only concern the theories defending the unconditional moral obligation; and when an obligation has a conditional character dilemmas do not appear either in theory or in practice. Excluding the possibility that it is the theory as such that generates moral dilemmas does not exclude experiencing them. It is so because from the fact that there is a resolution to a dilemma it does not follow that the subject is able to find it. When putting the question about the unsolvability of dilemmas however, we consider their objective, and not subjective dimension. Authors rejecting the possibility of existence of moral dilemmas on the ground of ethical theory do not claim that the subject never faces situations having features of moral dilemmas; we face them as result of the fact that we are limited. MacIntyre remarks that only an ideal moral subject, perfectly well knowing all morally relevant facts could never experience a dilemma situation. Since in reality we do not have such ideal knowledge we understand certain situations as dilemmas. MacIntyre also notices an analogy between theoretical conflicts in sciences (e.g. in physics – between the quantum and corpuscular theories of light) and conflicts of moral nature. Contradictions that appear in the field of sciences do not result from the scientists' conscious ignoring facts. Their appearance is accompanied by the conviction on the part of the scientists that there is a right solution to the problem – otherwise they would not search for it. It is not science as such that generates contradictions; if we perceive them it results from our insufficient knowledge of the surrounding reality (MacIntyre 1990, 376). Then the drama of the subject who is in a situation of moral dilemma would – *per analogiam* – consist in the fact that the subject is aware that one of the rival options should be rejected, but he does not know which one; hence making the right choice is questionable. Moral dilemmas would be – according to the above – *secundum quid* dilemmas, and the *quid* refers here to the subject's morally relevant discernment. In claiming this, MacIntyre is not original; as early as the thirteenth century, St Thomas Aquinas wrote about the possibility of the subject's entanglement *secundum quid* and *simpliciter*.

In Donagan's interpretation, Aquinas' introduction of a double possibility of the subject's entanglement (*perplexus secundum quid* and *perplexus simpliciter*) is necessary to defend the consistency of the theory. In the discussion of the kinds of entanglement the starting point for Aquinas would be examples given by St Gregory the Great in *Moralium Libri* (Gregorius Magnus 1878, 657–658). The two first ones are concerned with people who inconsiderately took an oath – one of them of keeping the secret, and the other one of obedience – on the strength of which they would have to behave in a way they considered morally vile. The third example concerns a clergyman who by way of simony took charge of the congregation, and realizing his offence he has the choice: either to continue vilely taking care of the faithful, or to abandon them. However, both alternatives seem inadmissible to him. The situation the clergyman has found himself in was to be – as Donagan (1977, 144–145) interprets St Thomas' thought – a kind of *secundum quid* entanglement, that is entanglement conditioned by some ignorance or illegality (MacIntyre does not mention this last element) as opposed to *simpliciter* entanglement that is not subject to such conditioning, that is it appears as result of lack of consistence of ethical judgments. St Thomas positively rejects the possibility of the existence of the *simplicite* type of entanglement on the ground of the theory. Aquinas also decidedly rejects the suggestion that examples of *secundum quid* could infringe the consistency of an ethical theory (S. Thomae de Aquino 1984, I–II, q. 19, a. 6 ad. 3; S. Thomae de Aquino 1972, q. 17, a. 4). A moral system is inconsistent only when it allows for situations in which without his own fault the subject may avoid committing one evil act only by committing another one – also evil, which is recognized by some authors as a characteristic feature of moral dilemmas.

The way Donagan interprets the *secundum quid* and *simpliciter* entanglement is not completely faithful to St Thomas' thought. First of all because in the quoted fragments of his writings Aquinas does not give his opinion on consistency of the ethical theory, but asks the question about how far is the erroneous moral awareness (a mistaken mind) binding for us? This does not mean that in Donagan's interpretation there are no right intuitions. Thomas states clearly that when a mind's or a conscience's mistake results from the subject's lack of knowledge that is his own fault, the subject is not in a dead-end situation (S. Thomae de Aquino 1984, I–II, q. 19, a. 6, ad. 3). Hence we are entitled to assume that even when someone intensely claims that he has found himself in a dilemma, this does not mean that he faces a real (ontological) moral dilemma. The entanglement of a man who has an erroneous moral awareness is not an utter entanglement (*simpliciter perplexus*), but entanglement in some aspect (*perplexus secundum quid*).

> quod ille qui habet conscientiam faciendi fornicationem, non est simpliciter perplexus, quia potest aliquid facere quo facto non incidet in peccatum, scilicet conscientiam erroneam deponere; sed perplexus secundum quid, scilicet conscientia erronea manente. (S. Thomae de Aquino 1972, q. 17, a. 4)

Disentangling this "aspect" is decisive for the possibility of resolving a situation that shows the features of a dilemma. However, as long as the subject is mistaken, he may be convinced that he faces the necessity of choosing an action that agrees with his conscience but does not respect God's commandment. Further St Thomas claims that the same deed cannot be both good and evil in the moral aspect (S. Thomae de Aquino 1984, I–II, q. 20, a. 6) which seems to exclude situations of real, that is unsolvable moral dilemmas.

Secundum quid entanglement, in accordance with the examples given by St Gregory, does not also have to be connected only with mistaken conscience (as it is the case in St Thomas). If a clergyman illegally holding his office faces a choice between further, vile, holding the office on the one hand and leaving the faithful on the other, and both the possible options seem inadmissible to him, this is result of the illegality committed earlier, and not of an error of the conscience. Hence it would follow that in a situation interpreted as a moral dilemma one may find himself either without his own fault – when e.g. someone making him take an oath conceals his true intentions from him, or through his own fault – when as a consequence of the wrong decisions made earlier we have found ourselves in a situation that can be disentangled depending on giving the decisions up, and we have to make our new decision "here and now", without the possibility of simply withdrawing from the mistakes that had been made. Do situations of this type appear in bioethical debates?

9.3 Examples of Dilemmas in the Bioethical Debate

Referring the above analyses to the context of bioethics assumes understanding bioethics as an ethical subdiscipline (of applied ethics) that is a detailed study of the general assumptions of the theory. Emphasizing this is important also because today a lot of authors suggest understanding bioethics not as a subdiscipline of ethics, but a kind of "social discourse" (Irving 2014). In a multicultural, pluralist society the discourse is assumed to be subject to the principles of democracy that are supposed to be helpful in working out an agreement. Speaking about any moral dilemmas on the ground of so understood bioethics is pointless. Nobody who has dealt even a little with bioethics can deny that it is a peculiar field where a variety of views and positions clash. The question "what does bioethics say to this?" practically cannot be unambiguously answered today. It depends on what bioethics is meant. If the only aim of the bioethical discourse is to be a presentation of various views, lack of an unambiguous answer will be something most natural; however, if in a discussion we ask the question "what does bioethics say to this?" we expect prescriptive answers, we assume a possibility of accepting a particular position. A multitude of theories and positions does not exclude accepting a definite opinion in a discussion; and what is more: it is only when we are able to clearly specify our position that our participation in a discussion that is supposed to lead to normative decisions becomes sensible. Rejecting bioethics as a discourse does not decide yet what character the assessments and norms functioning within it will have – this is because practicing bioethics within a theory means accepting general rudiments of a definite theory, and these – as we have already noticed – function today within two basic paradigms: the consequentialist one and the non-consequentialist one (Chyrowicz 2015, 75–79). In practice, this means that the norm that is fundamental for bioethical analyses, the one that orders respecting life and excluding destroying it, will be justified in various ways, having sometimes a conditional and sometimes an unconditional character. Moral, and more precisely bioethical dilemmas, will be mentioned only when the norm "thou shalt not kill" will be given the status of an unconditional one; in the opposite case one should speak about difficult, or even dramatic choices, about the legitimacy of which the subjects, however, would not have any essential doubts. Supporters of the unconditional character of the norm "thou shalt not kill" also tend to call difficult choices dilemmas. This is incompatible with the definition of the dilemma, but it strengthens the gravity of the situation. Do real moral dilemmas that cannot be resolved appear in the bioethical debate?

Let us take the example of a doctor who is convinced that abortion is an evil. The cases of abortion do not include situations in which the baby's developing organism does not have any chances for surviving and, at the same time, it is a serious threat to the life of the mother. For example, a pregnancy during which uterine cancer develops is a case of such a situation. A medical intervention consisting in saving the life of the mother will be then inevitably connected with removing the developing human life at the stage of the embryo or the fetus, although it is not the primary aim of the operation; that is, we are not dealing here with classical cases of abortion, if we understand abortion as a medical intervention aiming at killing the unborn baby. The doctor may not be reproached with devoting the life of the baby for the life of the mother, if the life of the baby cannot be saved. Hence the doctor may not have any doubts about the justness of saving the life of the mother, but he may have problems with assessing the chances of the survival of both the mother and the baby. This is important, since as long as there is a chance that both the mother and the baby can survive, a medical intervention is not morally unambiguous. If in such a situation the doctor says "I have a dilemma", it will not basically be a moral dilemma, but an epistemic one. The doctor lacks knowledge of medical character, and not of moral one. If he could predict the further progress of the disease he would not have a problem with making a decision about performing or refraining from an operation of hysterectomy (surgical removal of the uterus). Uncertainty is one of the elements of acting closely connected with the context of moral dilemmas, however, moral uncertainty, in which the doctor ponders over the admissibility of the very intervention as such (this also can happen), should be distinguished from uncertainty of the epistemic nature, when the doctor has doubts if he should perform the operation "here and now". Probably a lot more situations can be indicated in which doctors call the choices they face dilemmas only because they lack sufficient knowledge: they do not know how the patient's organism will react to the given dose of the medicine, it is difficult to predict how the patient will bear an operation... If these types of situations do not deserve the name of moral dilemmas it is not because doctors find it easy to make decisions, but because the reason why the choice is called dramatic is not strictly of moral nature; it is such only in an indirect way. The doctor is convinced that human life – if only this is possible – should be saved, and so he does not question the moral norm; he has a problem assessing the situation. For this type of dilemma it is characteristic that they disappear when knowledge is gained, that is along with the progress in the field of broadly understood biomedicine. In this type of situation, knowledge has moral significance in the sense that decisions are dependent on it (on having or not having it). Admittedly, the ultimate reason for making a decision is the value of human life and the norm "do not kill/ respect life" protecting it; however, this norm is in no way questioned, it also does not collide with any other one. In the field of bioethics knowledge or lack of it refers to even more complex situations.

Achievements of contemporary biotechnology allow us to gain information about the genetic condition of a particular person, which proves to have considerable influence on the decisions to be made. Genetic screening may – in its supporters' opinion – prevent moral dilemmas connected with decisions about abortion in the case of diagnosing in conceived babies such serious genetic disorders as Tay-Sachs disease or Lesch-Nyhan syndrome. Probably part of the parents who are told that they are carriers of the lethal genes will abandon their efforts to have their own biological offspring. Incidentally, calling the decisions about abortion moral dilemmas also seems to be a misunderstanding. For those who defend human life from the moment of conception this is no dilemma; neither a serious disease nor a physical or mental handicap are reasons for excluding someone from their right to have their life protected. Supporters of the position according to which the above mentioned diseases are connected with such psychophysical quality of the organism with which it is much better not to be born, also will not define decisions about abortion as dilemmas. They will defend the conditional character of the norm "thou shalt not kill", and such a character – as has been stated above – beforehand excludes dilemmatic situations. Coming back to the significance of genetic information – gaining knowledge about genetic predispositions of our organism or of the organisms of our children, which predispositions with a definite degree of likelihood will lead to a disease in the future, will probably significantly influence decisions about our life. The higher degree of probability of the occurrence of a disease is, the stronger the reason to state: "I have a dilemma over how to plan my future life", but this is an epistemic, and not a moral dilemma. We recognize this type of dilemmas as unsolvable as long as we do not have a chance to gain the necessary information. Hence such dilemmas have a temporary character, and gaining the necessary knowledge is at the same time their resolution.

In the bioethical dispute moral dilemmas also appear; and they have the already mentioned character of the *secundum quid* entanglement. One of them is especially connected with reproductive medicine: what can be done with cryopreserved human embryos that no one is interested with anymore? The embryos were brought to life by one of the methods of assisted reproductive technology. Since the process of acquiring embryos is connected with the previous hormone treatment of the woman, and the efficiency of the method is not one hundred per cent, acquiring a greater number of embryos is supposed to make stronger the chance to conceive a baby. In the case where the first implantation fails, another transfer of embryos is performed without the need of another hormone treatment. It happens that the first transfer results in successful implantation and a birth of a baby, and spare embryos turn out to be unnecessary. Thawing the unnecessary embryos is equal to destroying them, but there are also other options: the use of them for research or for acquiring stem cells. They may also be kept in liquid nitrogen *ad calendas graecas*, but this does not seem sensible and does not solve the problem. If someone recognizes that embryos have a normative status proper for a human person (this is one of the positions in the dispute over the ontological status of embryos), and so their life should be protected, they cannot find "here and now" a fair answer to the question about what should be done with them, because whatever decision they will make: to use the embryos for research, or to thaw them, their life will not be saved. It would be best if they were not cryopreserved at all; but we have an accomplished fact here. In this case the *secundum quid* entanglement consists in the problem of possible use of embryos that are not needed by anybody which appears because of an earlier decision about their cryopreservation that, in the opinion voiced by supporters of the normative status of embryos, was a mistake, and in their adversaries' conviction, it was a medical procedure necessary for the efficiency of methods of assisted reproductive technology. Only the former say it is a dilemma. It is true that they could say that from the beginning they opposed the cryopreservation of embryos and now it is not their problem, but this would be an expression of nonchalance and lack of responsibility. The suggested solution in the form of adopting embryos also does not seem to be a satisfactory solution. Admittedly, it is a chance for them to survive, which generally should satisfy opponents of the destruction of embryos, but at the same time it generates quite new problems. Firstly, it is hard to imagine that this type of adoption should not be preceded by preimplantation diagnostic, through which only the best embryos will be selected for adoption. It may also be supposed that in the situation when prenatal diagnosis performed already during the pregnancy shows serious defects of the developing fetus, the adoptive parents will not have a problem with making the decision about abortion – whatever one says, it is not their biological child. It is true that it is better when at least some embryos will not be destroyed, but for someone who is of the opinion that the life of every embryo is valuable, adoption is not an ultimate solution to the problem.

Apart from the epistemic and *secundum quid* dilemmas also ones appear in the bioethical debate that we face not because of lack of knowledge or of earlier mistakes, but of biological anomalies. Perhaps one of the most difficult cases discussed in bioethics in recent years was the one of separating the Siamese twins Mary and Jodie. The girls were joined at the abdomen with Jodie being the stronger sibling who was sustaining the life of Mary. Mary's nervous, respiratory and vascular systems were poorly developed. She could not live without Jodie whose vascular and respiratory systems had a double duty to sustain her own life and the life of her sister. Doctors thought the infants would die within 3–6 months at best. Separation of the twins was possible, however, it was certain that Mary would not survive the operation. Separating her from her "host-sister" was tantamount to her death, because the girl's heart and lungs were not strong enough to make her capable of independent survival. Hence the saving of Jodie was connected with the inevitable killing of Mary, and refraining from any intervention meant, according to all predictions, the death of both girls. The girls were separated, and Jodie survived the operation. The doctor who performed the operation claimed that there was not the slightest doubt that performing it was just, that is he did not consider the situation a dilemma. However, the girls' parents, who did not agree to the operation, did see a dilemma in it, and the decision was made by the court of justice. Who was right, the doctor or the parents? One may ponder over the question of whether the doctor should have waited to perform the operation until the moment when one of the girls had no chances of surviving; then we would be dealing with the situation in which of two endangered human beings we are saving the one we can save. Perhaps then the parents would agree to the operation without having the feeling that they were giving the life of one girl in order to save the life of the other. The case of Mary and Jodie is an example of a dilemma in which helplessness against biological anomalies is confronted with the possibilities given by medicine. In the situation when there are no chances for both the endangered persons to survive, and saving the life of one of them is only possible at the cost of depriving the other one of the chances to survive, consequentialists who accept the conditional character of the norm "thou shalt not kill" will refer to the ultimate outcome, without recognizing the situation as a dilemma, and without denying that the choice is difficult. Deontologists who are convinced that the norm "thou shalt not kill" is unconditional, will first ask if the medical intervention does not mean an action that could be considered to be deliberate killing; if not, they will agree to the saving of the life of the one of the beings entangled in this dramatic situation whose life can be saved. This may be Jodie, or a woman with whom ectopic pregnancy has been diagnosed, or a mother who cannot be saved during the childbirth together with the still unborn baby.

An example of biological anomaly indicated above as a basis of the situation defined as a dilemma is close to the case of pregnancy endangered with a developing tumor. The reason why these cases have been quoted as examples of different types of dilemmas is "medical uncertainty". If in the case of a cancer disease of the mother there may be a chance of both the mother and the baby surviving, on which chance the doctor should make intervention conditional, the quoted example of Siamese twins or ectopic pregnancy are dead end situations from the point of view of medicine. Are the presented cases "dead end situations"? The suggested solutions deny it, albeit it is not certain that they are satisfactory. In the case of dilemmas of the *secundum quid* type the indicated solution will be the most acceptable one among the basically unacceptable options. In our conclusion, let us refer then to theoretical reasons which prove to be helpful in resolving dilemmas.

9.4 Theoretical Strategies of Resolving Dilemmas

As we have already remarked, dilemmas do not appear on the ground of theories but in practice, in a definite context of acting. Biomedical practice is the context of the above mentioned examples. Recognizing them as dilemmas is connected with the fact that the decision-maker who has norms accepted in the domain of a given theory cannot cope "here and now" with deciding which of the rival options should be given priority. Hence if dilemmas do appear in practice, recognizing them as such results from accepting definite assumptions that are proper for a given ethical (bioethical) theory (Chyrowicz 2008, 393–401). A debate about resolving dilemmas seems reasonable exactly within a theory, and not outside it, and this is because of a few reasons that I will try to explain below.

1. Since interpreting a dilemma as a real (unsolvable) one is connected with an unconditional character of duties coming into conflict, this unconditionality should be in some way accounted for. This task – not an easy one! – rests with the defenders of deontological ethical theories, including the deontological approach to bioethics. Justification of the unconditional character of the moral duty to act outside the ethical theory seems to be an undertaking both risky and impossible to carry out.

2. A moral dilemma is not a standard situation, so experience and moral intuition that is typical of us cannot directly prompt what decision is right. The choice of the right option of acting – if it is to be a rational and justified choice – ought to be preceded by a reliable analysis of the rival moral demands, which will allow us to verify them and possibly recognize one of them as apparent only, or as worse than the other. An answer to the question "ought I really to do a or b" belongs to the theory in the sense that outside it we cannot indicate ultimate reasons why we ought to do anything in a dilemmatic situation. Although the rule of synderesis (good is to be done and pursued, and evil is to be avoided), the general formulation of justice (give each person what he or she deserves) or the "golden rule" (that which you hate to be done to you, do not do to another) are accepted irrespective of the ethical theory, they are too general to give immediate aid in a particular situation of a dilemma. It is only the ethical theory that explains what should be understood as materially specified good, and whatever others (as well as we) deserve, that is what kind of goods we should protect both with respect to us and to others.

3. It would be naïve to think that in debates over resolving moral dilemmas unambiguous answers will be given. If in the debates that are now being held, e.g. about the possibility to use embryos for acquiring stem cells in the perspective of working out therapies of diseases that have been incurable until now different answers are given, it is because they are formulated on the basis of different anthropological and ethical principles. Hence the ethical theory proves to be the field where both dilemmas are resolved and other than the suggested resolutions are verified.

4. The role of the ethical theory both in interpreting and in resolving moral dilemmas proves to be so significant that situations interpreted as real moral dilemmas on the basis of some theories, according to other ones will not be dilemmas at all; an example being the operation of separating the Siamese twins, Mary and Jodie, will be (at least initially) examined as a dilemma in deontological moral theories, and consequentialist theories will not see a special moral problem in separating the sisters.

5. Dilemmas that we encounter and towards which we have to take a definite position sometimes have many layers, in the meaning that it is difficult to decide about situations that result from previous mistaken or wrong decisions. These are dilemmas of the secundum quid type whose example is the above mentioned case of the cryopreserved embryos. Whatever action the subject will take in this type of situations, it will be a choice of the action he does not fully accept in the moral aspect. It is in this context that the reference to the "necessary evil" is made. May an attempt to justify the subject's doing "the necessary evil" in such situations be made? I think it may, however, with the reservation that the accepted solution will not be called just, that is, appropriate to the normative order considered to be just. It is not a prima facie just solution. It may be an optimal resolution "here and now", but only thanks to this it will not become just, and its justification does not excuse the subject from making painstaking efforts to – as far as it is possible – restore moral values to their deserved place.

6. Facing unprecedented situations the subject more than once acts "in good faith", which – provided he does not reject the theory – does not have to mean a purely subjective judgment. It will not be one if he makes all efforts to recognize subjective-neutral reasons, i.e. the objectively binding order of norms, both in relation to the one, and to the other option of acting.

References

Chyrowicz B (2008) O sytuacjach bez wyjścia w etyce. Znak, Kraków

Chyrowicz B (2015) Bioetyka: Anatomia sporu. Znak, Kraków

Donagan A (1977) The theory of morality. University of Chicago Press, Chicago

Gowans CW (1987) The debate on moral dilemmas. In: Gowans CW (ed) Moral dilemmas. Oxford University Press, New York

Irving D (2014) What is bioethics? In: Koterski JW (ed) Life and learning X: proceedings of the tenth University Faculty for Life conference. University Faculty for Life, Washington, DC, pp 1–84

Kant I (1995) Die Metaphysik der Sitten. In: Kant I (ed) Die Religion innerhalb der Grenzen der blosen Vernunft. Die Metaphysik der Sitten, Werke, bd 5. Könemann, Köln

MacIntyre A (1990) Moral dilemmas. Phil Phenomen Res 1(Supp Fall):367–380

Magnus G (1878) Moralium Libri sive Expositio in Librum B. Job xxxii 20. In: Migne JP (ed) Patrologia Latina, vol 76. Garnier, Paris

Mill JS (2001) Utilitarianism. Batoche Books, Kitchener

Thomae de Aquino S (1894) Summa theologica. Ex Typographia Forzani Et S, Romae

Thomae de Aquino S (1972) Opera omnia iussu Leonis XIII P.M. edita, t. 22, 2: Quaestiones disputatae de veritate … Qq. 8–20, Editori di San Tommaso, Romae
Williams B (1985) Ethical consistency. In: Williams B (ed) Problems of the self: philosophical papers 1956–1972. Cambridge University Press, Cambridge
Williams B (2006) Ethics and the limits of philosophy. Routledge, London

Part II
The Moral, Legal and Social Challenges of Reproductive Genetics

Chapter 10
Reproductive and Therapeutic Cloning

Henning Rosenau

10.1 Reproductive Cloning

Everything seems straightforward with reproductive cloning. The risks – particularly those of a moral-ethical nature – are deemed to be so high that this process is outlawed worldwide (Taupitz 2003, 221and 226).

10.1.1 Cloning Banned in Germany by Section 6(1) of the Embryo Protection Act (Embryonenschutzgesetz – ESchG)

This is clearly what the German legislature had in mind when banning the cloning of human beings. It is not possible to interpret the corresponding penal provision of Section 6(1) of the Embryo Protection Act in any other way. However, this legislative intent was not perfectly implemented, and there is a real debate about whether the legal provision really covers all conceivable processes today, and prohibits them subject to penalties.

10.1.1.1 The Characteristic of Identical Genetic Information

The first difficulty arises in that with the cell transfer method, what develops is not another human with the same genetic information – as posited in Section 6(1) of the Embryo Protection Act – but the transfer of mitochondria and organelles from the

H. Rosenau (✉)
Martin-Luther-University of Halle-Wittenberg, Halle an der Saale, Germany
e-mail: henning.rosenau@jura.uni-halle.de

© Springer International Publishing AG, part of Springer Nature 2018
M. Soniewicka (ed.), *The Ethics of Reproductive Genetics*,
Philosophy and Medicine 128, https://doi.org/10.1007/978-3-319-60684-2_10

cytomembrane of the donated egg only develops into an embryo with almost similar genetic material (Keller 1998, 447, 448; Schroth 2002, 170, 172). The organism that develops is never an absolute clone, but can only be more than 99% identical with the donor of the cell nucleus.

This seems to be a quite pedantically meticulous interpretation of the norm. But it should be noted that the biotechnical rules are centred on the provisions of the Embryo Protection Act of 13 December 1990 (Federal Law Gazette, BGBl 1990, 2746). The Embryo Protection Act classifies various actions and processes with and concerning the embryo or the germ line cells as improper misuse, and prohibits such under threat of criminal punishment. This criminalization, which is already formally manifested by the fact that the prohibitions of the Embryo Protection Act are classified as criminal acts (whoever does *x* shall be punished with imprisonment or a fine), means that the Embryo Protection Act is a purely criminal statute. There are historical reasons for this. When the Act was passed in 1990, under Germany's federal system the legislature did not have competence to adopt laws in the field of reproductive medicine. The Embryo Protection Act was thus formulated as a purely criminal statute based on general federal legislative competence in criminal matters pursuant to Article 74(1) no. 1 of the Basic Law of the Federal Republic of Germany (*Grundgesetz* – GG). It was not until 27 October 1994 that the legislature was given additional competence under Article 74(1) no. 26 of the Basic Law (BGBl 1994, 3146), which set out a clear federal competence to legislate in the areas of reproductive medicine, genetic technology, and organ transplantation. This means that Embryo Protection Act is interpreted as a criminal statute and is subject to the strictly analogous prohibition of Article 103(2) of the Basic Law, under which an act may be punished only if it was defined by a law as a criminal offence before the act was committed (*nulla poena sine lege*). An interpretation under the Embryo Protection Act is thus only permissible if it is strictly permitted in the wording of the corresponding provision. It is not possible to apply the Embryo Protection Act to circumstances beyond the actual wording of the Act (BVerfGE 71 [1985], 108, 115, 92 [1995], 1, 12), even if such an approach would be supported by other recognised interpretive topoi such as systematic interpretation, historic interpretation on the basis of the intention of the legislature, or objective-teleological interpretation (see: Puppe 2008, 64 sqq., 66 sqq., 78 sqq. and 80 sqq.; Larenz 1991, 344 on relationship between four approaches).

In consequence, some people argue that the wording already does not apply when the 99% conformity is subject to Section 6(1) of the Embryo Protection Act, i.e. where there is only *almost* identical genetic information. Under this interpretative approach, there would be no prohibition of cloning in Germany! (Schroth 2002, 170, 172).

With the unambiguous sense and purpose of Section 6 of the Embryo Protection Act, this approach will not be used due to the analogous prohibition. However, it also seems pedantic to say that the legislature has not been referring to the same (*derselbe*) genetic information (v. Bülow 1997, 718, 719), when a difference is drawn between absolutely identical and almost identical. Genetic information that is almost identical cannot be described with the word *gleich* (same), and the prevailing opinion assumes that minimal differences in DNA can still be reconciled with the wording of Section 6(1) of the Embryo Protection Act (ESchG) (Schreiber 2011, 891, 902; Witteck and Erich 2003, 258, 259; Günther et al. 2008, par. 6 marginal note 2; Hilgendorf 2001, 1147, 1160), because relevant cell nuclei that are also important for the development demonstrate an identical DNA (Günther et al. 2008, par. 6 marginal note 6). However, this interpretation is open to challenge.

10.1.1.2 The Characteristics of an Embryo

Another problem is that the prohibition on cloning only applies when an embryo is created.

Embryos are legally defined in Section 8(1) of the Embryo Protection Act thus:

'bereits die befruchtete, entwicklungsfähige menschliche Eizelle vom Zeitpunkt der Kernverschmelzung an' (already means the human egg cell, fertilised and capable of developing, from the time of fusion of the nuclei [BGBl 1990, 2746].)

But with cloning using the Dolly method, the egg is not fertilized, and so there is no fusion of the nuclei. Therefore, it is questionable whether an embryo is created during the cloning of an individual.

The legal definition with its controversial German word *bereits* (already) can be understood in two ways. The first is that an embryo is any human egg cell capable of developing and that it is *already* an embryo from the time of conception. This interpretation would include the embryo arising from the transfer of the nucleus: Any human egg cell capable of development would be an embryo, including the fertilized egg cell.

Yet Section 8(1) of the Embryo Protection Act can also be interpreted as applying only to an embryo resulting from the fusion of nuclei, but that the embryo should be directly protected from the moment of the fusion of the nuclei: an embryo is a fertilized egg cell and it is *already* protected from the moment of fusion of the nuclei (Witteck and Erich 2003, 258, 259; Gutmann 2001, 353, 355). Whether this rather stretched understanding of the word *bereits* (already) is more appropriate than if the word *auch* (also) is applied to the first interpretation (Gutmann 2001, 353, 355) does not settle the matter in dispute. As the discussion shows, both interpretations are semantically possible and do not exceed the possible meaning of the words as an absolute limit of the analogous prohibition. Therefore, the sense and purpose of the legal provisions give an indication for the understanding of the norm. As the rationale of the norm supports an extensive prohibition of cloning, the definition in Section 8(1) of the Embryo Protection Act need not be regarded as absolutely definitive. In consequence, an embryo within the meaning of Section 8(1) of the Embryo Protection Act is also a human egg cell capable of development that was transferred into the cell nuclei by way of cell transfer (Hilgendorf 2001, 1147, 1162).

As cloning logically excludes a classical fertilization – namely the fusion of a male with a female haploid set of chromosomes to create a complete genotype of the new individual – and as recourse to an embryo as a necessary intermediate step is not possible for all cloning processes, the term 'embryo' cannot require a prior fertilization as a precondition. Otherwise, a person created by such a method[1] would never have reached the embryo stage, which seems to be rather incongruous (see also: Taupitz, in: Günther et al. 2008, par. 8 marginal note 54 with further evidence).

But the methodological expediency of defining an embryo differently in Section 6 and in Section 8(1) of the Embryo Protection Act seems rather dubious (Müller-Terpitz, in: Spickhoff et al. 2014, par 6 marginal note 3; v. Bülow 1997, 718, 720; Schreiber 2011, 891, 902; Witteck and Erich 2003, 258, 259). Quite apart from the fact that Section 8(1) of the Act expressly states that "For the purposes of this Act, an embryo means…", it would not be very logical in a short statute with only 13 sections for the legislature to give a clear definition but not to apply its core meaning. However, as shown, this extrapolation is not necessary.

10.1.1.3   Need for Reform

However, the whole discussion shows that it is better to regard the characteristics set out in Section 6 of the Embryo Protection Act as unfortunate and in need of reform (for clarification, see: Schreiber 2011, 891, 903; Schroth 2002, 170, 172). This applies in particular as extremely questionable cloning processes would clearly not be covered by the express prohibition under Section 6(1) of the Embryo Protection Act. It would be easy to change the genetic material of the human being to be cloned before the transfer, in order to exclude genetic disabilities or illnesses or other flaws of the donor, and therefore optimizing the duplicate (compare v. Bülow 1997, 718, 720 sq.). This would mean that the clone would not have the same but significantly changed genetic material, meaning that Section 6(1) of the Embryo Protection Act would clearly not be applicable (Taupitz 2001, 3433, 3434; Hilgendorf 2001, 1147, 1160). One possible way out of this complicated legal position would be for Germany to sign the Additional Cloning Protocol to the Convention on Human Rights and Biomedicine (CETS 1998a – dated 12 January 1998, came into force on 1 March 2001). This clearly prohibits simple cloning without genetic manipulation. Under Article 1(2) of the Protocol, genetic identity makes reference to sharing of the same nuclear gene set, which would make the differing German questions of interpretation redundant. Germany has as yet been unable to take this step, as the Cloning Additional Protocol is only available to the current 35 Member States who have signed the Convention on Human Rights and Biomedicine. Germany has so far declined to sign the Council of Europe's Convention.

However, Section 6 of the Embryo Protection Act is absolutely clear on one matter: the cloning of an entire human being is also forbidden in Germany.

---

[1] That the entity resulting from such a cloning process would be a human being, and that the protection of the Basic Law (GG) would also apply to them, is generally recognised. (Klonbericht Bundestag printed matter, BT Drs 13/11263 [1998], 20).

10.1.2 The Legally Protected Rights of the Cloning Prohibition

Although the prohibition on cloning of human beings is unchallenged, and attracts moral indignation worldwide, the reasons for this prohibition are unclear. What is the legal interest being protected by the penal provisions of Section 6(1) of the Embryo Protection Act? How can we classify the archaic revulsion generated by the idea of cloned likenesses (Habermas 1998, 13) into a legal category?

The Christian theological view argues that human beings are playing God and meddling with the process of creation when we interfere in the natural processes of procreation and conception (Gröner 1991, 293, 296). The assertion is that mankind has already been doing this for thousands of years, and that biotechnology is a culturally historical practice. When a farmer breeds and improves new sorts of livestock, or a gardener propagates new plants, this interferes with natural evolutionary processes, Schreiber and Rosenau 1998, 395), and this argument is used to counter modern intervention with genomes. It is possible that there will be irreversible damage, and the argument is an ideological approach.

However, it must be countered that with the current state of research into cloning, the risk of malformations in cloned human beings is so high that using such processes on human beings seems to be out of the question at the moment (Gruss 2003, N1). In fact, similar methods used on mice, rabbits, goats, pigs and cows have shown that reprogramming of transferred somatic cells in an egg cell into a full organism works sufficiently well only in exceptional circumstances. Instead many deformed and damaged animals result from the cloning attempts, with degenerated immune systems and reduced life expectancies. An attempt to clone human beings under these poor research conditions would be irresponsible. Carrying out clinical research would already be outlawed under the applicable rules of the Helsinki/ Tokyo Declaration of the World Medical Association, due to a failure to meet the risk-benefit assessment under Section 16 of the Declaration. The benefits of cloning experiments are heavily outweighed by the risks and impact on the test person – the clone. Any medical experiments would fail to meet medical ethical standards.

But this does not mean that there should be a prohibition on reproductive cloning (explicitly: Taupitz 2002, 449, 452). With the current pace of medical developments, it is likely that such difficulties will be overcome in the future (Kahn 2002, 103, 107). Similar to the debate about xenotransplantation – where the danger of new viruses and infectious illnesses being transferred to patients seemed not to be controllable, thus prompting suggestions that clinical trials should be postponed (comp.: Beckmann et al. 2000, 292; Schreiber 2003a, 315, 322; Jungeblodt 2001, 67, 73–76 and 116; Beckmann and Müller-Ruchholtz 2002, 196, 200 sq.) – the endangerment argument only supports a moratorium on cloning, and not for a prohibition in principle. Therefore, the argument to be followed is that which focuses mainly on the concept of human dignity (comp.: Hilgendorf 2001, 1147, 1152 sq.), whereby the dignity of both the original being and the clone is to be taken into consideration.

The rights of the *original being* could be affected in that he or she could no longer have the feeling that they are unique (Joerden 1999, 79, 85). Indeed, it would be monstrous to think that at any time various versions of oneself could return in the future. But there would only be a breach of human dignity when the rights of the individual over their genetic information could be ignored without their approval. One would imagine that the impetus for cloning would come from the cell donors themselves, whether from some utopian idea of immortality, utilitarian considerations of old-age provision, or to maintain their family tree. If the original persons were to approve the cloning, their rights would not be contravened (Joerden 1999, 79, 86).

This argument also cannot be countered by the assertion that the respect for human dignity recognised in Article 1(1) of the Basic Law of the Federal Republic of Germany (GG) could not be waived (Witteck and Erich 2003, 258, 261; Herdegen 2016, Art. 1(1) marginal no. 98). For often the violation of the human dignity is only constituted when there is interference with the rights of the person against their free will. A medical intervention does not constitute a violation of human dignity per se, but only when, for example, a medical intervention is made against the will of the affected party. Cloning is also classified in the same manner: the cloning violates the dignity of the original human being only when the process takes place without their agreement. This genetic uniqueness is difficult to invoke (against this: Witteck and Erich 2003, 258, 262), because human beings cannot be reduced to their genetic code and millions of identical twins speak against such a view.

Thus the problem focuses on the *created cloned being*. On the whole, the assumption is that the dignity of this being is affected by its cloned existence (comp.: Brohm 1998, 197, 204; Ach et al. 1998, 223 et sqq. For another view Gutmann 2001, 353, 370 et sqq.; Joerden 1999, 79, 85). The clone would regard itself as a mere repetition of the original being and would be regarded by contemporaries as a curiosity, as a simple blueprint. The clone would be compared to the original being and measured against that person, and must live with this burden (Jonas 1985, 188 et sqq.). It is not the case that the clone must live the exact same life as the original being (Hilgendorf 2001, 1147, 1156). The genetic predisposition is much exaggerated. Human beings are not just the sum of their genes. This is demonstrated by looking at the development of numerous identical twins, which can vary greatly. The psychological burden – in addition to the physical dangers of cloning itself (Gutmann 2001, 353, 364) and other burdens that remain unexplained, such as a tendency to age more quickly – is quite real.

The real lack of human dignity here, namely the disregard of the subjective quality of the clone, is more clearly characterized by a further consideration. No one may have command over another, and make them the object of their wishes. But the clone is such an object. The clone's characteristics and abilities are not the work of chance: they are the result of another's intent. It must be admitted that even a person conceived and born naturally can be regarded as the result of various intentions of their parents. But there is a residual uncertainty and indeterminacy that differentiates a person conceived normally from a clone. Consequently, the genetic match is less relevant for human dignity (compare Glauben 1997, 305, 307). The genetic uniqueness is also not present in identical multiple births (Gutmann 2001, 353, 370). The determination of genetic specifications by another represents the interference identified in Article 1(1) of the Basic Law (GG) (Trute 2001, 385, 395; Siep 1998, 5, 13; Honnefelder 2013, 183, 190; see also: CETS 1998b, sec. 3).

One further problem must be clarified. How can the cloning of a person impact on the human dignity of the clone when this person does not yet exist? At the time of creation, the clone does not exist and cannot therefore have its human dignity impacted by the act of creation as such (Joerden 1999, 79, 85; von der Pfordten 1998, 213, 215).

It is correct that constitutional theory tells us that a person need not exist – or more correctly need no longer exist – in order to be afforded human dignity. The entirely undisputed post-mortal right of personality of deceased persons is based on rights of human dignity BVerfGE 30 [1971], 173, 193 sq.).

The same approach can also give answers to whether a life that does not yet exist can logically suffer a breach of its human dignity. As there is a post-mortal human dignity, then there is also a pre-natal protection. The protection of human dignity is afforded to the potential future person, and at the same time projected onto their future life. The pre-natal treatment, which would be regarded as a breach of human dignity in a post-natal setting, is thus covered by Article 1(1) of the Basic Law (GG) (Neumann 1998, 153, 160; Brohm 1998, 197, 204; Frommel 2002, 39, 42; Herdegen 2016, Art. 1(1) marginal note 98).[2] Thus this is a type of pre-natal protection of human dignity. But at this point it should be noted that the legally protected interest underlying Section 6(1) of the Embryo Protection Act (ESchG) is the human dignity of a potentially cloned person.

10.2 Therapeutic Cloning

A much more controversial issue is whether therapeutic cloning is acceptable from a biopolitical viewpoint.

10.2.1 The Term 'Therapeutic Cloning'

With therapeutic cloning, even the term itself has already been the subject of some dispute. Some would like to call it "research cloning" (Höfling 2001, 277, 287). The word "therapeutic" is justified by the aims linked to this type of cloning. There is no intention that a human being should be born as a result of this cloning. Instead, researchers are interested in the embryonic stem cells, which can be extracted about four days after fertilization of the blastocyst, and are then themselves no longer

---

[2] Herdegen (2016, Art. 1 GG marginal note 63 and 98) sees the cloned embryo after nidation falling under the protection of Art. 1(1) GG and must admit that in light of the nidation perspective represented here, there may be a pre-effect. Thus there is no pre-conditional requirement of a transcending claim of non-creation, but a transcending claim of no creation in this manner – i.e. on no creation by cloning.

totipotent. This means they are no longer capable of developing into a full human organism, but as "all-rounders" have the ability to develop into each cell type of all three germ layers – that is to develop into all 210 cell types of the human body (Taboada 2003, 129, 133). One can refer to the pluripotency of these embryonic stem cells. This might allow for nerve cells to be cultivated that might provide treatment or a cure for patients with Alzheimer's or Parkinson's disease, hematopoietic cells for leukemia sufferers, or even for whole organs to be cultivated from embryonic stem cells. In reproductive cloning, the aim of the process is cloning. But in therapeutic cloning, the aim is to develop treatment possibilities opened up by the use of cloned egg cells.

The intent behind the technical process is also important for the words chosen to describe it. *Reproductive cloning* refers to the process intended to duplicate a human being. The aim is to create another person. But *therapeutic cloning* is aimed at providing medical treatments. The aim of the research is to develop possible medical treatments by means of human tissues conceived by tissue engineering. It is designed to help people who are ill.

However, the current denomination of this kind of research as therapeutic is subject to hefty criticism that it is dishonest (comp.: Winnacker 2003, 6; Höfling 2001, 277, 278; Mieth 1999, 224, 233, suspects language politics and persuasive strategies lie behind this). Effective therapies are a long way off. Therapeutic cloning is not even an established method for animal experiments. The use of the term suggests a value that is currently baseless. It would be more correct to merely refer to *research cloning*.

The counter-argument that the experimental stage is not yet concluded, and that human testing is still a long way off,[3] is correct but falls short. It makes no sense to use terms today that will be irrelevant in a few years,[4] but which themselves fail to provide the necessary clarity and unequivocality. The term 'research cloning' also hides much more than it makes clear in the processes under discussion. In particular, what is currently called 'reproductive cloning' could be labelled mere research cloning. The proviso that there is currently no real basis for a practical application for humans is applicable just as for therapeutic cloning.[5] Here too there are numerous problems to be overcome. Cloning experiments presently lead to many anomalies and abnormalities, and 'normal' clones are practically not possible. This is linked to the fact that up to 5% of genes that carry hereditary factors important for embryonic development

---

[3] Some treatment successes have been noted with animal experiments, such as for varieties of diabetes in mice (see: Reich 2003, 106).

[4] The uncertainty of future therapeutic use is also not a persuasive argument against (basic) medical research (comp.: Beier 2003, 73, 74). In contrast, the uncertainty about the additional value of a new type of treatment is often an approval condition for clinical studies. If it is already clear that either the current standard treatment or the new method brings better results, then that alternative should be pushed from the beginning (see: Schreiber 1983, 13, 15). Uncertainty about a therapeutic approach is thus by itself not an acceptable argument in favour of or against a new research approach.

[5] The claim by the American Raëlian sect that the first cloned baby had been born at Christmas 2002 is generally regarded as mere media allegation, not least because of the refusal to allow genetic testing of the alleged child (see: FAZ 2002, 1; Reich 2003,101; Ganten 2003, 25).

cannot be correctly regulated. The epigenetic regulatory mechanisms are not yet fully understood (comp.: Herrler et al. 2003, 84, 90; Jaenisch 2003, N1). The low level of cloning efficiency is recognized: per organism 0.5–5% of implanted oocytes result in live-birth animals. In the case of Dolly the sheep, nucleus transfer was carried out on around 300 egg cells and 300 embryos were created (Rohwedel 2002, 18, 26).

The descriptors 'reproductive' and 'therapeutic' have entered the public domain, and they clearly state the function of the process they describe much more clearly than the term 'research cloning'. As these terms are also used internationally, the use of this terminology is justified and should not be abandoned (Rosenau 2006, column 230, 231 sq.).

10.2.2 The Dispute About the Constitutionality of Therapeutic Cloning

Whether therapeutic cloning should be allowed is a matter of strong debate. The United Nations General Assembly passed a Declaration on this matter on 8 March 2005, but the voting showed that there are deep divisions between states on this matter. Only 84 nations – less than half of all Members of the UN – approved a total prohibition. Germany has not signed the Convention.

In Germany, some people regard creating a cloned embryo with the aim of using this for research and therapy purposes as a contravention of the instrumentalization ban under Article 1(1) of the Basic Law (GG). This raises in particular the question of the instrumentalization of embryos. The consumptive research would objectify the embryos and use them exclusively for another purpose, namely the treatment of patients (Starck 2001, 59; Herdegen 2002, 1, 10). In contrast to stem cell creation from embryos left over after regular in-vitro fertilization processes, in the case of therapeutic cloning the embryos are only created in order to generate stem cells. Therapeutic cloning demonstrates such instrumentalization that it could be regarded as breaking taboos. This lies in the "functionally reduced creation of human embryos …, in order to *use* them directly as research material" (Höfling 2001, 277, 283).

The background to this is that with therapeutic cloning doctors want to create stem cells with genetic material identical to that of the patient. The replacement tissues created from the patient's stem cells have the same genetic information as the patient, and under any treatment the autologous transfer would not likely result in rejection of the new tissues (Schindehütte et al. 2002, 67, 77; Rohwedel 2002, 18, 25). Such transfers would probably be more successful than transplantation of donated organs.

There is no assumption of a contravention of Article 1(1) of the Basic Law (GG). The determination that from the beginning the embryos created for research purposes are not intended to be allowed to develop into human beings is essential for ascertaining a breach of the human dignity provision. Under such circumstances, a breach of human dignity cannot be projected on to a person to come into being in the future. There is no forward application of the human dignity protection. With therapeutic cloning, by definition no life will result from the embryo. Due to the fact that the focus is on creating tissues rather than a person, no contravention of human dignity can be assumed (Rosenau 2004, 135, 152; Hilgendorf 2001, 1147, 1157; Dederer 2002, 1, 23; Ipsen 2001, 989, 996). As the perspective of development into a human being is missing, the argument about contravention of dignity fails (Herdegen 2016, Art. 1 (1), marginal notes 99 and 107). There is no future person, and therefore no self-will to protect. The biological development is halted in good time – long before nidation.

This evaluation corresponds to the current legal position in Europe, whereby Article 3(2)(d) of the Charter of Fundamental Rights of the EU prohibits the reproductive cloning of human beings. Article 18(2) of the Council of Europe's Convention on Human Rights and Biomedicine goes no further. Article 18(2) states:

"The creation of human embryos for research purposes is prohibited."

However, it must be taken into consideration that the definition of embryo is not uniform among the Member States. As some countries differentiate between the pre-embryo up to the 14th day of development and the later embryo, the Convention on Human Rights and Biomedicine cannot be interpreted as prohibiting therapeutic cloning (Taupitz 2001, 3433, 3437; Haßman 2003, 18; for the assumption of a prohibition see: Kopetzki 2002, 15, 61; Herdegen 2016, Art. 1 (1), marginal notes 96; Holznagel 2003, 75, 78). As said, this is in line with the evaluation of Article 3(2)(d) of the Charter of Fundamental Rights of the EU (OJ C 303 dated 14 December 2007, 1 et sqq.), which expressly prohibits reproductive cloning but not therapeutic cloning. It does not allow for the creation of human embryos for research purposes. But there is no clarification of whether that includes an embryo in vitro. There is no consensus at this stage among the Member States of the Council of Europe about the nature and legal position of an embryo (ECtHR 2005, 727, 731; see: Jofer 2014, 387).

10.2.3   Use of Induced Pluripotent Stem Cells (iPS Cells) as an Alternative

The dispute about the prohibition also of therapeutic cloning is likely to become irrelevant soon. A new option of generating induced pluripotent stem cells (iPS cells) from any simple adult body cell will make the process of therapeutic cloning unnecessary in the medium term.

In 2006, with a combination of the four genes c-Myc, Klf-4, Oct-4 and Sox-2 – which had been introduced into somatic cells by means of a retrovirus – Shinya Yamanaka succeeded in returning these cells to a pluripotent state (Takahashi and Yamanaka 2006, 663 et sqq.).

In 2007, this reprogramming of mature differentiated somatic cells into the pluripotent state was also carried out with human cells (Yu et al. 2007, 1917 et sqq.). It was also discovered that the c-Myc gene, which is known to be oncogene (potentially causes cancer) can be replaced (Takahashi et al. 2007, 861 et sqq.).

Since this time, scientists have been able to effectively create human stem cells themselves, so that this incredibly important cell type will no longer be dependent on human embryos for research and possible future therapies.

As well as learning about the development of living organisms, the sense and purpose of this research is to refine these cells into various differentiated somatic cells and use the resulting tissues for therapeutic purposes. So therapeutic cloning and iPS cells are moving in the same direction. They allow for further differentiations, including blood cells, liver cells, abdominal cells and even brain cells, into which one can add various different transcriptor genes (Hochendlinger and Plath 2009, 512 with further evidence).

With current scientific processes, iPS cells can be developed into completely viable living beings by means of a certain biotechnical process called tetraploid complementation assay.

This has led to consideration about regarding such a combination with a tetraploid blastocyte as creating the necessary conditions within the meaning of Section 8(1)(2) of the Embryo Protection Act (ESchG). Under this, iPS cells would also have the potential to develop into an individual human being and would therefore be subject to the definition of an embryo (Minwegen 2011, N1). In effect, they would not only be pluripotent, but actually totipotent.

However, this is not convincing. For one thing, under these premises it is almost impossible to sensibly differentiate between the protection afforded to cells worthy of fundamental rights protection and those that do not require such protection (Minwegen 2011, N1). "A skin cell is not an embryo." (Kersten 2004, 550) In addition, this approach is countered by factual arguments. The legislature regarded necessary further conditions to comprise necessary medical and biological conditions required by a naturally conceived embryo to develop into a human being. Here, the iPS cell must be combined with a blastocyte in order to develop further. But in doing so, the pluripotent cell loses its identity as an organism and becomes part of another chimeric organism. Thus we cannot speak of an iPS cell having the individual characteristics under the conditions necessary to develop into an individual as set out in Section 8(1)(2) of the Embryo Protection Act (ESchG).

The iPS cell lacks the potential to develop into an individual, which means that it cannot at least reach the stage of nidation. The fact that such a stage can be brought about in conjunction with a tetraploid blastocyte has no effect on this argument. The induced pluripotent cell is like an embryo, but is not quite an embryo. It is not an embryo within the meaning of the Embryo Protection Act (ESchG) (Rosenau 2015, 233, 253; also: Höfling, in: Prütting 2013, par 8 ESchG marginal note 7; Kersten 2004, 549 sq.; Taupitz, in: Günter/Taupitz/Kaiser 2008, par. 8 ESchG marginal note 62; *idem* 2008, 107, 138). Research with iPS cells thus seems to be a legally acceptable form of human biological research, and offers an alternative to therapeutic cloning.

10.2.4 The Legal Prohibition of Therapeutic Cloning
in Germany

The ordinary legal position in Germany is simply stated. As long as the legal term of human dignity remains under discussion and can be interpreted in different ways, the legislature – in Germany this is the *Bundestag* and the *Bundesrat* – is called upon to define protection of human dignity, but it enjoys some broad discretion (Taupitz 2001, 3433, 3440; Heun 2002, 517, 523 sq.; Faßbender 2003, 279, 280). The legislature has the discretion to prohibit or limit embryo research, but overall has decided on a restrictive approach. This approach is constitutionally permitted, but not required by the constitution (compare Kloepfer 2002, 417, 422). It would also be possible for Germany to approve therapeutic cloning.

The legislature clearly penalized therapeutic cloning in Section 2(1) of the Embryo Protection Act (ESchG) because the removal of embryonic stem cells from the blastocyte means that the embryo is not used in the intended manner (compare Höfling 2001, 277, 278; Günther et al. 2008, par. 2 marginal note 30). The criminal prohibition on cloning under Section 6(1) of the Embryo Protection Act (as discussed above) is also applicable, because the cell nucleus transfer method is used to clone an embryo within the meaning of Section 8(1) of the Embryo Protection Act. All cloning processes are thus prohibited and penalised in Germany, including therapeutic cloning.

10.3 The Status of the Embryo

Whether this legal position is sustainable, or should be amended (a proposal is included in: Gassner et al. 2013, 8 and 67 et sqq.: "Proposal 2: Any action carried out with the intention of facilitating the birth of a human being whose nuclear genome is identical with that of another living or deceased person shall be prohibited"), is mainly down to whether or not the embryo is already classed as a human being. This is where the status debate comes into play. The issue about whether the embryo enjoys the protection of fundamental rights to life can be answered with one short question: Is the embryo 'life' within the meaning of Article 2(2) sentence 1 of the Basic Law (GG)?

10.3.1 Human Life After Fusion of the Nuclei

This argument is very strong and operates on the basis of the so-called PIC argumentation. The human life comes into being as soon as the egg and sperm cell fuse. The embryo has the potential (P) to develop into a person; it has individuality and identity (I); and it develops with continuity (C) into a fully-fledged person. Yet all three points can be offset by weighty counter-arguments.

10.3.1.1 Potentiality

An embryo can potentially become a person: a conclusion on the *status ad quem* must be derived from the *status quo*. They embryo is given the protection of a fundamental right to which it would also later be entitled when born.

The potentiality approach is less plausible as it does not explain why cells in the pre-nucleus stage – and in Germany there are around 25,000–30,000 instances of such cells being cryo-conserved – should also not be entitled to full protection of fundamental rights. With these cells, where the sperm has already penetrated the formed oocyte to be fertilised, but cell fusion has not yet taken place, the genetic identity has already been established. If development would continue uninterrupted, an embryo would have developed, and there can be no evidence of increased potentiality (Heun 2002, 517, 520; Merkel 2001a, 480).

Drawing a conclusion of *status ad quem* from the *status quo* is also less than logically persuasive. Prince Charles has the potential to become the King of England. But that does not mean that we should today (when this article was written) accord him the status of the English monarch.

More persuasive is that the potentiality of the embryo remains fictitious unless important circumstances come to pass (Schreiber 2003b, 157, 163 sq.). An embryo in cellular culture has no real chance of realising its attributed potentiality of becoming a human being unless it can find its way into a uterus. According to Schlink, the categorical difference is huge: Before nidation, the embryo lives only in its potentiality, and does not actually live its potentiality until after nidation (Schlink 2000, 17). The fictive potential does not become real potential until after nidation and it is the realization of the future status that justifies the attribution of the fundamental rights of life.

Finally, the discovery of iPS cells fully exposes the irrationality of the potentiality argument. The iPS technology allows any sort of human cells – such as the skin cells of grown-up humans – to be reprogrammed as an iPS cell. Tetraploid embryo complementation then allows this cell to form a totipotent entity that can develop into a human being. This means that every skin fragment we shed each day has the potential to develop into a human being. Our skin fragments evidently are not regarded as human beings. The potentiality argument leads to absurdity.

10.3.1.2 Individuality or Identity

From the moment of fusion of the nuclei there is a new, genetically defined individual life in existence (Classen 2002, 141, 143; Höfling 2001, 277, 281; regarding argument for and against PIC see: Damschen and Schönecker 2003). However, this statement is only sustainable with respect to the genetic identity. Until nidation or individuation from the 13th to 14th day of development of the embryo,[6] there is a

---

[6] Individuation – namely the cessation of divisibility and when it is no longer possible for identical multiple embryos to form – is concurrent with, or happens shortly after, nidation (comp.: von

possibility of multiple embryos. Identical twins are genetically identical. But it seems highly questionable to link fundamental rights of life protection to this identity. There is no doubt that each such twin has a separate right to life. This protection extends not to the twins together but to each individual and identical being (Heun 2002, 517, 521). The individuality is not fixed until the possibility of cell division has ended. Before that, with the destruction of a zygote it is not possible to say which life is being destroyed (Merkel 2001a, 496). But genetic identity is an insufficient criterion for attribution of human qualities, as the human being is not just the sum total of their genes (Merkel 2003a, 23, 39 sq.).

The attempt to save the individuality argument in this way – by separating individuality (which is present) and singularity (that is absent prior to nidation) (Höfling 2001, 277, 281; similarly Beckmann 2003a, 97, 100, who does not regard division into twins as characterising an individual; Beckmann 2003b, 9 et sqq.) is a sophisticated contortion. It misunderstands that the human being is not created once their genes are determined, so that identical twins can also be different individuals and are such from the moment of nidation (Heun 2002, 517, 521).

10.3.1.3 Continuity

The continuity argument is linked with the idea that the human development process is a continuing smooth process without any sharp demarcations. Any differentiation within this continuum is arbitrary, so the protection of fundamental rights must extend to all stages, including embryonic development (Starck 1999, art. 1, marginal note 18; Höffe 2002, 111, 138).

Rightly so, the logical-formal objection of the sorites paradox is raised. The lack of noticeable demarcation does not negate the necessity and possibility of making differentiations (Heun 2002, 517, 520). When individual grains of sand are piled up into a heap, at exactly what point does the heap come into existence? At this point one can differentiate between heaps and grains.

More important is the objection that continuity can only arise if the nidation of the egg cell in the uterus is accompanied by a very important basic condition. The fertilized egg cell by itself does not develop into a person. It is the symbiosis with the mother that sets off embryogenesis and opens up the way to further development (Taupitz 2002, 111, 113; Taupitz 2002b, 10 et sqq.). This demonstrates a very relevant demarcation that in a non-arbitrary fashion allows for a legal differentiation of the phases before and after nidation.[7]

Loewenich 2002, 43, 47; Giwer 2001, 67) and should therefore be regarded equally for legal assessment – as in the Federal Constitutional Court (*Bundesverfassungsgericht*) (BVerfGE 39 [1975], 1, 37).

[7] The acceptance of watersheds in embryonic development is gaining in acceptance abroad. There is an assumption of a pre-embryonic phase that ends about 14 days after conception. During this period, the embryo is often categorised as a pre-embryo. The differentiation is proposed in the *Warnock* Report (1984, 66) and underlies both the English Human Fertilisation and Embryology Act (1990) and the Spanish Ley (1988, Art. 4).

10.3.1.4 Constitutional Requirements

Reliance on the above argumentation topoi became necessary because the Basic Law (GG) leaves the status of the embryo open. There was no parliamentary majority in favor of including an unborn life under Articles 1 and 2 of the Basic Law when it was being drawn up (Schreiber 2003b, 157, 162; Merkel 2003b, 151, 154; therefore the 'mothers and fathers of the Basic Law' have considered this issue, according to: Beckmann 2003a, 97, 101).

The question has still not been clarified by a constitutional court in Germany. The authentic interpreting body of the Basic Law (GG) – the Federal Constitutional Court (*Bundesverfassungsgericht*) – has expressly not made a decision on the fundamental rights of embryos outside the mother's uterus (as noted by: Berghäuser 2015, 111; Merkel 2001b, 493, 495). Both abortion decisions with respect to time limits refer in their headnotes to the "unborn" (BVerfGE 88 [1993], 203) life or to the life "developing in the womb" (BVerfGE 39 [1975], 1). The second abortion decision dated 28 May 1993 cites General Prussian Law from the year 1794 (ALR). It states (in translation): "General human rights also apply to unborn children, *from the time that they are conceived*" (ALR 1794, Sec. 10 I 1, highlights by author) which expressly contradicts that a decision has been made as to whether human life begins with the fusion of egg and sperm cells, *even when* this knowledge is supported by medical anthropology (BVerfGE 88 [1993], 201, 251). Of course, as *obiter dicta* the additional statement is not legally binding. If the Federal Constitutional Court were to consider in vitro embryonic rights at some point in the future, there is no requirement for it to follow this reasoning.

Therefore, the oft-cited doctrines of the Federal Constitutional Court (BVerfG) do not apply to the embryo at the cell-stage. These doctrines state that the (no longer divisible!) life "during the process of growth and development does not develop into a human being but as a human being" (BVerfGE 88 [1993], 203, 252, comp.: 39 [1975], 1, 37) and that the development process ("in any case, from the 14th day after conception"!) is a continuing process "that indicates no sharp demarcation and does not allow a precise division of the various steps of division of human life" (BVerfGE 39 [1975], 1, 37, comp.: 88 [1993], 203, 267). Unfortunately, this is not often observed (*pars pro toto* Benda 2001, 2147, 2148).

10.3.1.5 Inconsistencies

The view that human life begins with the fusion of sperm and egg cannot be cogently justified. It also leads to indisputable contradictions.

The protection of life has already been removed in that the use of contraceptive intrauterine devices and other nidation inhibitors is permitted (Taupitz 2002, 111, 113). These devices destroy fertilized eggs – which according to the prevailing concept are already classified as human beings. Now in these cases it could be argued that the conflict created by pregnancy is already present in the female body, even though it must be admitted that the conflict situation is not properly felt by the woman (Taupitz 2002, 111, 113).

There is another everyday situation where embryos that are no longer in utero, but would be just as worthy of protection as the in vitro embryos, are destroyed. Around 70% of all embryos created in a natural way by fusion of egg cells and sperm do not nidate or do not develop fully due to being swept out of the uterus by monthly bleeding. This natural process is often caused by a defect in the embryos. But this functional process in the female uterus can be regarded as a trigger, whereby fully intact embryos are flushed out of the female body. Under this critical viewpoint, these would be human beings that would fall under the protection of the Basic Law. But these embryos are left to their own fate. The idea of selecting and saving viable embryos is as unappetizing as it is absurd, and no one would seriously consider such consequences of this attribution of the fundamental right to life (Lüderssen allocates this determination to the indisputable premises in: 2002, 209, 210; see also *idem* 218: Green 2001, 55). But it also demonstrates the overall inconsistency of the argument.

The approach is not certain about the consequences. Some regard the right to life as absolute, and refuse to consider a balanced restriction in favor of other rights (Höfling 2002, 34, 39). Such approaches are exemplified by the pointed statement by *Deutsch* that in Germany an embryo is protected absolutely up to the point where it may be aborted (Deutsch 1991, 721, 724).

Others rely on the argument that Article 2(2) sentence 3 of the Basic Law (GG) allows for fundamental rights to life to be interfered with only pursuant to a statute. This evidently shows that the right to life can be offset against other rights (Taupitz 2001, 3433, 3437). Interference with the fundamental right to life is as such not unconstitutional. The general restrictions of fundamental rights must be observed with reference to moral barriers – especially the principle of proportionality. It should also be taken into account that interference with life always means full abandonment of the protection of the right to life, and with such all-or-nothing constellations an exemption can only be considered in exceptional circumstances. At the same time, many believe that the destruction of embryonic life is not disproportionate to the purpose of embryonic stem cell research, namely research into the development of treatments for severe illnesses (Kloepfer 2002, 417, 421). The main argument is that the intensity of the protection of the right to life in the different development stages of human beings may vary. The concept is of a graduated prenatal or tiered protection of life (Dreier 2002, 4, 5). The fact that abortion is permitted shows that an unborn life may also be sacrificed in favor of other legal values (Taupitz 2001, 3433, 3437; Kloepfer 2002, 417, 420). Thus it is also permissible to expect that the unborn life of an embryo will also attract restricted rights of protection and that its interests may be sacrificed in favor of high-value medical considerations. The derived interpretation on the basis of a tiered protection of the fundamental right to life seems less consistent. All cases founded on a restriction of the fundamental right to life under Article 2(2) sentence 3 of the Basic Law (GG) are of a different nature. They concern police shooting in emergency circumstances, or private self-defense and emergency aid – cases where intervention is needed in the fundamental rights of the person from whom the attack or disorder emanates. The embryo to be used here is not attacking or causing disorder in the rights of third parties (which Tauplitz concedes in: 2001, 3433, 3427 footnote 32). It is not the source of any danger.

In a legal system where the prevailing opinion is that rescuing life by mandatory blood donation from a third party is incompatible with the most important value principles of the legal community (Wessels et al. 2015, marginal note 473 with further evidence), it is difficult to argue for intentional destruction of life not for the purposes of saving a life but just for basic research. Those who argue that embryos should have a right to life *ab initio* run into the difficulties caused by the principle of prohibition of excessiveness when they wish to allow such destruction for scientific purposes (for a tiering of the right to life, see: Hoerster 2003, 529, 530). The attribution of a right to life would thus lead to absurdity.

10.3.2 Human Life from the Point of Nidation

However, the boundary for the beginning of human life within the meaning of Article 2(2) sentence 1 of the Basic Law (GG) lies elsewhere – from the point of nidation. This is the last important development stage prior to the birth that is an important *conditio sine qua non* for the realization of human potential, for the determination of the individual – in short for becoming a human being. The successful interdependency after the nidation into the uterus of the maternal organism allows the embryo to develop fully and be born. Without this nidation in the uterus of the mother there would be no living mammal and no human being (Schindehütte et al. 2002, 67, 78; Nüsslein-Volhard 2001, I; falsely: Jofer 2014, 368 sq.).

The genetic program is fixed from the fusion of the nuclei, but an in vitro process never results in a human being. It is not sufficient to have a larger Petri dish, or to turn the fertilized egg cell into a homunculus using alchemistic distillation and chemical analysis to facilitate putrefaction – a type of decomposition process caused by moist warmth. What Theophrastus Paracelsus regarded as possible in 1537, in the *De Natura rerum* attributed to him, is unthinkable. He said (in translation) that "a human could be born outside the womb and be born in a natural way" (citation after: Schöne 2002, 47, 51). The genetic program itself cannot create the human being – the symbiosis with the mother's body is required (Nüsslein-Volhard 2001, I; Reich 2003, 26). Such an intensive interdependency is required alongside the full genetic code – it is irreplaceable and indispensable (Nüsslein-Volhard 2003, 24). The process leading towards humanization is not complete until nidation, meaning that nidation has a quite distinct quality than other necessary development stages (Haßmann 2003, 101, 103; Rosenau 2003, 761, 773; Schmidt-Jortzig 2003, 33. On the objection that nidation is also a process that can stretch out over several days and does not take into account that conclusion of nidation must be regarded as definitive, see: Beckmann 2003a, 97, 100. For the necessity of differentiating between passive and active potentiality, see: Maio 2002, 175, 178 sq., who already regards the moment of fertilisation as active potentiality), as it is that step at which passive potentiality turns into active potentiality. A non-nidated embryo cannot develop into anything. The benchmark for the attribution must be whether the starting point for the protection of the right to life under the Basic Law (GG) – and that is central for human life – can be half way secured. Not until that is established can the precursors to human life under Article 2(2) sentence 1 of the Basic Law be attributed, as it is not until then that we can speak of something developing "as a human being" (BVerfGE 88 [1993], 203, 252). As Lüderssens has said (in translation): "The degree of inescapability is not reached until the nidation in the mother intended to accept and provide sustenance for the embryo" (2002, 210, 221).

Human life should not be confused with mere biological existence (according to: Höfling 2001, 277, 281). Otherwise, cells removed from a patient and then grown in cultures would also constitute human life, but these circumstances clearly do not fall under the protection of Article 2(2) sentence 1 of the Basic Law (GG) (Heun 2002, 517, 519). This question has been decided with respect to the determination of brain death as the time of death of a person. That cells outside of the brain can then continue to live, and can be kept alive, in no way justifies that the brain-dead person is alive. There is something familiar about the debate about brain death. As with the debate about when life comes into being, the arguments posited are often emotional, ideological or moral theological, although here a type of life piety has gained the upper hand.

No doubt an emotional component has its place in deciding normative points of contention, and in weighing up results with balanced judgments of everyday morals we speak of coherentist moral theories (Merkel 2002, 155, footnote 203 based on John Rawls). This was what the great experimental physicist and polymath Georg Christoph Lichtenberg from Göttingen recognized, though he did not label it so, when he talked of 'ratifying his findings while they were still quite warm.' The example scenario provided by Merkel and originally by Annas, is illustrative. Fire breaks out in a biotechnical laboratory containing live in vitro embryos and an unconscious infant. A rescuer arriving at the very last minute can save either the infant or the embryos. No one doubts that the rescuer would save the infant (Annas 1989; Merkel 2001c, 37, 38, *idem* 2002, 151 et sqq.). But this scenario does not bring us much further. It only answers the question of whether the right to life of embryos is inferior to that of human beings. It does not answer the question of whether in principle embryos should have a fundamental right to life. In order to ratify this proposition – that human life does not come into being until nidation – the Merkel example would have to be modified. Suppose the laboratory was on fire, but there were only ten embryos there in a petri dish. Would anyone consider doing everything possible in order to save those ten embryos? The reader should consider what they would do in such a situation. I think, intuitively, that no one would do anything. So our intuition speaks against the assumption that an embryo constitutes a human being.

There are still some objections to the solution put forward here. It is argued that the determination of the beginning of life at the point of nidation is a classic naturalistic fallacy, that is the logical impossible derivation of a normative evaluation from a fact – here the fact that over 70% of all fertilised eggs fail to nidate in natural processes and are rejected by the body before nidation (Härle 2003, 63, 64; Maio 2002, 175, 182). Quite apart from the fact that without recourse to biological circumstances it is not possible to decide (Rosenau 2003, 761, 775 with further evidence; that the law must always make determinations that are somewhat arbitrary, see: Pawlowski 2002, 71, 74 sq.) when human life begins, this allegation is neutralized by the fact that it can be turned exactly on its head. The normative determination that life begins with the fusion of egg and sperm cells is linked to a natural factor – and incidentally is also supported by an assertion of natural continuity.

It is also argued that it is only a question of time until the maternal uterus is superfluous and an embryo can be created and born using a uterine machine (Höffe 2002, 111, 139). Whether such development of an in vitro fertilized cell outside of the womb will be technically possible is not known at present. It is called ectogenesis (Hilgendorf 2002, 387 et sqq.). But even if this undesirable development, the argument remains that the embryo requires a uterus-like environment in which to grow to a person (Lüderssen 2002, 209, 219). Even in the future, the embryo will not be able to develop of its own accord.

Finally, it is argued that the in vitro embryo is the subject of various research interests and is also more vulnerable than an embryo in vivo, and that it consequently most deserves the protection of the Basic Law and therefore should not fall outside its scope of applicability (Duttge 2003, 411, 412; Laufs 2003, 40 and 42; Wolbert 2003, 453, 455; Fink 2000, 210, 212, Sönnecken 2002, 80). In light of the protective role of the law, the fusion of the nuclei should be the earliest point at which the protection of the embryo could be reasonably justified. But this ignores the fact that such an argument goes in circles. The fact does not justify why the embryo should be protected at all, and the argument is derived from the result, merely justifying the need for protection with an assertion of required protection.

10.3.3 *Pre-vital Protection of Human Dignity for the In Vitro Embryo*

In addition, it is also incorrect that before nidation the embryo is not afforded any fundamental rights protection. It does not follow from the decision that the right to life under Article 2(2) sentence 1 of the Basic Law (GG) applies to the embryo after nidation. The right to life and human dignity must not be linked (Heun 2002, 517, 518, who admittedly never wants to set the right to life later than human dignity). The sponsorship of human dignity and the right to life must not automatically be considered together, but can apply at different times. "Where human life exists, human dignity is present to it", said the Federal Constitutional Court (in translation) (BVerfGE 39 [1975], 1, 41). This does not mean that the sentence can be turned on its head. It is incorrect and inaccurate to say that where no human life exists, there is no human dignity. There is no question that it applies to deceased persons, whose human dignity also extends beyond death (Starck 1999, art. 1 GG, marginal note 17; Taupitz 2002, 111, 113). Similarly, post-mortal rights of personality should also be recognized for prenatal protection of dignity for the human life that is created with nidation. This is also the case here. We recall the discussion about the legal value of the prohibition of cloning. As there, the embryo is afforded the forward protection of human dignity (similar: Jofer 2014, 412 et sqq.).

Others think that considering life as starting at the point of nidation goes too far. Parallels are drawn to the concept of brain death. If the cessation of all brain functions means the end of human life, it would only be consistent to apply the same principles to the start of brain functions (Hoffman 1990, 115, 119; Sass 1989, 160 et sqq.). Some place the start of life thus defined at around the 57th day after conception, others stipulate the 30th (Merkel is thus cited by: Schreiber 2002, 231, 240) or the 35th day (comp.: Starck 1999, art. 1 GG, marginal note 18) after conception, as at this time the neural canal has developed, thus paving the way for the start of brain activity (Sass 1989, 160, 172 sq.). It is the brain that first allows for a subjective awareness of an own identity and provides the basis for the faculties of self-determination and reason that qualify as the human condition (Merkel 2001a, 503 sq.; with the respect to the attribution of human dignity following: Heun 2002, 517, 522).

It is doubtful whether if, as is the case with brain death, it will be possible to determine a halfway unshakeable starting point for brain activity from a biomedical perspective (Schlingensiepen-Brysch 1992, 418, 420). But the brain concept falters above all on the fact that self-awareness and the ability to self-determine are not necessary conditions for the attribution of the human condition. Otherwise, a comatose patient existing without consciousness would also not enjoy a right to life and human dignity. The same applies to those suffering from anencephaly (comp.: Starck 1999, art. 1 GG, marginal note 17), where a large part of the brain is missing from birth. Putting the start of life back to a point when some sort of brain activity commences or until brain cell material forms is therefore less than convincing.

10.4 Conclusion

So human life starts with nidation. Before, the embryo *in vitro* is not life within the meaning of Article 2(2) sentence 1 of the Basic Law (GG).

References

Ach J et al. (1998) Hello Dolly?: Über das Klonen. Suhrkamp, Berlin
Annas G (1989) At law: a French homunculus in a Tennessee court. Hast Cent Rep 19(6):20–22
Beckmann J-P (2003a) Wachsendes Lebensrecht? Erwiderung zu Dreier. Zeitschrift für Rechtspolitik 36:97–101
Beckmann J-P (2003b) Selbstbestimmung durch Mutmaßungen über den Sterbewillen. Zeitschrift für Biopolitik 4:1–15
Beckmann JP, Müller-Ruchholtz W (2002) Xenotransplantation of cells, tissues or organs: scientific achievements, ethical justifiability and legal permissibility. Transplantationsmedizin 3:196–202
Beckmann JP et al (2000) Xenotransplantation von Zellen, Geweben oder Organen. Springer, Berlin/Heidelberg
Beier HM (2003) Forschungsverbot zum therapeutischen Klonen – Spiegel einer hohen oder einer irrigen Moral? Reproduktionsmedizin 19(2):73–75

Benda E (2001) Verständigungsversuche über die Würde des Menschen. Neue Juristische Wochenschrift 30:2147–2150

Berghäuser G (2015) Das Ungeborene im Widerspruch: Der symbolische Schutz des menschlichen Lebens in vivo und sein Fortwirken in einer allopoietischen Strafgesetzgebung und Strafrechtswissenschaft. Duncker & Humblot, Baden Baden

BGBl (1990) Gesetz zum Schutz von Embryonen [Embryo Protection Act], 13 December 1990, Bundesgesetzblatt Part I, pp 2746–2748

Brohm W (1998) Forum: Humanbiotechnik Eigentum und Menschwürde. Juristische Schulung:197–205

BVerfGE Entscheidungen des Bundesverfassungsgerichts (settled case law of the Federal Constitutional Court), edited by the members of the Federal Constitutional Court, cited after volume and year, Mohr Siebeck, Tübingen

CETS (1998a) Additional protocol to the convention for the protection of human rights and dignity of the human being with regard to the application of biology and medicine, on the prohibition of cloning human beings dated 12 January 1998

CETS (1998b) Explanatory report on additional cloning protocol of the Council of Europe

Classen C (2002) Die Forschung mit embryonalen Stammzellen im Spiegel der Grundrechte. Deutsches Verwaltungsblatt 117:144–145

Damschen G, Schönecker D (2003) Der moralische Status menschlicher Embryonen: pro und contra Spezies-, Kontinuums-, Identitäts- und Potentialitätsargument. de Gruyter, Berlin

Dederer H-G (2002) Menschenwürde des Embryo in vitro? – Der Kristallisationspunkt der Bioethik-Debatte am Beispiel des therapeutischen Klonens. Archiv des öffentlichen Rechts 127:1–26

Deutsch E (1991) Embryonenschutz in Deutschland. Neue Juristische Wochenschrift 12:721–724

Dreier H (2002) Stufungen des vorgeburtlichen Lebensschutzes. Zeitschrift für Biopolitik 1(2):4–11

Duttge G (2003) Review of Böckenförde-Wunderlich, Präimplantationsdiagnostik als Rechtsproblem. Juristenzeitung 58:411–412

ECtHR (2005) Recht des ungeborenen Lebens auf strafrechtlichen Schutz bei Schwangerschaftsabbruch wegen ärztlichen Verschuldens – Vo/Frankreich. Neue Juristische Wochenschrift 11:727–735

Faßbender K (2003) Der Schutz des Embryos und die Humangenetik: Zur Verfassungsmäßigkeit des neuen Stammzellengesetzes und des Embryonenschutzgesetzes im Lichte des einschlägigen Arzthaftungsrechts. Medizinrecht 5:279–286

Fink U (2000) Der Schutz des menschlichen Lebens im Grundgesetz – zugleich ein Beitrag zum Verhältnis zur Menschenwürdegarantie. Jura 22: 210–216

Frommel M (2002) Die Menschenwürde des Embryos in vitro. Kritische Justiz 35(4):411–426

Ganten D (2003) Gezielte Zellvermehrung und spezifische Zelltransplantation – Ein Plädoyer für die Stammzellforschung und eine Klärung der Begrifflichkeiten. Zeitschrift für Biopolitik, ppp 25–33

Gassner U, Kersten J, Krüger M, Lindner JF, Rosenau H, Schroth U (2013) Fortpflanzungsmedizingesetz: Augsburg-Münchner-Entwurf. Mohr Siebeck, Tübingen

Giwer E (2001) Rechtsfragen der Präimplantationsdiagnostik: eine Studie zum rechtlichen Schutz des Embryos im Zusammenhang mit der Präimplantationsdiagnostik unter besonderer Berücksichtigung grundrechtlicher Schutzpflichten. Duncker & Humblot, Berlin

Glauben P (1997) Die Ethik setzt den Maßstab. Deutsche Richterzeitung 45:306–308

Graul E, Wolf G (eds) (2002) Gedächtnisschrift für Dieter Meurer. de Gruyter, Berlin

Green (2001) Frankfurter Allgemeine Zeitung, 29 November

Gröner K (1991) Klonen, Hybrid- und Chimärenbildung unter Beteiligung totipotenter menschlicher Zellen. In: Günther H-L, Keller R (eds) Fortpflanzungsmedizin und Humangenetik – Strafrechtliche Schranken. Mohr Siebeck, Tübingen, pp 293–326

Gruss P (2003) Klone – Traum oder Albtraum? Frankfurter Allgemeine Zeitung, 14 May, p N1

Günther H-L, Taupitz J, Kaiser P (2008) Embryonenschutzgesetz, 2nd edn. Kohlhammer Verlag, Stuttgart

Gutmann T (2001) Auf der Suche nach einem Rechtsgut: Zur Strafbarkeit des Klonens von Menschen. In: Roxin C, Schroth U (eds) Medizinstrafrecht, 2nd edn. Boorberg Verlag, Stuttgart, pp 353–380

Habermas J (1998) Sklavenherrschaft der Gene: Moralische Grenzen des Fortschritts. Süddeutsche Zeitung, 17–18 January, p 13

Haßmann H (2003) Embryonenschutz im Spannungsfeld internationaler Menschenrechte, staatlicher Grundrechte und nationaler Regelungsmodelle zur Embryonenforschung. Springer, Berlin

Härle W (2003) "Lehrt Euch nicht auch die Natur...?" Was wir von der Natur (nicht) lernen können. In: Beyreuther K, Merkel R, Fuchs T et al (eds) Wider die Natur. Studium Generale. Sammelbände der Vorträge des Studium Generale an der Ruprecht-Karls-Universität, Heidelberg, pp 62–76

Herdegen T (2002) Humangenetik und die Steuerungskraft des Verfassungsrechts. In: Hillenkamp T (ed) Medizinrechtliche Probleme der Humangenetik. Springer Verlag, Berlin, pp 1–16

Herdegen M (2016) In: Maunz T, Dürig G (eds) Grundgesetz, Kommentar Rechtsstand: Mai 2016. München, CH Beck

Herrler A et al (2003) Epigenetische Kontrolle der Genaktivität. Reproduktionsmedizin 19:84–92

Heun W (2002) Embryonenforschung und Verfassung – Liebensrecht und Menschenwürde des Embryos. Juristenzeitung 57(11):517–524

Hilgendorf E (2001) Klonverbot und Menschenwürde – Vom Homo sapiens zum Homo xerox? Überlegungen zu § 6 Embryonenschutzgesetz. In: Geis M-E, Lorenz D (eds) Staat, Kirche, Verwaltung, Festschrift für Hartmut Maurer zum 70. Geburtstag. CH Beck, München, pp 1147–1157

Hilgendorf E (2002) Biostrafrecht als neue Disziplin? Reflexionen zur Humanbiotechnik und ihrer strafrechtlichen Begrenzung am Beispiel der Ektogenese. In: Eberle C-E, Ibler M, Lorenz D (eds) Der Wandel des Staates vor den Herausforderungen der Gegenwart: Festschrift für Winfried Brohm zum 70. Geburtstag. CH Beck, München, pp 387–413

Hochedlinger K, Plath K (2009) Epigenetic reprogramming and induced pluripotency. Development 136(4):509–523. doi:10.1242/dev.020867

Hoerster N (2003) Forum: Kompromisslösungen zum Menschenrecht des Embryos auf Leben? Juristische Schulung 43(6):529–532

Höffe O (2002) Gentechnik und Menschenwürde: An den Grenzen von Ethik und Recht. DuMont, Cologne

Höfling W (2001) Verfassungsrechtliche Aspekte des so genannten therapeutischen Klonens. Zeitschrift für medizinische Ethik 47:277–283

Höfling W (2002) Das Tötungsverbot und die Grenzen seiner Einschränkbarkeit aus verfassungsrechtlicher Sicht. Zeitschrift für Lebensrecht, pp 34–41

Hofmann H (1990) Die Pflicht des Staates zum Schutz des menschlichen Lebens. In: Schluchter E, Laubenthal K (eds) Recht und Kriminalität, Festschrift für Friedrich-Wilhelm Krause zum 70. Geburtstag. Heymann, Köln, pp 115–122

Holznagel B (2003) Perspektiven der Stammzellforschung und der Präimplantationsdiagnostik: Rechtliche Grenzen und Normen. In: Siep L, Quante M (eds) Der Umgang mit dem beginnenden menschlichen Leben. LIT Verlag, Hamburg, pp 75–86

Honnefelder L (2013) Ethische Aspekte in der gegenwärtigen deutschen Diskussion um die Stammzellforschung. In: Tauplitz J (ed) Das Menschenrechtsübereinkommen zur Biomedizin des Europarates – taugliches Vorbild für eine weltweit geltende Regelung? Springer Verlag, Berlin, pp 183–195

Ipsen J (2001) "Der verfassungsrechtliche Status" des Embryos in vitro, Anmerkungen zu einer aktuellen Debatte. Juristenzeitung 56(20):989–996

Jaenisch R (2003) Dolly war auch als Embryo ein Schaf. Frankfurter Allgemeine Zeitung, 25 Mai, p N1

Joerden J (1999) Wessen Rechte werden durch das Klonen möglicherweise beeinträchtigt? Jahrbuch für Recht und Ethik 7:79–85

Jofer P (2014) Regulierung der Reproduktionsmedizin. Nomos Verlag, Baden Baden

Jonas H (1985) Technik, Medizin und Ethik: Zur Praxis des Prinzips Verantwortung. Suhrkamp, Berlin

Jortzig E (2003) Rechtsfragen der Biomedizin. Nomos Verlag, Baden Baden

Jungeblodt S (2001) Rechtliche Aspekte der Xenotransplantation. In: Quante M, Vieth A (eds) Xenotransplantation: Ethische und rechtliche Probleme. Mentis Verlag, Paderborn, pp 67–116

Kahn A (2002) "Therapeutic" cloning and the status of the embryo. In: McLaren A (ed) Cloning. University of Virginia Press, Charlottesville

Keller R (1998) Klonen, Embryonenschutzgesetz und Biomedizin-Konvention: Überlegungen zu neuen naturwissenschaftlichen und rechtlichen Entwicklungen. In: Eser A, Schittenhelm U, Schumann H (eds) Lenckner: Festschrift für Theodor Lenckner zum 70. Geburtstag. CH Beck, München, pp 477–494

Kersten J (2004) Das Klonen von Menschen: Eine verfassungs-, Europa- und völkerrechtliche Kritik. Mohr Siebeck, Tübingen

Kloepfer M (2002) Humangentechnik als Verfassungsfrage. Juristenzeitung 57(9):417–428

Kopetzki C (2002) Grundrechtliche Aspekte der Biotechnologie am Beispiel des "therapeutischen Klonens". In: Kopetzki C, Mayer H (eds) Biotechnologie und Recht. Manz, Wien, pp 15–61

Larenz K (1991) Methodenlehre der Rechtswissenschaft, 6th edn. Springer, Berlin

Laufs A (2003) Auf dem Wege zu einem Fortpflanzungsmedizingesetz?: Grundfragen der artifiziellen Reproduktion aus medizinrechtlicher Sicht. Nomos Verlag, Baden Baden

Lüderssen K (2002) Der Schutz des Embryos und das Problem des naturalistischen Fehlschlusses: Skizze einer Zwischenbilanz. In: Graul E, Wolf G (eds) Gedächtnisschrift für Dieter Meurer. de Gruyter, Berlin, pp 209–236

Maio G (2002) Ist die Instrumentalisierung des Embryos moralisch zu rechtfertigen? Zur Ethik der Forschung an embryonalen Stammzellen. In: Oehmichen M, Kaatsch HJ, Rosenau H (eds) Praktische Ethik – in Klinik und Forschung. Lübeck, Schmidt-Roemhildt, pp 175–192

Merkel R (2001a) Früheuthanasie: rechtsethische und strafrechtliche Grundlagen ärztlicher Entscheidungen über Leben und Tod in der Neonatalmedizin. Nomos Verlag, Baden Baden

Merkel R (2001b) Grundrechte für frühe Embryonen? In: Britz G et al (eds) Grundfragen staatlichen Strafens: Festschrift für Heinz Müller-Dietz zum 70. Geburtstag. CH Beck, München, pp 493–521

Merkel R (2001c) Die Abtreibungsfalle. Die Zeit, 25 January, pp 37–38

Merkel R (2002) Forschungsobjekt Embryo. Deutscher Taschenbuch Verlag, München

Merkel R (2003) Grundrechte für frühe Embryonen? Normative Grundlagen der Präimplantationsdiagnostik und der Forschung an embryonalen Stammzellen. In: Beyreuther K, Merkel R, Fuchs T et al (eds) Wider die Natur. Studium Generale. Sammelbände der Vorträge des Studium Generale an der Ruprecht-Karls-Universität, Heidelberg, pp 23–39

Merkel R (2003b) Embryonenschutz, Grundgesetz und Ethik. In: Schweidler et al. (eds) Menschenleben – Menschenwürde. Lit Verlag, Berlin, pp 151–164

Mieth D (1999) Ethische Probleme der Humangenetik: Eine Überprüfung üblicher Argumentationsformen. In: Engels E-M (ed) Biologie und Ethik. Reclam Philipp Jun, Stuttgart, pp 224–256

Minwegen R (2011) Was meint ihr, wenn ihr vom Embryo sprecht? Frankfurter Allgemeine Zeitung 06 July, p N1

Neumann U (1998) Die Tyrannei der Würde. Argumentationstheoretische Erwägungen zum Menschenwürdeprinzip. Archiv für Rechts- und Sozialphilosophie 84:153–166

Nüsslein-Volhard C (2001) Der Mensch nach Maß – unmöglich. Süddeutsche Zeitung 1–2 December, p I

Nüsslein-Volhard C (2003) Wann ist der Mensch ein Mensch? C.F. Müller, Heidelberg

Pawlowski H-M (2002) Zu den Grundlagen der Bioethik: Verfassungsrecht oder Moral? Zum Verhältnis von Recht und Moral – nicht nur in Deutschland. Zeitschrift für Rechtsphilosophie 1:71–92

Prütting D (2004) Fachanwaltskommentar Medizinrecht. Hermann Luchterhand Verlag, München

Puppe I (2008) Kleine Schule des juristischen Denkens. Vandenhoeck & Ruprecht, Stuttgart

Reich J (2003) "Es wird ein Mensch gemacht": Möglichkeiten und Grenzen der Gentechnik. Rowohlt, Reinbeck

Rohwedel J (2002) Gewinnung und Verwendung gewebespezifischer und embryonaler Stammzellen. In: Hauskeller C (ed) Humane Stammzellen. Therapeutische Optionen – Ökonomische Perspektiven – Mediale Vermittlung. Pabst Science Publishers, Lengerich, pp 18–26

Rosenau H (2003) Reproduktives und therapeutisches Klonen. In: Amelung K (ed) Strafrecht, Biorecht, Rechtsphilosophie: Festschrift für Hans-Ludwig Schreiber zum 70. Geburtstag am 10, Mai 2003. C. F. Müller, Heidelberg, pp 761–781

Rosenau H (2004) Der Streit um das Klonen und das deutsche Stammzellgesetz. In: Schreiber H-L, Rosenau H, Ishizuka S, Kim S (eds) Recht und Ethik im Zeitalter der Gentechnik: deutsche und japanische Beiträge zu Biorecht und Bioethik. Vandenhoeck & Ruprecht, Göttingen, pp 135–168

Rosenau H (2006) Art. Biotechnologie. In: Heun W et al (eds) Evangelisches Staatslexikon. Kohlhammer, Stuttgart, columns 230–231

Rosenau H (2012) Ein zeitgemäßes Fortpflanzungsmedizingesetz für Deutschland. Nomos Verlag, Baden-Baden

Rosenau H (2015) Tip Hukuku Dergisi. Journal of Medical Law 4(8):233–253

Sass H-M (1989) Hirntod und Hirnleben. In: Sass H-M (ed) Medizin und Ethik. Reclam, Stuttgart, pp 160–183

Schindehütte J, Meyer B, Gruss P (2002) Woraus wir werden. Die Stammzellen und ihr Potential. In: Elsner N, Schreiber H-L (eds) Was ist der Mensch? Wallstein Verlag, Göttingen, pp 67–80

Schlingensiepen-Brysch I (1992) Wann beginnt das menschliche Leben? Zeitschrift für Rechtspolitik 11:418–422

Schlink B (2000) Aktuelle Fragen des pränatalen Lebensschutzes Überarbeitete Fassung eines Vortrages, gehalten vor der Juristischen Gesellschaft zu Berlin am 19. Dezember 2002. de Gruyter, Berlin

Schöne A (2002) 'Von vorn die Schöpfung anzufangen' – Goethes Homunkulus. In: Elsner N, Schreiber H-L (eds) Was ist der Mensch? Wallstein Verlag, Göttingen, pp 47–66

Schreiber H-L (1983) Rechtsprobleme bei Therapiestudien. Verhandlungen der Deutschen Krebsgesellschaft 4:13–15

Schreiber H-L (2003a) Xenotransplantation – rechtliche Aspekte. In: Grimm H (ed) Xenotransplantation: Grundlagen – Chancen – Risiken. Schattauer Verlag, Stuttgart, pp 315–322

Schreiber H-L (2003b) Das Menschenwürdeprinzip in der Diskussion über Embryonenforschung, Klonen und Präimplantationsdiagnostik. In: Bernsmann K, Ulsenheimer K (eds) Bochumer Beiträge zu aktuellen Strafrechtsthemen: Vorträge anlässlich des Symposions zum 70. Geburtstag von Gerd Geilen. Carl Heymanns Verlag, Köln, pp 157–170

Schreiber H-L (2011) Recht als Grenze der Gentechnologie. In: Schünemann B et al (eds) Festschrift für Claus Roxin zum 70. Geburtstag am 15, Mai 2001. de Gruyter, Berlin, pp 891–904

Schreiber H-L, Rosenau H (1998) Biotechnik: Rechtlich. In: Korff W, Beck L, Mikat P (eds) Lexikon der Bioethik, vol 1. A-F. Gütersloher Verlagshaus, Gütersloher, p 395

Schroth U (2002) Forschung mit embryonalen Stammzellen und Präimplantationsdiagnostik im Lichte des Rechts. Juristenzeitung, pp 170–179

Siep L (1998) Zue ethischen Problematik des Klonens. Jahrbuch für Wissenschaft und Ethik 3:5–14

Sönnecken I (2002) Die Nidation als Zäsur im Rechtsschutz menschlichen Lebens. Shaker, Aachen

Spickhoff A (ed) (2014) Medizinrecht, 2nd edn. CH Beck, München

Starck C (1999) Kommentar. In: von Mangoldt H, Klein F, Starck C (eds) Das Bonner Grundgesetz, vol 1. F. Vahlen Verlag, München

Taboada P (2003) Stammzellenforschung und Menschenwürde. In: Schweidler W (ed) Menschenleben – Menschenwürde. LIT Verlag, Münster, pp 129–150

Takahashi K, Yamanaka S (2006) Induction of pluripotent stem cells from mouse embryonic and adult fibroblast cultures by defined factors. Cell 126(4):663–676

Takahashi K et al (2007) Induction of pluripotent stem cells from adult human fibroblasts by defined factors. Cell 131(5):861–872

Taupitz J (2001) Der rechtliche Rahmen des Klonens zu therapeutischen Zwecken. Neue Juristische Wochenschrift 47:3433–3434

Taupitz J (2002) Therapeutisches und reproduktives Klonen aus juristischer Sicht. Z Arztl Fortbild Qualitatssich 96:449–455

Taupitz J (2002) Stammzellen: Rechtliche Aspekte. GenTechnik & Recht:10–16

Taupitz J (2003) Rechtliche Regelung der Embryonenforschung im internationalen Vergleich. Springer, Berlin

Trute H-H (2001) Forschung an humanen Stammzellen als Ordnungsproblem des Wissenschaftsrechts. In: Hanau P (ed) Wissenschaftsrecht im Umbruch: Gedächtnisschrift für Hartmut Krüger. Duncker & Humblot, Berlin, pp 385–395

von Bülow D (1997) Dolly und das Embryonenschutzgesetz. Deutsches Ärzteblatt 94(12):718–719

von der Pfordten D (1998) Klonierung als Manipulation. In: Ach J et al (eds) Hello Dolly?: über das Klonen. Suhrkamp, Berlin, pp 213–219

von Loewenich V (2002) Ab wann ist der Mensch ein Mensch. Biologische Gesichtspunkte. Annales d'histoire et de philosophie du vivant 7:43–49

Yu J et al (2007) Induced pluripotent stem cell lines derived from human somatic cells. Science 318(5858):1917–1920

Wessels J, Beulke W, Satzger H (2015) Strafrecht: Allgemeiner Teil, 45th edn. C.F. Müller, Heidelberg

Winnacker E-L (2003) Heilversuche mit Stammzellen in wenigen Jahren denkbar. Süddeutsche Zeitung 25–26 January, p 6

Witteck L, Erich C (2003) Straf- und verfassungsrechtliche Gedanken zum Verbot des Klonens von Menschen. Medizinrecht 21(5):258–262

Wolbert W (2003) The accusation of double standard in the debate on embryonic stem cell research. DMW – Deutsche Medizinische Wochenschrift 128(9):453–456

Chapter 11
Germline Gene Therapy in the Era of Precise Genome Editing: How Far Should We Go?

Peter Sýkora

11.1 Introduction

For decades, geneticists have followed the ethical rule "do not interfere with germline genes in humans" and for a very good reason which nobody understands better than geneticists themselves – it would be extremely irresponsible to change the human germline with the technologies used for the creation of genetically modified organisms. Apart from the risk rate aspect, more arguments have been raised including ethical, religious, social, political ones against germline gene engineering in humans – the risk of destruction of the essence of the human, the genetic fundamentals of those properties which make a human, as well as jeopardizing the fundamentals of democratic institutions, playing God, reviving the practices of eugenics, creating a new genetic aristocracy ("genobility"), and new forms of social discrimination which would lead so far as the genocide of the human species (for overview: Silver 1997; Resnik et al. 1999; Frankel and Chapman 2000; Annas et al. 2002; Stock 2003; Rasko et al. 2006; Mehlman 2012; Knoepfler 2015). In 24 countries (including France, Germany, UK, Italy, Canada) germline gene modification in humans has been prohibited by law; in other countries like China, Japan, India and USA only by guidelines (Ishii 2015).

But the situation has recently dramatically changed and the genetic taboo was broken twice in 2015. First, in February by the decision of the British Parliament to change the law prohibiting germline intervention in humans in order to allow fertility clinics to carry out mitochondrial donation/replacement techniques which alter the germline. Second, in April, by of the appearance of a publication written by Chinese scientists in which they described the first-ever experiment using a newly emerging technology, CRISPR-Cas9, for editing germline genes in non-viable human embryos.

P. Sýkora (✉)
University of Sts. Cyril and Methodius, Trnava, Slovak Republic
e-mail: petersykora111@gmail.com

© Springer International Publishing AG, part of Springer Nature 2018
M. Soniewicka (ed.), *The Ethics of Reproductive Genetics*,
Philosophy and Medicine 128, https://doi.org/10.1007/978-3-319-60684-2_11

The first event was preceded by a several-year long discussion in the UK in academic circles as well as in public (HFEA 2015). The second event of 2015 provoked an avalanche of new discussions on the consequences brought about by the new technologies of genome editing, particularly the rapidly developing CRISPR-Cas9 technology among scientists, bioethicists, in both professional and public affairs organizations. It began with a meeting of a small group of scientists and bioethicists in Napa, California in January 24, which initiated a global discussion headed by David Baltimore culminating in the International Summit on Human Gene Editing in December 1–3, in Washington, D.C.; it was organized by the National Academy of Sciences and the National Academy of Medicine's Human Gene-Editing Initiative and co-hosted with the Chinese Academy of Sciences and the U.K.'s Royal Society. In his opening remarks David Baltimore, the initiator of the summit and chair of the organizing committee said:

> We are taking on a heavy responsibility for our society because we understand that we could be on the cusp of a new era in human history. Today, we sense that we are close to being able to alter human heredity. Now we must face the questions that arise. How, if at all, do we as a society want to use this capability? (Olson 2016, 1)

In this study I focus solely on the safety aspects associated with the potential germline gene therapy, while leaving aside the moral, ethical and societal aspects of inheritable genetic modifications in humans, which are no less important and extensively discussed in professional writings.

I will demonstrate that the new emergent genome editing technologies are taking us very rapidly into an era wherein it will be possible to edit genes without any unwanted offtarget changes, which makes the germline gene therapy a potentially safe technology. However, only for the cases where a healthy wild-type DNA sequence would be restored by the precise DNA repair of faulty genes, since the contemporary knowledge of human genome complexity is still very limited and the risk of unintended harmful consequences is high.

11.2   Gene Modifications in Humans

Modifying genes in an organism does not automatically mean that these changes are also hereditary and will be transferred to the offspring. There are two different ways of intervening in the genetic information in multicellular organisms. Either with DNA of specialized cells of the body (so-called somatic cells), or with DNA in sperms and eggs, or with DNA in early embryos, or with DNA of embryonic stem cells from which the reproductive cells are formed (so-called germline cells).

In the first case, intervention in genetic information is not hereditary. When an individual with the altered DNA in somatic cells dies, his altered DNA will also cease to exist as the genetic change will not affect the DNA in the germline cells that give rise to the next generation. In contrast, in the second case in which the DNA in germline cells is altered, the genetic alteration is hereditary as it is transferred to all subsequent generations. These cases are most often referred to as *germline gene modification, germline gene intervention, germline genetic/gene engineering*, and most recently *germline gene/genome editing*.

Not all genes in the cell are located on the chromosomes in the cell nucleus. In human cells, 37 genes out of the total of approximately 23,000 are located in the mitochondria. The U.K.'s Parliament approved in 2015 two techniques of mitochondrial replacements, the pro-nuclear transfer and the maternal spindle transfer, when at the end faulty mitochondria with mutant mtDNA in mother's ova are replaced with healthy mitochondria carrying unmuted mtDNA from donor ova. However, this form of germline gene modification is not a part of recent genome editing debate which has focused only on nuclear gene manipulations.

A half century of debate on the ethics of gene manipulations in humans in terms of their safety, moral justification, and the potential social impact has led to a gradual establishment of an ethical matrix with four alternatives created by the crossing of two ethical demarcation lines, around which the whole debate revolves (Anderson 1989). The first demarcation line is the differentiation between manipulations with human genes with the aim of repairing them (gene therapy) and manipulations with physiologically normal genes with the goal of improving them (genetic enhancement); its drawback in terms of practical policy is the fact that this line is blurred, with no clear-cut distinction between therapy and enhancement in general. Another demarcation line is the differentiation between interventions into somatic cell genes and germline genes. This red line has the advantage of being sharp because, from the biological point of view, somatic cells can be clearly distinguished from germline cells.

What is generally considered to be an ethically non-problematic alternative is somatic cell gene therapy – genetic interventions in somatic cell genes. Before 2012, about 2000 pre-clinical and clinical trials of somatic cell gene therapy took place worldwide and it is on the way to becoming a standard therapy method in the near future (Ginn et al. 2013). The first remedies based on somatic gene therapy have already been approved by the relevant institutions – Gendicine in China and Glybera in the European Union.

The remaining three alternatives are in various degrees controversial. The most controversial is germline gene enhancement while somatic gene enhancement is less so. Germline gene therapy is also considered controversial despite the fact that it is a therapy, as there is the question of an informed consent of future generation persons, as well as the concern about the slippery slope towards germline gene enhancement. Despite the fact the phrase could have negative connotations some scientists suggest to speak openly in case of altered germline in humans about "genetically modified humans" or "GM humans" (Knoepfler 2015).

11.3 A Specter of GM Humans Has Been Haunting Public Opinion for Decades

Several months before the Chinese experiment with editing the human embryo genome was published in the Springer Open access on-line scientific journal *Protein and Cell* in April 2015, a small group of elite experts in genome editing met in Napa, California, on January 24 in response to the unofficial information on such experiments in China, to discuss the risks associated with using the new CRISPR-Cas9 technology in biomedicine. The summary of their meeting was later published (March 19) in the scientific journal *Science* (Baltimore et al. 2015). The speed with which the scientific community responded to the first rumors on editing the human embryo genome may seem surprising and lead to the assumption of an exaggerated or even hysterical reaction from scientists, before it is examined in a wider historical context. Only then can such a response from scientists prove fully justified. They sent out a joint message to society that despite the ubiquitous and Hollywood-generated Frankenstein image of the "mad scientist", they can take responsibility. We have witnessed a biotechnological revolution associated with the development of genetically modified organisms (GMO). If we realize that today there exists, in a considerable part of society, especially in Europe, a negative attitude even towards GM plants it is not difficult to imagine what resistance might develop in society if information emerges that scientists have the technology to very easily create GM people.

Moreover, the risk that lay society is unable to understand the difference between editing genes in somatic cells, which creates changes not transferred to further generations, and inheritable germline gene editing, and therefore will respond to gene editing in humans by its overall ban, was pointed out in the open letter in the prestigious *Nature* by the scientists using genome editing for advanced somatic cell gene therapy. According to them, "[K]ey to all discussion and future research is making a clear distinction between genome editing in somatic cells and in germ cells." Therefore, they propose that

> [A] voluntary moratorium in the scientific community could be an effective way to discourage human germline modification and raise public awareness of the difference between these two techniques. (Lanphier et al. 2015, 411)

What is interesting is that it was precisely the appeal for a better biological education of the public, in an attempt to save the promising genetic research for the potential treatment of genetic diseases from an overall ban, that the whole debate on gene manipulations in humans started at the beginning of the 1970s. At that time it was about sheer speculation based on the most recent revolution knowledge in biology in 1960s, such as deciphering the genetic code and discovery of how to introduce foreign genes into the cells of micro-organisms by means of the virus (later, this principle was used for somatic cell gene therapy).

The public learned about the upcoming biological revolution from the popular science bestseller by the British author and journalist G.R. Taylor "The Biological

Time Bomb" (Taylor 1968) and the extensive cover story "Man into Superman: the promise and peril of the new genetics" published by TIME magazine in April 1971. It was here that the question was raised of whether biologists are sufficiently responsible and whether their research should not be supervised by the public, as modern genetics give the man an unprecedented possibility to biologically "reshape himself", and thus gain control over his own biological evolution. In this respect, the article refers to the biblical story of the forbidden fruit from the tree of knowledge by which the man will become God the Creator. By the way, it was in this article in the Time magazine that the newly proposed term "bioethics" (Jonsen 1998, 27) was used in the media for the first time ever.

Bernard Davis in his study "Prospects for Genetic Intervention in Man" published at the end of 1970 in Science magazine (perhaps the first ever study on this topic), responding to the attempts to ban all genetic interventions, including the cure of hereditary diseases, reminds us of how the "politically based attack on science, Lysenkoism, utterly destroyed genetics in the Soviet Union" (Davis 1970, 1283). Davis does not argue that the project of the genetic improvement of humans is unethical, but rather he argues that from the scientific point of view it makes no sense and is impossible to realize it, so the concerns of the public are futile and are a result of its ignorance of genetics.

Forty-five years later, Time magazine, in its July 4th, 2016 issue, came up with a cover story dedicated to the huge potential of CRISPR-Cas9 genome editing technology for genetic engineering. The article considers that the impact of CRISPR technology on the democratization of gene engineering technology could be similar to that of the shift from big mainframe computers to personal computers had on the democratization of IT technologies. The misgivings about abuses of CRISPR technology are mentioned in the article in connection with the creation of super humans (designer babies in today's terminology), as well as in connection with its potential abuse by hostile states or terrorist groups to create a genetic weapon, as was also raised in the aforementioned TIME cover story of 1971. According to Time magazine, a killer mosquito that transmits a deadly disease, or a DNA-damaging virus, that could infect human cells and decimate the population, would constitute such a weapon. It is one of the reasons, as the article states, why, at the beginning of this year, the Director of National Intelligence James Clapper put gene-editing methods like CRISPR on a list of mass-destruction threats (Park 2016).

11.4 Taboo Breaking Experiments with Genome Editing in Human Embryos

In the taboo breaking experiment from April 2015, Chinese scientists from Sun Yat-sen University in Guangzhou used human tripronuclear zygotes obtained from local fertility clinics in order to repair β-thalassemia mutations in human β-globin (HBB) gene with the help of CRISPR-Cas9 gene editing technology (Liang et al. 2015).

Tripronuclear zygotes (3PN) are polyspermic zygotes with an extra set of chromosomes which occur in frequency about 2–5% during in vitro fertilization, and are known to be unable of further development in vivo, so there is not even a theoretical possibility that such a genetically manipulated embryo would develop into a child. The main goal of the experiment was to fill the gap in our understanding of whether CRISPR-Cas9 genome editing, already successfully used for gene editing in monkey embryos and human cells, would also work in early human embryos.

In a paper published a year later, another group of Chinese scientists (Kang et al. 2016) has conducted a similar experiment for the purposes of evaluating the CRISPR-Cas9 technology for the introduction of precise genetic modifications in early human 3PN embryos. In this study, rather than repair of genetic mutation, the goal was to introduce the Δ32 mutation precisely at the CCR5 locus and for good reason. The CCR5Δ32 allele is a naturally occurring 32-base-pairs deletion which introduce a premature stop codon into a gene coding CCR5 membrane receptor for HIV virus on T cells. Due to a nonfunctional receptor, the HIV virus is not able to enter into T cells. Individuals homozygous for this mutation are resistant to HIV infection. The idea to knock-out CCR5 gene in T cells as a strategy for somatic cell gene therapy of HIV infection is now in fact the most advanced use of new genome editing technologies (see below) for treatment. Until October 2015 several Phase 1 clinical experiments using autologous T cells with CCR5 gene knocked-out with Zinc Finger Nuclease were successfully completed and entered Phase 2. The experiments seem to support pre-clinical experiments on animal models and human tissue cells that T cells with ex vivo mutated CCR5 gene transplanted back would lead to a decrease in HIV viral loads. Recently, other genome-editing techniques, TALENs, meganucleases and CRISPR-Cas9 were also applied for knocking-out CCCR5 gene in T cells (see for review: Maeder and Gersbach 2016).

The Chinese experiments on human 3PN embryos resulted in efficient and targeted generations of CCR5Δ32 mutations in four of the 26 total samples, but they also created offtarget indel mutations at the same locus. However, we have to see the Chinese experiments in a broader context of an already discussed idea (e.g. Harris 2007). Why not introduce CCR5Δ32 mutation into the germline and thus make all of the following generations of humans genetically resistant to HIV, of course when the technology would be safe enough and ethically accepted? The Chinese scientists themselves in their paper have pointed out that clinical application of the kind of experiments they have been doing is premature (Liang et al. 2015), that question of dangerous side effects as off-target and mosaicism has to be resolved first and more comprehensive understanding of the CRISPR-Cas9 gene editing mechanisms is necessary. In their view "it is foreseeable that a genetically modified human could be generated" and therefore any attempt to do it "needs to be strictly prohibited until we can resolve both ethical and scientific issues" (Kang et al. 2016). However, the use of genome editing for germline modification in principle has not been excluded in the paper.

11.5   Precise Genome Engineering Revolution

Chinese scientists used in their experiments on 3PN human embryos the CRISPR-Cas9 technology, which is now the most popular genome editing technology (Hsu et al. 2014; Ledford 2015a). There are four emerging genome editing platforms which have been recently developed on a "path to clinic" (Corrigan-Curray et al. 2015): zinc finger nucleases (ZFNs), transcription activator-like effector nucleases (TALENs), meganucleases, and clustered regularly interspaced short palindromic repeats associated with a endonuclease Cas9 (CRISPR-Cas9).

What makes a difference between these gene editing platforms and recombinant DNA techniques is targeting. In contrast to classical genetic engineering, which uses restriction enzymes and ligases, genome editing technologies can introduce targeted changes into DNA with single base-pair precision. Restriction enzymes cut DNA at a specific short DNA sequence or nearby, but there are many randomly distributed restriction sites in a long molecule DNA. The targeted cut has to be selected out from a huge number of random cuts. In contrast, genome editing platforms use programmable nucleases, which means that they can be designed to cut a DNA at a specific DNA sequence.

All four endonucleases generate double-strand breaks (DSBs) in DNA which induce two types of naturally occurring cellular mechanisms of DNA repair – the nonhomologous end joining (NHEJ) repair pathway and the homology-dependent repair (HDR). NHEJ repair can be used in experiments for creation of targeted mutations in genome because it introduces small sequence insertions or deletions (so called *indels*). HDR repair pathway is able to incorporate a new DNA sequence synthetized by the user into a target DNA break. In this way a complete new sequence can be inserted into a specific site of DNA or a mutant sequence can be replaced by a correct wild type sequence (Maeder and Gersbach 2016). Similarities with a cut-and-paste function in word editing software is what gave the name to this technology.

ZNFs, TALENs and meganucleases recognize target DNA sequence directly by a protein-DNA interaction. To program those endonucleases for a specific target DNA sequence one has to specifically design the endonucleases. For each target sequence a new nuclease has to be designed in a rather complicated way. In contrast, the Cas9 endonuclease does not recognize target sequence but so called guide RNA to which Cas9 is attached. A sequence in well-defined proximity to target sequence is recognized by short single nucleotide guide RNA sequence on RNA-DNA complementarity interaction. Compared to the designing of a endonuclease, which is a protein, the synthesis of short complementary RNA is easy, cheap and can be ordered in commercial companies. It is also very easy to use. For example, a simple co-injection of guide RNA, Cas9 nuclease (or mRNA coding Cas9 nuclease) and short DNA template (if the goal is to insert a specific sequence) into a single cell zygote (or any cell) is all that is needed for the introduction of targeted DNA change.

It is so easy, effective and cheap that DIY biologists are eager to use it for their citizen-science/amateur genome editing projects (Ledford 2015b). CRISPR-Cas9 platform seems to be the most promising and most widely used emerging genome editing platform in biotechnology. It has already been successfully used in genome editing experiments in more than 50 different organisms, including microbes, plants, animals, and human and it is generally considered to be a revolutionary new tool in biotechnology and biomedical research, "the biggest game changer to hit biology since PCR" (Ledford 2015a; Hsu et al. 2014).

Genome editing has recently been successfully used in more than 30 preclinical studies of somatic cell gene therapy and few phase I clinical trials with the use of ZFNs and TALENS (for recent review see: Cox et al. 2015; Maeder and Gersbach 2016). For example, for already mentioned CCR5 knocking-out gene clinical trials entering phase II, as well as for experiments with gene corrections of β-globin mutations causing sickle-cell disease or β-thalassemia, and other hematologic disorders like X-SCID, ADA-SCID, RS-SCID. In case of Duchenne muscular dystrophy gene function correction by targeted sequence insertions or reading frame restoration by sequence deletions has been used, in case of cystic fibrosis gene correction in stem cells.

Wide public attention was devoted to the first case of a life being saved by a gene editing procedure, namely that of a 1-year-old girl dying from leukemia after all conventional treatments failed (Le Page 2015). Scientists with the use of TALENS edited the genome in T-cells from a healthy donor in order to disable a gene receptor so the recipient's immune system does not reject them when they are used for bone marrow transplants into the patient.

Recently, on June 21, 2016, an advisory committee at the US National Institute of Health (NIH) approved the phase I clinical trials using CRISPR-Cas9 genome editing technique for a somatic cell gene therapy (Reardon 2016). A similar somatic cell gene therapy experiment has been approved by the West China Hospital ethics review board for a team of scientists from Sichuan University. At the same time British scientists from the Francis Crick Institute have been granted by the Human Fertilisation and Embryology Authority (HFEA) permission to deactivate genes in leftover embryos from IVF clinics genome with the use of CRISPR-Cas9 technique to study early embryo development (Knapton 2016).

11.6 Genome Editing Technology Easy to Use

The revolutionary benefits of CRISPR-Cas9 for inheritable genetic modifications lie not only in that it is able to create targeted germline gene modifications, but also in the fact of how simple it is to work with this technique. To achieve targeted modifications in the genome with classical recombinant DNA techniques by random mutations and selection had to be done through a complex and lengthy procedure which could not be practically applied to primates including humans. In frequently used technique of pre-genome editing, a foreign DNA is entered by means of a vector (usually an altered adenovirus or HIV virus) into embryonic stem (ES) cells cultivated in vitro. Since the vector with foreign DNA inserts randomly into chromosomes, it is necessary with the help of antibiotic and metabolic markers to select out all of the cells with offtarget insertions and pick out those few ES cells with targeted insertions. A nucleus is then transferred from those targeted ES cells into an enucleated zygote which is implanted into the mother. In offspring developed from such embryos all somatic cells have targeted gene modification including germline cells, so the targeted gene modification is transferred to future generations (Capecchi 2000).

In an earlier version of the method (Capecchi 1994), before the somatic cell nuclear transfer method (the Dolly cloning method) was discovered, the protocol was even more complicated. Targeted ES cells had to be injected into a blastocyst stage embryo first, and then implanted into the mother. Only chimeric offspring were received this way and they had to be later mated and then their offspring selected (identified by DNA analysis) for homozygotes in a target modified gene.

With the CRISPR-Cas9 technique, transgenic mice, rats and finally non-human primates as animal models of human diseases have already been achieved. Targeted germline gene editing in cynomolgus monkeys (*Macaca fascicularis*) by Niu et al. (2014) was particularly appreciated since transgenic mice and rats with their simple nervous system cannot be used as animal models of neuropsychiatric disorders. On the other hand, the creation of a generation of monkeys genetically engineered to have Parkinson's, Alzheimer's or autism has been criticized from an animal welfare perspective (Coghlan 2016).

It is important to point out that the modification of the human germline is far from being the only research goal of genome editing in early human embryos. With the help of precise gene modifications, like gene knockout, it is possible to obtain fundamental knowledge about what genes are needed for the healthy development of an early embryo. However, Chinese genome editing in human embryos can be seen rather as the next logical step in a long journey from the first recombinant DNA bacteria and transgenic animals, through knockout mutations in transgenic mice models and transgenic non-human primates, towards future transgenic humans.

11.7 How Far Should We Go in Germline Gene Modifications in Near Future?

The reason why the scientific community began immediately to raise the alarm over the Chinese experiments with genome editing in human embryos is that it is basically a proof-of-principle experiment, thus breaking the technological barrier which for decades had successfully prevented any ideas of germline gene modifications in humans from being realized. These experiments of Chinese scientists clearly demonstrated that genome editing technology can also function in the cells of the early human embryo (that e.g. NHEJ and HDR repair mechanisms are functional in them), and that by means of this they can induce targeted modifications in germline genes in humans.

As we have already shown, the traditional gene engineering technologies used to create GM plants and GM animals are unacceptable in higher primates. Unacceptable in the same way are methods successfully used in humans for somatic gene therapy. In both cases the reason is the same – sites of genome modifications are not under control. In the 2003–2005 clinical trials of somatic gene therapy of boys with SCID-X1 (a severe combined immunodeficiency, known as "bubble boy syndrome") 5 of the 25 treated children developed leukemia. Later analysis confirmed that the development of leukemia was a consequence of offtarget insertions of retroviral vector and transgene into the genome which activated some proto-oncogenes and deactivated a tumor suppressor gene in these patients (Hacein-Bey-Abina et al. 2008). Had it been a case of germline gene instead of somatic cell gene therapy, it would have resulted in the creation of a new inheritable leukemia disease.

One of the very frequent objections to germline gene therapy is that it is unnecessary since genetic diseases can be eliminated in future generations by means of IVF combined with preimplantation genetic diagnostics (PGD). It is true for most cases of affected or unaffected heterozygous parents carrying one of about 4000 monogenic diseases identified yet, when we wish to avoid of devastating genetic diseases in progeny. But it does not work when one of the parents is homozygous for a dominant disease gene. Then all embryos are affected and PGD is not applicable. The only way to eliminate a disease gene from the progeny of such a couple would be germline gene therapy. In general they are very rare, but some of them, like familial hypercholesterolemia, myotonic dystrophy, Huntington's disease, neurofibromatosis, polycystic kidney disease have a higher incidence in specific geographic areas (Nussbaum et al. 2007, 126–127). PGD is also inapplicable in the case of common diseases with a polygenic component, like heart disease, Alzheimer's, schizophrenia and obesity, because the number of allele combinations involved far exceeds the number of IVF embryos available for PGD selection. The multiplexable functioning of the CRISPR platform and its ability to change many genes simultaneously has been already demonstrated in animals (the simultaneously targeted three distinct genome loci in pigs in order to generate an animal Parkinson's disease model – Wang et al. 2016) which makes the genome editing technology applicable also in cases of polygenic inheritance in the future.

Another very common safety objection is that the specific risk of the germline gene intervention is just too high because it would affect many individuals (without their consent). In fact, it would permanently change the genome in all subsequent descendants and therefore gene therapy should exclusively focus on somatic cell gene interventions when a risk-benefit balance can be achieved in a similar way as in other standard cases of therapy (not to mention the fact that a patient's informed consent can be obtained in this gene therapy in contrast to germline therapy).

However, a germline gene intervention also has an advantage over somatic cell gene therapy (Capecchi 1994). In developmental genetic diseases when damage in embryogenesis has already happened with irreversible harmful consequences, somatic cell gene therapy is useless. In contrast to somatic gene therapy, when the transgene is introduced to target tissue, germline gene therapy might be no more, but less complicated, since targeted gene modification in the early embryo will be at the right place at the right time, so there will be no problem with delivering a transgene into a cell, cytotoxicity, stability, expression or an adverse immune reaction.

Therefore a number of scientists have begun to consider the possibility of separating germline gene therapy from the risks associated with inheritable intervention in the germline. In other words, to get inheritance of germline modifications under control, not only for safety reasons, but also for the fact that any genetic improvement that is optimal now will be very likely "outmoded" within the next 20–30 years (Capecchi 2000, 38). In this regard, several strategies have been considered, e.g. putting transgenes on an artificial chromosome, or inserting transgenes in cassettes at docking sites on chromosomes together with a special regulatory gene which controls of transgene expression. A regulatory gene could be, for example, a gene for an ecdysone receptor (ecdysone is an insect hormone not occurring in humans). Unless the hormone ecdysone is injected into the body, the transgene remains inactive in cells. In such a way, the transgene can be either activated when it is needed (e.g. transgene with anticancer function) or a transgene can be turned off by offspring if they do not wish it to function (Campbell and Stock 2000).

The bottom line according to Capecchi is that "whatever procedures we might adopt for human germline therapy, they should, at the very least, be reversible. Fortunately, this can be accomplished" (Capecchi 2000, 8–40). What he had in mind was to exploit selective deletion mechanisms via a site-specific recombinase known as CRE. Now, genome editing technologies provide more effective tools for a precise reversion of gene modification to a previous state. For example, scientists with the use of CRISPR-Cas9 technology recently succeeded in precisely removing the entire HIV-1 provirus integrated in genome of human T-cells (Kaminski et al. 2016). In the recent proposal of guidelines for CRISPR germline editing therapies it is suggested that any preclinical research proposal has to include a companion specific reversal mechanism and multigenerational animal models of increasing complexity (Evitt et al. 2015).

At the present time the CRISPR-Cas9 and other genome editing technologies also create, together with targeted modifications, several off-target changes in genome. This is the main reason why scientists have now unanimously called for a moratorium on any attempt to use genome editing of the human germline for clinical applications. On the other hand, there has been rapid progress in improving the precision of the CRISPR-Cas9 platform. Recently, just in the last few months, several new studies have been published to demonstrate different strategies of how to reduce the off-target effects caused by the fact that Cas9 endonuclease can cleavage sites that are not fully complementary to the RNA guide (Slaymaker et al. 2016; Paquet et al. 2016). Even they succeeded in engineering CRISPR-Cas9 in such a way that it is able to edit the DNA sequence directly at the level of single base-pair at the specific target without a double-strand DNA backbone cleavage and a donor DNA (Komor et al. 2016).

We are quickly approaching a new era of precise genome editing, which means editing without off-target effects. Does it mean then that in this situation decisions on germline gene interventions will be taken solely on ethical and social grounds? Definitely not. Off-target effects are not the only purely technological risk associated with interventions in the human genome. The risk results from the fact that even targeted and well-meant interferences can have unwanted and dangerous consequences, if they are interfering in complex natural systems which we are yet to fully understand. This risk was clearly pointed out by the President's Bioethics Committee Report "Beyond Therapy" (Kaas et al. 2003), which, in this respect, speaks of applying the "precautionary principle" and the need to learn a lesson from environmentalists who know how dangerous it can be to interfere in natural systems that are "highly complex and delicately balanced as a result of eons of gradual and exacting evolution" (Kaas et al. 2003, 287–288). Fukuyama, in his book *Our Post-Human Future*, connects this risk to the possibility of fatal damage to the human essence, brought about by well-meant small interferences, because we still do not even know what the human essence is (he calls it "Factor X"), or how it is genetically endowed; what we know though is that it is a complex whole in terms of the complex system theory. For such complex systems it follows that even a small change might lead to catastrophe ("Butterfly effect") by destroying a well-adjusted balance (Fukuyama 2002, 170–173). However, such interference, as also asserted in both the Presidential Bioethics Committee Report and Fukuyama's book, is not meant to be therapy, but rather human enhancement, the endeavor to somehow improve humans, especially by biotechnology.

This attempt to argue for the protection of the human genome against any enhancement, especially a genetic one, has been ridiculed by Harris (2007) as naïve and unrealistic. The complexity of the human genome portrayed as "a seamless web", a "house of cards" has been criticized by Buchanan as a pure assumption made "from the armchairs" and ignoring "biological sciences which do not support it" (Buchanan 2011). He has argued that the human genome is not a fragile complex system, but a quite robust system as it has been demonstrated by successful experiments with transgenic animals. However, later developments in biology seem to confirm that the complexity of the human genome is much higher than previously expected. According to a systematic study of the human genome conducted recent years by ENCODE (Encyclopedia of DNA elements) project, the international consortium of more than 400 scientists in 32 labs worldwide, the human genome is not a collection of independent genes or their modules, but a complex network in which genes, along with regulatory elements interact in overlapping ways (Maner 2012; Djebali et al. 2012). More than 80% of the human genome, until recently considered to be parts of the genome with no function (so-called junk DNA), are estimated to be regulatory sequences which control the activity of 23,000 human genes (which in turn equal to only 1,5% of the genome). Even some well-known advocates of human enhancement and post-humanism like Nick Bostrom and Allen Sandberg accept that long term evolutionary "wisdom of nature" is embedded into the complexity of the human genome, and it would be easy to destroy it unless one follows the evolutionary heuristic approach they have proposed (Bostrom and Sandberg 2009). Anyway, what should be highlighted is that the precautionary "environmentalist argument" by the Presidential Report and by Fukuyama, based on the complexity of human genome, is an argument against germline gene enhancement, not against germline gene therapy.

Gene therapy, as it has evolved within the development of somatic gene therapies in recent years, today includes a broad variety of gene manipulations with a therapeutic aim, and the correction of inherited monogenic diseases (like cystic fibrosis) represents only 22.4% of them. By far the most clinical trials of gene therapy (64.4%) have been aimed at the treatment of cancer (transfer of tumor suppressor p53 gene into cells). Among other diseases approved for gene therapy clinical trials are cardiovascular diseases, infectious diseases (like hepatitis B and C, adenovirus, influenza), neurological diseases (like Alzheimer's, epilepsy, Parkinson's), ocular diseases and inflammatory diseases (Ginn et al. 2013).

It is very important to be aware that the kinds of gene therapy approved as safe enough for somatic gene therapy could be very risky as germline gene therapy. With one exception – when it is about a gene correction, when a mutant, disease causing DNA sequence, is repaired by a targeted genome editing to a healthy, wild-type DNA sequence. All other kinds of gene therapies, as mentioned above, are in fact alternations of a genome, although for therapeutic aims. However, such therapeutic genome alternations might bring the risk of unintended harmful consequences if the complexity of gene interactions in the human genome is not very well understood.

11.8   Conclusion

So, how far should we go with germline gene interventions in humans in the era of precise genome editing technologies? I believe that recent empirical biological studies revealing the very complex functional structure of the human genome, which we are only beginning to understand, strongly support the argument against any alternation of germline genome, whether for therapeutic or enhancement aims. At the present level of knowledge about the complexity of the human genome we can hardly go further than it is the precise DNA repair of faulty genes, with no offtarget effects. However, I believe that reversal mechanisms should be demanded as a *conditio sine qua non* for all germline gene interventions, including gene corrections.

Acknowledgements   This work was supported by the Slovak Research and Development Agency under the contract No. APVV-0379-12.

References

Anderson WR (1989) Human gene therapy: why draw a line? J Med Phil 14:681–693
Annas GJ, Andrews LB, Isasi RM (2002) Protecting the endangered human: toward an international treaty prohibiting cloning and inheritable alterations. Am J Law Med 28(2/3):151–178
Baltimore D, Berg P, Borchan M et al (2015) A prudent path forward for genomic engineering and germline gene modification. Science 248(6230):36–38

Bostrom N, Sandberg A (2009) The wisdom of nature: an evolutionary heuristic for human enhancement. In: Savulescu J, Bostrom N (eds) Human enhancement. Oxford University Press, Oxford, pp 375–416

Buchanan A (2011) Beyond humanity? Oxford University Press, Oxford

Campbell J, Stock G (2000) A vision for practical human germline engineering. In: Stock G, Campbell J (eds) Engineering the human genome. Oxford University Press, New York, pp 9–16

Capecchi MR (1994) Targeted gene replacement. Sci Am 270(3):52–59

Capecchi MR (2000) Human germline gene therapy: how and why. In: Stock G, Campbell J (eds) Engineering the human genome. Oxford University Press, New York, pp 31–42

Coghlan A (2016) First monkey genetically engineered to have Parkinson's created. New Scientist on-line (15 June 2016, update 21). https://www.newscientist.com/article/mg23030784-200-first-monkey-model-of-parkinsons/. Accessed 10 Jul 2016

Corrigan-Curray J, O' Reilly M, Kohn DB et al (2015) Genome editing technologies: defining a path to clinic. Mol Ther 23(5):796–806

Cox DBT, Platt RJ, Zhang F (2015) Therapeutic genome editing: prospects and challenges. Nat Med 21(2):121–131

Davis BD (1970) Prospects for genetic intervention in man. Science 170:1279–1283

Djebali S, Davis CA, Merkel A et al (2012) Landscape of transcription in human cells. Nature 489:101–108

Evitt NH, Mascharak S, Altman RB (2015) Human germline CRISPR-Cas modification: toward a regulatory framework. Am J Bioeth 15(12):25–29

Frankel MS, Chapman AR (2000) Human inheritable genetic modifications. Assessing scientific, ethical, religious, and policy issues. American Association for the Advancement of Science, Washington, DC

Fukuyama F (2002) Our posthuman future. Farrar, Straus and Giroux, New York

Ginn SL, Alexander IE, Edelstein ML et al (2013) Gene therapy clinical trials worldwide to 2013 – an update. J Gene Med 15:65–77

Hacein-Bey-Abina S, Garrigue A, Wang GP et al (2008) Insertional oncogenesis in 4 patients after retrovirus-mediated gene therapy of SCID-X1. J Clinic Investig 118(9):3132–3142

Harris J (2007) Enhancing evolution. Princeton University Press, Princeton

HFEA (2015) Statement on mitochondrial donation. http://www.hfea.gov.uk/9606.html. Accessed 27 Jul 2016

Hsu PD, Lander ES, Zhang F (2014) Development and application of CRISPR-Cas9 for genome engineering. Cell 157:1262–1278

Ishii T (2015) Germ line genome editing in clinics: the approaches, objectives and global society. Brief Funct Genomics. doi:10.1093/bfgp/elv053

Jonsen AR (1998) The birth of bioethics. Oxford University Press, New York

Kaas L, Balckburn E, Dresser RS et al (2003) Beyond therapy: biotechnology and the pursuit of human improvement. The President's Council on Bioethics, Washington, DC

Kaminski R, Chen Y, Fischer T et al (2016) Elimination of HIV-1 genomes from human T-lymphoid cells by CRISPR/Cas9 gene editing. Sci Rep 6:22555. doi:10.1038/srep22555

Kang X, He W, Huang Y et al (2016) Introducing precise genetic modifications into human 3PN embryos by CRISPR/Cas-mediated genome editing. J Assist Reprod Genet 33:581. doi:10.1007/s10815-016-0710-8

Knapton S (2016) British scientists granted permission to genetically modify human embryos. The Telegraph. http://www.telegraph.co.uk/science/2016/03/12/british-scientists-granted-permission-to-genetically-modify-huma/. Accessed 8 Jul 2016

Knoepfler P (2015) GMO sapiens. World Scientific Publishing Co., Singapore

Komor AC, Kim YB, Packer MS et al (2016) Programmable editing of a target base in genomic DNA without double-stranded DNA cleavage. Nature 533:420–424. doi:10.1038/nature17946

Lanphier E, Urnov F, Haecker SE et al (2015) Don't edit the human germ line. Nature 519:410–411

Ledford H (2015a) CRISPR, the disruptor. Nature 522:20–24

Ledford H (2015b) Biohackers gear up for genome editing. Nature 524:398

Le Page M (2015) Layla's gene-editing legacy. New Sci 228(3047):10–11

Liang P, Xu Y, Zgang X et al (2015) CRISPR/Cas9-mediated gene editing in human tripronuclear zygotes. Protein Cell 6(5):363–372

Maeder ML, Gersbach CA (2016) Genome-editing technologies for gene and cell therapy. Mol Ther 24(3):430–446

Maner B (2012) The human encyclopedia. Nature 489:46–48

Mehlman MJ (2012) Transhumanist dreams and dystopian nightmares. The John Hopkins University Press, Baltimore

Niu Y, Shen B, Cui Y et al (2014) Generation of gene-modified cynomolgus monkey via Cas9/RNA-mediated gene targeting in one-cell embryos. Cell 156:836–843

Nussbaum RL, McInnes RR, Willard HF (2007) Thomson and Thomson genetics in medicine, 7th edn. Saunders Elsevier, Philadelphia, pp 126–127

Olson S (ed) (2016) International summit on human gene editing: a global discussion. Committee on Science, Technology, and Law; Policy and Global Affairs; National Academies of Sciences, Engineering, and Medicine. National Academies Press (US), Washington, DC. http://www.nap.edu/catalog/21913/international-summit-on-human-gene-editing-a-global-discussion. Accessed 10 Jul 2016

Paquet D, Kwart D, Chen A et al (2016) Efficient introduction of specific homozygous and heterozygous mutations using CRISPR/Cas9. Nature 533:125–129. doi:10.1038/nature17664

Park A (2016). A new technique that lets scientists edit DNA is transforming science and raising difficult questions. TIME 188(1). July 04. http://time.com/4379503/crispr-scientists-edit-dna/. Accessed 10 Jul 2016

Rasko JE, O'Sullivan GM, Ankeny RA (2006) The ethics of inheritable genetic modification. Cambridge University Press, Cambridge

Reardon S (2016) First CRISPR clinical trial gets green light from US panel. Nature News (22 June). doi:10.1038/nature.2016.20137

Resnik DB, Steinkraus HB, Langer PJ (1999) Human germline gene therapy: scientific, moral and political issues. Medical intelligence unit 9. R.G. Landes Company, Austin

Silver LM (1997) Remaking eden. Avon Books, New York

Slaymaker IM, Gao L, Zetsche B et al (2016) Rationally engineered Cas9 nucleases with improved specificity. Science 351(6268):84–88. doi:10.1126/science.aad5227

Stock G (2003) Redesigning humans. Houghton Mifflin Company, Boston

Taylor GR (1968) The biological time bomb. Thames and Hudson, London

Wang X, Cao C, Huang J et al (2016) One-step generation of triple gene-targeted pigs using CRISPR/Cas9 system. Sci Rep 6:20620. doi:10.1038/srep20620

Chapter 12
Gene Editing in Human Embryos.
A Comment on the Ethical Issues Involved

Iñigo De Miguel Beriain and Ana María Marcos del Cano

12.1 Introduction

In 2015, a research team at Sun Yat-sen University in Guangzhou, China, led by Junjiu Huang, reported the first attempt to genetically modify a human embryo (Liang et al. 2015). This initiative was followed by a second experiment carried out by another Chinese team (Kang et al. 2016), in which the genes of human embryos were edited to try to make them resistant to HIV infection. At the same time, a British team at the Francis Crick Institute was permitted by the Human Fertilisation and Embryology Authority to use clustered regularly interspaced short palindromic repeats (CRISPR)–Cas9 technology in embryos for early-development research, an initiative that was strictly forbidden just 5 years ago (Callaway 2016). It seems, therefore, that trends in bioethics have changed dramatically. Indeed, embryo gene editing is becoming increasingly popular and will play a key role in the coming years, if unpredictable scientific issues do not conspire to impede it.

The main question is: should we enjoy this possibility or should we worry about it? At first glance, it seems that gene editing in human embryos might be excellent news, since this new technology might allow human beings to avoid some of our most terrible diseases, which could be eradicated from the very beginning. Moreover, we could even guarantee better genes for future human generations, thanks to the progressive modification of the gene expressions that are involved in the predisposition to certain diseases, or even consider human cognitive or moral enhancement as an available option. Finally, in the long term, its applications could dramatically change our lifestyle, insofar as gene editing could delay or arrest ageing, an effect

I. De Miguel Beriain (✉)
University of the Basque Country (UPV/EHU), Bilbao, Spain
e-mail: idemiguelb@yahoo.es; inigo.demiguelb@ehu.eus

A.M. Marcos del Cano
Universidad Nacional de Educación a Distancia, Madrid, Spain

© Springer International Publishing AG, part of Springer Nature 2018
M. Soniewicka (ed.), *The Ethics of Reproductive Genetics*,
Philosophy and Medicine 128, https://doi.org/10.1007/978-3-319-60684-2_12

that has already been achieved in mice (Bartke et al. 2001), offering the prospect of humans living longer, without loss of memory or frailty.

However, it would clearly be absurd to not mention that gene editing in human embryos also involves a number of issues that require deep analysis. The concerns, indeed, are twofold: the first is connected with the safety of the technology, and the second relates to some ethical issues. Regarding the first, one must keep in mind that CRISPR manipulation includes a number of technological issues that require urgent answers, including both off-target and on-target, but unwanted, effects that might cause severe harm to the patients and their descendants (Lanphier et al. 2015). Off-target mutations might cause severe side effects in the patient that could only be checked by clinical trials. Changes in tumor suppressor genes, for example, can cause cancer (Daesik et al. 2015). Even worse, in the case of human embryos, gene changes will inevitably affect the patent's germ line, and thus, they will be perpetuated via his/her descendants, perhaps provoking a change in the human genome (Thompson 2015).

Concerns directly linked to our common ethical beliefs are even more worrying, due to the simple fact that technical and safety issues have the potential to be resolved over time by further research and advances, while moral considerations will probably continue to be the focus of public debate (Hinxton Group 2015). It is precisely these kinds of dilemmas that this chapter will deal with. Indeed, in the following pages we will carefully discuss the most challenging ethical issues regarding gene editing in human embryos, with the aim of making some essential clarifications on this extremely sensitive topic. To this purpose, we will focus on four main issues: the necessity of this technology and the risk issues involved; the embryo loss involved; the alteration of the human genome and human identity; and the enhancement/eugenics issue.

12.2 The Necessity of This Technology and the High-Risk Arguments

The necessity of gene editing in human embryos has been strongly debated. Indeed, some scholars deny it on the basis that all the clinical applications in human embryos that this technology might provide us with are already available. To give an example, it is often argued that gene editing for disease prevention can already be performed thanks to preimplantation genetic diagnosis (PGD) (Lander 2015). Therefore, there are no good reasons to continue with these practices. However, this is quite a weak argument due to a number of facts. First of all, it is simply untrue that existing technologies can produce the same benefits as embryo gene editing. To continue with the example, some hereditary diseases cannot be avoided simply via PGD. As a result, couples need egg donations to ensure healthy offspring whilst gene editing would allow people to have their own healthy biological children

(Savulescu et al. 2015). Nevertheless, the most important argument against the futility claim comes from another angle. As the Hinxton Group pointed out,

> while much of the focus of public discussions of human genome editing has been on potential clinical applications, the immediate and perhaps most exciting uses of this technology are in basic scientific research… These distinctions are important to make clear that, even if one opposes human genome editing for clinical reproductive purposes, there is important research to be done that does not serve that end. That said, we appreciate that there are even categories of basic research involving this technology that some may find morally troubling. Nevertheless, it is our conviction that concerns about human genome editing for clinical reproductive purposes should not halt or hamper application to scientifically defensible basic research (Hinxton Group 2015)

Therefore, it is quite simple to deny that (at least) basic research on human embryos holds no value at all. On the other hand, closing this door might constitute a serious attempt against the beneficence principle, insofar as it would render substantial benefits in terms of health care (Savulescu et al. 2015).

The high risk that gene editing carries is a much stronger objection against this technology, as public discussion has demonstrated. Soon after the first Chinese experiment was published, a group of scientists published a letter in *Nature* asking for a moratorium on their application by stating that "genome editing in human embryos using current technologies could have unpredictable effects on future generations. This makes it dangerous and ethically unacceptable. Such research could be exploited for non-therapeutic modifications. We are concerned that a public outcry about such an ethical breach could hinder a promising area of therapeutic development, namely making genetic changes that cannot be inherited" (Lanphier et al. 2015).

These claims are, of course, reasonable, in the sense that no one is willing to expose a human being to unnecessary risks. However, it is also true that risk considerations require some subtle considerations that are not usually made. First of all, we need to make a distinction between basic science and clinical application, in so far as both of them entail totally different consequences. If we are to make clinical use of gene modification in embryos that will be transferred to a uterus, precautionary measures must be maximized. It is difficult, however, to hold that such initiatives must also be compulsory when we think about embryos that will not be transferred, but destroyed. One might think that this distinction is not so easy to make. Indeed, the European Group of Ethics wrote that "because of the blurring lines between basic and applied research, some also call for a moratorium on any basic research involving human germline gene modification until the regulatory framework is adjusted to the new possibilities" (European Group on Ethics 2015). However, if we are not to adhere to the slippery slope argument (which we will discuss later), there seems to be good reason to believe that embryo transfer draws an essential boundary between basic research and clinical application (Thomson 2015). To make a simple example, alleged risks to the human genome provoked by a change in the embryo's germ line can only be assumed if those embryos were transferred into a uterus, as otherwise, the modification will never be transmitted to any descendants. Moreover, we should keep in mind that even in the case of clinical application of the technique, there are a number of reasons to believe that definitive bans are too radical an approach (Hinxton Group 2015). Indeed, it is expected that basic research will severely reduce the risks that human gene editing might pose in the future, limiting off-target effects and providing us with a better picture of on-target events that have unintended consequences. If this were the case, then we should reconsider any previous regulation, so as to allow new paths to be followed.

To sum up, we adhere to the general principle stated by Baltimore et al., who suggested that "as with any therapeutic strategy, higher risks can be tolerated when the reward of success is high, but such risks also demand higher confidence in their likely efficacy" (2015). Zero risk, indeed, does not exist, and good practices and safety should not be confused with acceptability of risks (those judgements are ultimately personal, and perceptions also differ among researchers) (Lunshof 2016). Therefore, we strongly support the idea that if further research reduces the risks involved, then the reasons for the prohibition of gene editing in human beings will vanish. On this basis, we could at least reconsider the possibility of authorizing the clinical use of gene therapies in embryos, even if it involves a modification to our descendants' genome, assuming that it will give them relevant benefits (Isasi and Knoppers 2015).

Finally, we would like to end these paragraphs by mentioning a risk-related issue that is not usually considered when thinking of the risks of gene editing, even if it seems to us extremely important. In our opinion, in terms of risk prevention, banning gene editing in human embryos would not be wise if we consider the whole picture of the situation. New gene editing technologies are relatively cheap to use, and since they introduce edits at specific locations within targeted genomes in a way that is unprecedented in its ease of use and accuracy, their results are extremely accessible. This simply means that a general ban would only create "black markets" or lead to biotechnological tourism practices. Indeed, a worst-case scenario presents itself in such circumstances: some years ago, the Sunshine Project documented nearly a dozen possible uses of genetic science for biological warfare purposes, including the creation of ethnicity-specific pathogens (Sunshine Project 2003).

The situation has become even more worrying at the present time, due to current political circumstances. Moreover, even if we leave aside the terrorist attack threat (which seems quite improbable), we must assume that accidents occur when manipulating life. It does not seem so difficult to think about a scenario in which someone working in a lab might commit a significant mistake with terrible consequences. As Greely has mentioned, "someone could, with luck, change the genome of a huge population of mosquitoes, or weeds, in a very short time" (Skerret 2016). But this might create a huge, unpredictable disaster beyond the scope of the countermeasures we rely on nowadays. It is certainly due to all of these reasons that security agencies have already been alerted to the dangers entailed in gene editing (Regalado 2016). Our point is, therefore, quite simple. If we are to think in terms of risk prevention, then it looks like we should support the fast development of basic science in gene editing, since we need to gain a deep knowledge of how to mitigate the consequences of an incident or action intended to cause harm through the use of these technologies. Thus, it would be quite absurd to advocate a general moratorium that would paralyze "official" science, while the characteristics of gene editing make it impossible to control its use in "clandestine" science. On the other hand, we should try to strengthen the monitored use of gene editing, even in human embryos, if we want to gain the knowledge that we will need precisely to reduce the harm caused by an incidental or man-made biological disaster.

12.3 Gene Editing and Embryo Loss

A second and extremely important ethical issue in human embryo gene editing is that this technology is still far from being completely developed. Indeed, its adequate refinement will need an impressive number of experiments, which will surely involve hundreds of embryos that will not survive them. Therefore, it is quite easy to imagine that people considering human embryos as human dignity holders have something to say against it (Foht 2016). However, there are a number of reasons that converge to make this initial criticism quite weak. First of all, it is must be highlighted that gene editing research will be developed using discarded human embryos; that is, it will not imply the creation of human beings for research purposes at all. Therefore, it is plausible to build a response to this first criticism not only by rejecting the embryo's moral value, but also on the basis of the so-called "discarded-created distinction". According to this criterion, it is immoral to create human embryos for the sole purpose of being used in research, but it is not immoral to use surplus embryos, as they were created for laudable reasons and it was misfortune that altered their destiny. It was not the proposed research which sustained their creation or provoked it in any way at all. Thus, and according to the "nothing is lost principle" (Outka 2009; Pennings and Van Steirteghem 2002; Prieur et al. 2006; Zoloth 2002), even if we consider human embryos as equivalent to adult human beings, it would be morally acceptable to make use of them for the benefit of science. However, one might argue that the "nothing is lost principle" is not a definitive argument, but an objectionable approach that has received criticism for a number of reasons (Devolder 2005, 2013). For example, some authors have stated that even if we share the moral roots of this reasoning, it could never be applicable to the case of spare embryos, as they are not about to die: they will die if somebody takes the decision to destroy them. Otherwise, they could remain frozen forever (Brock 2013). On the other hand, other scholars have pointed out that once we decide to use embryos for research purposes, they become mere means, no matter why they were originally created. Therefore, these intermediate positions are inconsistent: one could maintain that the creation and use of embryos is morally acceptable, or not, but should not provide different answers in both cases (Robertson 1999).

This is not, of course, the right moment to solve this complex discussion, but it seems reasonable to point out that there is an alternative way to gain knowledge on the technique without facing the issues involved in the discarded/regarded discussion. We are referring to the simple fact that we do not really need to make use of "real" embryos in order to continue with the research on gene editing. For example, the first Chinese team implementing gene editing in human embryos took special care over the selection of the embryos which were indeed "non-viable" embryos. It seems crucial to highlight that from an ethical, and especially from a legal, point of view, it is totally different to make use of viable as opposed to non-viable embryos, even if from a scientific point of view they could both be useful, especially if we end up calling "embryos" entities that are not capable of developing into a human being. This alternative is aligned with the path traced by the EU High Court on 18 December 2014 (Case C-364/13). On that occasion, the High Court stated that "in order to be classified as a 'human embryo', a non-fertilized human ovum must necessarily have the inherent capacity of developing into a human being" (point 28) and "consequently, where a non-fertilized human ovum does not fulfil that condition, the mere fact that that organism commences a process of development is not sufficient for it to be regarded as a 'human embryo', within the meaning and for the purposes of the application of Directive 98/44" (point 29). According to this reasoning, we consider that any ovum that lacks an inherent potential to develop into a human being (for instance, due to defective mitochondrial DNA) should not be considered as a human embryo, but as an embryo-like creature. But if this were true, research on those creatures should be considered as perfectly acceptable according to the EU legal and (most probably) ethical standards, they would fit the same from a scientific point of view, and their results could be the object of a patent.

Finally, it is worth mentioning that some scholars have stated that gene editing in human embryos might indeed mitigate concerns about the discarding of embryos in the long term, due to a simple reason: if we were to improve the quality of the embryos, then we could dramatically reduce the number of spare embryos. Therefore, if we consider the human embryo to be a valuable being, we should support the use of this new technology (Savulescu et al. 2015). However, this argument has been refuted by some others, who consider that this would be difficult to achieve, since the IVF industry already routinely discards enormous numbers of human embryos. Thus, it is unlikely to adopt new, risky, unproven methods for editing disease-causing genes in embryos when it can simply selectively discard those embryos (Foht 2016).

12.4 Embryo Editing and the Sanctity of the Human Genome

A third argument against gene editing in human embryos comes from the idea that the human genome needs to be preserved as it is intrinsically valuable. This is, in general, the position held by all those who adhere to the so-called defence of the sanctity of the human genome or the "playing God" argument. According to them, we should never introduce any change to the human genome, in so far as it is "sacred" (Rifkin 1983), as it was created by God or because it is the main basis of human dignity (Kass 2004), as the UNESCO Declaration on the Human Genome and Human Rights stated. Therefore, if we alter it, we would be affecting the nature of the human being involved (Annas 2005; Habermas 2003).

If we are to support this type of belief, it is clear that all types of germ line gene editing (including embryo gene editing, of course) should be banned. However, there are a number of reasons to suggest that this could hardly be done. First of all, it goes against some of the practices that are extended in health care. For example, an alteration of the germ line can occur quite easily, even if it is not intended at all, as is the case with chemotherapy (Isasi et al. 2016). As a consequence, all those who receive this treatment are advised to abstain from having children during its application and for some time afterwards. However, if we were to consider our genome to be "sacred" material, then we should directly ban chemotherapy. This is what being "sacred" really involves. But it is hard to find someone who supports this conclusion in any way. Quite the opposite, we build our reasoning on the basis of the risk/benefit principle. However, in doing so we are recognizing that the human genome is not "sacred" at all, but is a good whose value could be comparable with other goods. However, this conclusion denies the assumption that its mere change is intrinsically immoral, opening the door to a risk/benefit analysis.

Moreover, even if one does not endorse this refutation, he/she should be mindful that human genome preservation does not directly imply a general opposition to germ line editing, because a gene alteration does not necessarily mean an alteration to the human genome. Suppose, for instance, that we modify someone's genome, by changing the expression of a gene responsible for Huntington's disease to its normal, healthy expression. Such an intervention would certainly modify the subject's (and his/her descendants') genome, but not the human genome at all (at least, if we manage to avoid off-target changes), since the modification would not introduce any novelty into the human gene information pool. This subtle distinction has been addressed extremely well by Japanese bioethicist Tetsuya Ishii, who wrote that "the functional correction of a small mutation in the embryo via HDR along with a short DNA template appears to be acceptable because this form of genetic modification can leave a wild-type gene, which is in a natural genetic state, and would fall outside of one of the ethical objections against germ line gene modification: transgression of the natural laws. The copying of a naturally occurring variant via HDR along with a short DNA template might be considered to be natural" (Ishii 2015).

Therefore, we must be aware that the concept of changes in *an* individual genome and changes in *the* human genome are not necessarily to be equated with one another. It is true that no one can introduce changes to the human species genome without altering someone's genome (since species are no more than the sum of their members), but it is also true that someone might change the genome of a single individual or even an extended group of human beings (if his/her germ line is affected, for instance) without changing *the* human genome. Therefore, changing *a* human embryo genome is not the same as changing *the* human genome. This is why "there could be cases of genetic enhancement when this practice would not alter human nature, and as such, should not be morally prohibited" (Morar 2015). To sum up, we must bear in mind that a change in the human genome does not come from a mere germ line modification, but necessarily involves the introduction of genuine novelty to the human gene pool. Therefore, a general ban on human embryo editing on the grounds of the defense of the human genome is clearly unjustified, even if we really believe in the sanctity of the human genome.

12.5 Embryo Editing and Human Identity

As is commonly known, the current ruling norms on gene editing permits some interventions based on this technology, but on the condition that they must not provoke an alteration in the embryo's identity. For instance, the EU Directive on biotechnological inventions (Directive 98/44/EC, 1998) states in its article 6(b) that "processes for modifying the germ line genetic identity of human beings shall be considered unpatentable". Similarly, it is worth mentioning that Regulation (EU) No 536/2014 of the European Parliament and of the Council of 16 April 2014 on clinical trials on medicinal products for human use, and repealing Directive 2001/20/EC states (article 90) that "no gene therapy clinical trials may be carried out which result in modifications to the subject's germ line genetic identity". These clauses constitute an interesting riddle. As Rosario Isasi et al. have noted, genetic identity "has yet to be defined, and we need to look for an approach to genome editing that can lead toward compromise or consensus" (2016). Indeed, who really does know when gene editing alters an embryo's identity? One might think that the cited regulation alludes to a wider sense of identity, the condition that makes us be who we are, that is, no one but us. However, who can define that essential condition in an in vitro embryo?

At first glance, it seems obvious that the issue is hard to solve. We could, for instance, consider that any intervention altering the embryo's DNA would constitute an alteration to its human identity. However, this does not work well with the clauses included in the EU regulation just cited. Indeed, if this were the case, the inclusion of the word "identity" would be totally superfluous in so far as any change in the germ line would be forbidden. On the contrary, the wording in the clauses suggests that even in the case that a germ line modification is procured, we could still trace distinctions between alterations that affect identity (that would be forbidden) and those which do not (that would be allowed). Nevertheless, where should we draw the line?

Even if it is difficult to make a suggestion, we dare to state that gene modifications related to health care or disease prevention could hardly be included in this catalogue because disability can never be considered to be a part of an embryo's identity. Therefore, any modification in gene expression related to a concrete embryonic disability should not be considered as an alteration to its identity. Or, to go one step further, even if we were to concede that disability is indeed a part of someone's identity, we should not protect that identity if it hinders a positive action that benefits the offspring's health. This is, indeed, the conclusion to which most of the scholars arrived when commenting on the famous case of Duchesneau and McCullough and their deliberate intention to have a deaf child. Even if we agree with Savulescu's proposal that a "couple have the right to procreate with whomever they want" (Savulescu 2002), it seems undeniable that once the embryo has been created, his own interests should prevail over his parent's interests. Therefore, we cannot allow them to interfere with medical interventions on the basis of the preservation of the offspring's identity, because disability (even if somehow considered a moral good) should never be considered a higher good than health (at least, if he/she is not able to state the opposite). Indeed, Savulescu also seems to be right when he states "imagine a couple has a child who is born deaf but who could hear if given a cochlear implant. They refuse. They clearly harm their child because that child is worse off (by remaining deaf) than it would otherwise have been (if it had the implant). There are legitimate grounds to interfere in such choices" (Savulescu 2008). If you change the word "cochlear implant" to "gene modification", the reasoning stays the same. And it is very good reasoning.

Of course, we are aware that in sustaining this criterion, we are stating that identity is not a superlative value, but a value that can be left aside when a superior value – such as health – is at stake. But this is not, in our opinion, a radical thought at all, especially if we keep in mind that adults – and nowadays even children – are often allowed to change some essential parts of their being, including their gender, if there are good reasons for. Therefore, it would be contradictory to hold that identity is a "sacred" good at the beginning of life but a changeable one as time goes by. A call for the embryo's consent does not seem to make a fundamental difference here, since we often make decisions to defend a child's health without asking for that consent, and no one thinks that we do something wrong in doing that. The child's health, indeed, seems to be a sufficiently important good so as to justify an intervention aimed to protect it, even in the strange case that we consider a medical condition to be part of his/her identity.

As a final conclusion, we consider that a general ban on gene editing affecting the embryo's germ line genetic identity is hard to understand and entails a number of questions that hinder its practical applicability, since no one knows what type of gene modifications might change someone's identity and which might not. However, we suggest that this clause can only be morally acceptable if we are not to include disabilities, or even predispositions to diseases, as a part of that genetic identity. Otherwise, we conclude that we would be imposing a moral good, such as identity, to another moral good, health, which holds greater importance, according to our vision.

12.6 Embryo Editing, Eugenics and Human Enhancement

Finally, we will consider one of the strongest claims against human gene editing: the claim that this technology might allow the new eugenics movement to put its ideas into practice, a final result that could bring humanity terrible consequences. This is not at all a feeble argument. First of all, we must keep in mind that the claim is far from unrealistic. Eugenics, as such, is a declared goal of the transhumanist movement, which has firmly supported gene editing as an effective tool to reach its objectives. Moreover, there do not seem to be reasons to doubt that there exists a real risk that this technology might finally be used for cognitive enhancement purposes. However, this could hardly be acceptable, since cognitive enhancement entails a number of significant moral issues (Center for Genetics and Society 2015; Douglas 2015; Mehlman 2012; Wilson 2007); for example, it defies equity if it is only available to those who can afford it, generally the richest part of the population. Furthermore, it might challenge autonomy (British Medical Association 2007; Farah et al. 2004), because if some people start making use of this possibility, everyone else would feel compelled to do the same, at least if they did not want their descendants to suffer the consequences of not having benefited from the procedure. In this sense we should not forget that cognitive conditions, such as intelligence, are often considered a comparative good. If a lot of people who share your capacities are enhanced, then you would become a handicapped human being, because you will have been reallocated to the lower part of table. Therefore, it would be wise to avoid any use of gene modification that is not strictly linked to human health (Carrol and Charo 2015). This point of view is, indeed, shared not only by most public opinion according to the most recent polls (Center for Genetics and Society 2015), but also by a vast number of scholars (Caplan et al. 1999) and, thus, it is important to guarantee that gene editing is not used for enhancement purposes. Therefore, it is needed to settle on an adequate regulation that discriminates between interventions aimed to improve human health and interventions that seek enhancement (Pollack 2015).

Instead, those adhering to the eugenics argument hold that in practice, it is not possible to make these distinctions. They hold that even if some applications of human gene editing might be acceptable from a moral point of view, they should be forbidden anyway because they would open the door to further unmoral applications of the technology, such as the aforementioned human enhancement and eugenics. In other words, they believe that we should ban germ line gene editing *as such* because if we do not so, we will be unable to stop building the road to a future world where eugenics would become real. This is, of course, a slippery slope argument, that could be expressed in a classical format: while there is nothing ethically wrong with germ line gene editing itself, if we pass a law making it permissible now, slowly and surely society will move towards a point where many will want to use it for unethical purposes, such as enhancement or eugenics. Moreover, it would even lead us to the division of human species into two different groups, human beings and enhanced human beings, a fact that is almost universally considered as morally unacceptable. Therefore, we should place a ban on each and every use of this technology.

Does this reasoning make real sense? Should we renounce all of the possible benefits of gene editing only because it will most likely bring us to a GATTACA world? The answer to this question requires a deep analysis of the argument. Keeping in mind its slippery slope essence, we must realize that in order to check the acceptability of the claim, we should concentrate on two fundamental points.

First of all, we should analyze whether the final consequence they oppose, in this case the eugenic use of gene editing, is really unmoral or not. Secondly, we should make a decision on whether the causal claim is as strong as the proponents of the argument state or, on the contrary, whether it is possible to construct effective firewalls to protect us from the unethical uses of germ line gene editing, while making their moral applications possible. If we concede that both of these premises are sustainable, then we must adhere to the conclusion of the argument, that is, the necessity to proceed to a general ban on gene editing in human embryos. However, if any of these premises is shaken, then we must reject the whole argument and keep on with the research. As we have already argued on the feasibility of the first premise – the immorality of the final result – we will now concentrate on the other one, that is, the causal claim.

The key point of the discussion, therefore, lies in the connection between the ultimately unmoral consequences that gene editing could bring us, and its use as an effective tool for addressing health care issues or reducing our exposure to certain diseases; that is, the real plausibility of the causal claim. If we consider that this connection is really undeniable, then we would have a convincing reason to ban gene editing, even if it might cause severe suffering for those who could benefit from its application in health care. If this is not the case, then we would be committing a serious and unmoral crime against those people, in so far as we would be depriving them of a treatment that could help them on the basis of a weak reasoning. Therefore, we need to ask ourselves, is it really impossible to place boundaries between the different sorts of applications of biotechnology?

In order to provide this question with an adequate answer, we consider it necessary to start by précising a very important issue. It is not those who deny the argument who need to demonstrate its weaknesses, but it is those who hold it who need to demonstrate the feasibility of the causal claim. This is due to a simple reason: as we consider human freedom to be a primary right, it must be the one proposing to settle some limits on who has to provide the reasons why. Therefore, in this concrete case, it is the opponents of gene editing in human embryos who have to demonstrate that it entails an unavoidable danger of ending up in eugenic practices. Of course, this is extremely hard to do, and it often happens that those proposing the argument are not willing to face the task. As Burgess wrote in his paper on the slippery slope arguments, "unfortunately, purveyors of the Great Argument rarely if ever work it into a detailed slippery-slope argument. They rest content with the sketchiest of formulations, leaving the detailed work to their opponents: we've shown you (sketchily) that it might happen; now show us (in detail) that it couldn't. But this is a fraud. The mere presentation of a slope does nothing to show that the onus of proof is on the reformer to demonstrate that a proposed change will not lead to disaster. This mistake, although of no great intellectual importance, sophistication or profundity in itself, is nonetheless encountered often enough to deserve analysis and diagnosis and it is clearly of great practical importance, for many have been persuaded of its soundness" (Burgess 1993).

On the other hand, it is worth mentioning that slippery slope arguments entail a considerable range of pessimism because they try to convince us of the impossibility of placing boundaries, separating effects. They neglect the utility of regulation as an effective tool to defend us against unwanted uses of technology, a belief that is, for instance, clearly embedded in a statement made by Marcy Darnovsky: "a regulatory line between traits construed as medical and those that are clearly enhancements would be impossible to draw or to hold" (2016). However, and again, they do not provide us with good reasons to support this statement, and facts seem to demonstrate the opposite. Consider, for example, the case of human cloning. One of the main arguments against the use of nuclear transfer for therapeutic purposes (so-called therapeutic cloning) was that, even if it was not unmoral as such, it should be banned because it would finally legitimize reproductive cloning. However, most regulatory authorities did not adhere to this slippery slope argument and decided to permit therapeutic cloning while banning reproductive cloning. Until now, the alleged consequences have not happened at all. Even if therapeutic cloning has been practiced for more than 15 years now, no human clones have been produced. Moreover, no attempts have been registered. Therefore, it is at least somewhat odd to state that a law can be ineffective in stopping some applications of biotechnology. If opponents to gene editing in human embryos consider that this case is totally different to that of stem cells, they should provide us with good reasons for that; to our knowledge, this is yet to transpire.

We understand, of course, that the boundaries between different purposes are extremely fuzzy and we will need to go further in the discussion on their limits, but this is precisely what governance is supposed to do. As Richard Hayes wrote in 2008,

> some have argued that the fact that it is difficult to draw bright lines regarding the therapy/ enhancement distinction means that no lines can be drawn. But this is a specious argument. Public policy is in large part a matter of drawing lines; we do it all the time. Putting our trust in commercial markets and the free play of human desire would unleash a genetic enhancement rat-race that could never be contained. The responsible alternative is to establish as a matter of law the clearest lines possible and a clear statement of intent, and delegate decisions over remaining gray areas – which typically impact fewer individuals – to accountable regulatory bodies" (2008)

Finally, it seems to us that these sorts of claims are somehow contradictory, because if they were true, that is, if regulation were really incapable of stopping unmoral practices, it would be difficult to understand their final goal: to impose a ban that would be as ineffective as any other. Indeed, as gene editing is already a technical reality and is already being applied, if the slippery slope argument is right, then a ban could hardly work and the most terrible consequences of this technology are surely to come. Instead, if we consider that adequate regulation might separate the different uses of gene editing in human embryos, only banning those that are considered unmoral, then we are recognizing somehow that the main assumption of the argument – the inevitability of future harm – is wrong.

To sum up, it seems that there are a number of strong reasons that go against the slippery slope to eugenics argument. Altogether, they recommend that we be extremely cautious about categorical affirmations denying the capability of regulation to set effective boundaries between different practices. Thus, it seems obvious that we must keep an eye open so as to prevent any possible use of gene editing that might bring about the excesses of eugenic practices. However, this could hardly justify a general ban on the use of gene editing in human embryos when one considers the benefits that are at stake and the moral imperative to pursue them.

Acknowledgments We would like to take this opportunity to acknowledge the support from the Spanish Ministry of Economy, award DER 2013-41462-R, Regional Government of Madrid, award H2015-HUM/3330 and COST Actions IS1303 Citizen's Health through public-private Initiatives: Public health, Market and Ethical perspectives (CHIP ME) and CA15105 European Medicines Shortages Research Network – addressing supply problems to patients (Medicines Shortages)

References

Annas G (2005) American bioethics: crossing human rights and health law boundaries. Oxford University Press, Oxford

Baltimore D, Berg P, Botchan M et al (2015) A prudent path forward for genomic engineering and germ line gene modification. Science 348(6230):36–38

Bartke A, Brown-Borg H, Mattison J et al (2001) Prolonged longevity of hypopituitary dwarf mice. Exp Gerontol 36:21–28. doi:10.1016/S0531-5565(00)00205-9

British Medical Association (2007) Boosting your brain power: ethical aspects of cognitive enhancements. British Medical Association, London

Brock DW (2013) Creating embryos for use in stem cell research. J Law Med Ethics 38(2):229–237

Burgess JA (1993) The great slippery-slope argument. J Med Ethics 19(3):169–174

Callaway E (2016) UK scientists gain licence to edit genes in human embryos. Nature. http://www.nature.com/news/uk-scientists-gain-licence-to-edit-genes-in-human-embryos-1.19270. Accessed 21 May 2016

Caplan AL, McGee G, Magnus D (1999) What is immoral about eugenics? West J Med 171(5–6):335–337

Carroll D, Charo RA (2015) The societal opportunities and challenges of genome editing. Genome Biol 16(242). doi: 10.1186/s13059-015-0812-0

Center for Genetics and Society (2015) About human germline gene editing. http://www.geneticsandsociety.org/article.php?id=8711. Accessed 2 May 2016

Daesik K, Sangsu B, Jeongbin P et al (2015) Digenome-seq: genome-wide profiling of CRISPR-Cas9 off-target effects in human cells. Nat Methods 12:237–243. doi:10.1038/nmeth.3284

Darnovsky M (2016) Should heritable gene editing be used on humans? Wall Str J Apr 10 2016. http://www.wsj.com/articles/should-heritable-gene-editing-be-used-on-humans-1460340173. Accessed 2 May 2016

Devolder K (2005) Creating and sacrificing embryos for stem cells. J Med Ethics 31(6):366–370. doi:10.1136/jme.2004.008599

Devolder K (2013) Killing discarded embryos and the nothing-is-lost principle. J Appl Philos 30(4):289–303. doi:10.1111/japp.12033

Douglas T (2015) The harms of enhancement and the conclusive reasons view. Camb Q Healthc Ethics 24(1):23–36. doi:10.1017/S0963180114000218

European Group on Ethics (2015) Statement on Gene Editing https://ec.europa.eu/research/ege/pdf/gene_editing_ege_statemen.pdf. Accessed 02 May 2016

Farah MJ, Illes J, Cook-Deegan R et al (2004) Neurocognitive enhancement: what can we do and what should we do? Nat Rev Neurosci 5(5):421–425

Foht B (2016) Experiments on human embryos offer little hope for curing genetic diseases. Natl Rev   http://www.nationalreview.com/article/430771/genetic-modification-embryos-morally-wrong-still. Accessed 2 May 2016

Habermas J (2003) The future of human nature. Polity, Malden

Hayes R (2008) Is there an emerging international consensus on the proper uses of the new human genetic technologies? Testimony given for the subcommittee on terrorism, nonproliferation and   trade.   http://www.geneticsandsociety.org/downloads/20080619_hayes_testimony.pdf. Accessed 2 May 2016

Hinxton Group (2015) Statement on genome editing technologies and human germ line genetic modification. http://www.hinxtongroup.org/hinxton2015_statement.pdf. Accessed 2 May 2016

Isasi R, Knoppers BM (2015) Oversight of human inheritable genome modification. Nat Biotechnol 33:454–455

Isasi R, Kleiderman E, Knoppers BM (2016) Editing policy to fit the genome? Science 351 (6271):337–339

Ishii T (2017) Germ line genome editing in clinics: the approaches, objectives and global society. Brief Funct Genomics 16(1):46–56

Kang X et al (2016) Introducing precise genetic modifications into human 3PN embryos by CRISPR/Cas-mediated genome editing. J Assist Reprod Genet 33(5):581–588

Kass LR (2004) Life, liberty and the defense of dignity. Encounter Books, San Francisco

Lander ES (2015) What we don't know? In: Commissioned papers for the International summit on gene editing. Washington, DC, 1–3 Dec 2015

Lanphier E, Urnov F, Haecker SE et al (2015) Don't edit the human germ line. Nature 519:410–411

Liang P, Xu Y, Zhang X et al (2015) CRISPR/Cas9-mediated gene editing in human tripronuclear zygotes. Protein Cell 6:363–372

Lunshof JE (2016) Human germ line editing – roles and responsibilities. Protein Cell 7(1):7–10

Mehlman MJ (2012) Will directed evolution destroy humanity, and if so, what can we do about it? St Louis Univ J Health Law Policy 3:93–122

Morar N (2015) An empirically informed critique of Habermas' argument from human nature. Sci Engin Ethics 21(1):95–113

Outka G (2009) The ethics of embryonic stem cell research and the principle of "nothing is lost". Yale J Health Policy Law Ethics 9(3):585–602

Pennings G, Van Steirteghem A (2002) The subsidiarity principle in the context of embryonic stem cell research. Hum Reprod 19(2):1060–1064

Pollack R (2015) Eugenics lurk in the shadow of CRISPR. Science 348(6237):871

Prieur MR, Atkinson J, Hardingham L et al (2006) Stem cell research in a Catholic institution: yes or no? Kennedy Inst Ethics J 16(19):73–98

Sunshine Project (2003) Emerging technologies: genetic engineering and biological weapons. http://www.sunshine-project.org/publications/bk/bk12. Accessed 2 May 2016

Regalado A (2016) Top US Intelligence Official calls gene editing a WMD threat. MIT Technology Review (February). https://www.technologyreview.com/s/600774/top-us-intelligence-official-calls-gene-editing-a-wmd-threat/. Accessed 2 May 2016

Rifkin J (1983) Algeny. Viking, New York

Robertson JA (1999) Ethics and policy in embryonic stem cell research. Kennedy Inst Ethics J 109:109–136

Savulescu J (2002) Deaf lesbians, "designer disability," and the future of medicine. Brit Med J 325(7367):771–773

Savulescu J (2008) Is it wrong to deliberately select embryos which will have disabilities? Practical Ethics. March 10 2008. http://blog.practicalethics.ox.ac.uk/2008/03/is-it-wrong-to-deliberately-select-embryos-which-will-have-disabiltites/. Accessed 2 May 2016

Savulescu J, Pugh J, Douglas T et al (2015) The moral imperative to continue gene editing research on human embryos. Protein Cell 6(7):476–479. doi:10.1007/s13238-015-0184-y

Skerret P (2016) Is do-it-yourself CRISPR as scary as it sounds? STAT. https://www.statnews.com/2016/03/14/crispr-do-it-yourself/#Lentzos. Accessed 2 May 2016

Thompson Ch (2015) Governance, regulation, and control: public participation. In: Commissioned papers for the International summit on gene editing. Washington, DC, 1–3 Dec 2015

Thomson Ch (2015) The human germline genome editing debate. Impact ethics. http://impactethics.ca/2015/12/04/the-human-germline-genome-editing-debate/. Accessed 2 May 2016

Wilson J (2007) Transhumanism and moral equality. Bioethics 21(8):419–425

Zoloth L (2002) Jordan's banks, a view from the first years of human embryonic stem cell research. Amer J Bioethics 2(1):3–30

Chapter 13
Geneticization and Bioethics: Ethical Dilemmas in Genetic Counselling

Ewa Baum and Jan Domaradzki

13.1 Genetic Essentialism

Many considerations in contemporary bioethics revolve around genetics and its impact on health and reproduction. The rapid increase in genetic knowledge and the development of biotechnology is changing our understanding of health, disease and risk (Petersen 2006), as well as social, and in particular parental, relations (Rapp 2000; Rothman 1993; Hallowell 1999). A consequence of these changes is the geneticization hypothesis, which has been discussed for almost three decades, and which Canadian researcher Abby Lippman has defined as "the ongoing process by which differences between individuals are reduced to their DNA codes, with most disorders, behaviours and physiological variations defined, at least in part, as genetic in origin," accompanied by "a process by which interventions employing genetic technologies are adopted to manage problems of health" (Lippman 1991, 19). According to some researchers, this assertion is not supported by social practice (Hedgecoe 1998; Arnason and Hjörleifsson 2007), but despite this, genetics today is the dominant paradigm of medicine where it is seen as the most adequate model to explain health and disease as well as norm and pathology (Strohman 1997; Rose 1995). Furthermore, genetic diagnosis provides new knowledge about diseases or vulnerability to these, hence it significantly impacts the very concept of disease, which is increasingly defined in terms of genetics. William E. Stempsey (2006) openly writes about "the geneticization of diagnostics," which generates a number of challenges and ethical problems that do not occur with traditional diagnostics.

Many researchers stress that the development of genetics reinforces thinking in terms of genetic essentialism, reductionism, determinism and fatalism. By providing new knowledge on the genetic causes of many diseases, personality traits and behaviors, genetics promotes a specific "molecular optics" which emphasizes the

E. Baum (✉) • J. Domaradzki
Karol Marcinkowski University of Medical Sciences, Poznań, Poland
e-mail: ebaum@ump.edu.pl; jandomar@ump.edu.pl

© Springer International Publishing AG, part of Springer Nature 2018
M. Soniewicka (ed.), *The Ethics of Reproductive Genetics*,
Philosophy and Medicine 128, https://doi.org/10.1007/978-3-319-60684-2_13

molecular basis of human existence. Nikolas Rose believes that by reducing the human self to molecular structures, genetics is becoming a source of a "new molecular ontology of life" (2007). Dorothy Nelkin and Susan Lindee write about a "DNA mystique" (1999), whilst Alex Mauron mentions a "genomic metaphysics" (2002). The reason being is that by describing the human genome as an unchangeable being independent of the body and environment, like the Aristotelian *eidos* or the Christian soul, it becomes perceived by many as a physical marker of life and the definitional essence of humanity. As a result, more and more individuals perceive themselves in genetic terms (Novas and Rose 2000; Greco 1993).

Such an essentialist view of humans is reinforced by numerous metaphors which describe genes and the DNA as a "profile," "portrait," "fingerprint," "personal bar code," "visiting card" or "passport", as well as a "unique code of existence" or a "human black box." Describing the DNA as the carrier of individuality suggests that the genome is the ontological foundation of a person's "I," that it describes the ego of the past and determines the future. Concurrently, genetic essentialism is based on the premise that a person is simply a set of genes which stores a complete manual for the construction of the human body, and that the genes are, in fact, responsible for life (Nordgren 2008, 252–266; Nordgren and Juengst 2009). Accordingly, genetic information is seen, by many, as the quintessential identity (Mauron 2002, 832). This view of people is manifested in seeing a growing number of behaviors, emotional states and personality traits through the prism of genes or in the search for the so-called "gene of humanity," like FOXP2, ASPM or MYH16, which enabled humans to separate from other primates (Dar-Nimrod and Heine 2011).

In the context of bioethics, it is important to note that genetic technology may affect the genetic identity of a person on several levels: (1) a range of reproductive biotechnologies, including preimplantation diagnosis, IVF or microinjection, can affect a person's identity understood as a continuity over time: as all these techniques determine, after all, which person is born; (2) through genetic engineering, genetics can affect a person's identity understood as a basic kind of being, as exemplified by interspecific hybrids; (3) finally, gene therapy and genetic enhancement make it possible to select an individual's specific traits and genetic or phenotypic properties (Zeiler 2007; Nordgren 2008).

Genetic essentialism also affects the public perception of risk. While formerly risk had a primarily external dimension, the present emphasis on the genetic causation of and vulnerability to various diseases makes it increasingly internal, and puts the responsibility for managing the risk on the individual (Hallowell 1999). This is so because genetic testing seems an effective tool to control the human interior and destiny, whereas to undergo this testing is becoming a new manifestation of the process of civilizing the body, as described by Norbert Elias (2000). Indeed, by promoting the special ethos of activism, the genetic discourse suggests that undergoing genetic testing is a manifestation of rationality, responsibility, altruism and solidarity with others, whilst genetic ignorance is defined as that of irrationality, selfishness and lack of accountability. Accordingly, individuals belonging to risk groups may experience pressure to learn about their risk and to take measures to manage it (Lippman 1991; Rapp 2000; Rothman 1993; Hallowell 1999).

However, since genetic testing may have a serious impact not only on the individual, but also on third parties, including offspring, spouses and relatives, the moralization of the genetic discourse produces a discussion on the individual's fundamental right to autonomy, self-determination and knowledge. While medical professionals are legally bound to respect the individual's right to not know, it is emphasized that genetic ignorance can be a source of harm to others (Domaradzki 2013, 2015).

13.2   Human Dignity

Both advocates and opponents of the right to genetic knowledge/ignorance cite the elementary value of human dignity, which has been one of the major concerns in the field of human rights, at least since the founding of the United Nations (Freeman 2007, 43–44). In the Preamble to the Universal Declaration of Human Rights, the UN recognizes dignity as "the foundation of freedom, justice and peace in the world" (Szawłowski 1982, 132–133). Furthermore, because respecting human dignity requires the individual to be treated as a goal and not a means in any undertaken ventures, in democratic countries dignity is one of the fundamental premises of the legal system, and as such, is legally protected (Świątkowski 2006, 168–169; see: Stanley 1981, 72). Every person, regardless of their social, physical or mental condition, has the right to expect positive treatment from others, and human dignity is violated when an individual is treated in a negative or degrading way by third parties.

It should be stressed that the notion of "human dignity" is used by supporters of biotechnological development as well as its critics, who see it as a threat to human significance, e.g. through the propagation of genetic engineering (Hołub and Duchliński 2008, 302–321). Both sides point out that while research should not be discontinued in the name of human dignity, one should not fall under the illusion that nothing but constant scientific advancement demonstrates our humanity. Hence, whilst highlighting the need for genetic or presymptomatic testing, they suggest that one should also consider other (non-scientific) circumstances, such as the assumption that genetic testing may considerably change social relations in those societies where they are used on a large scale (Cebrat and Cebrat 2012, 62).

Another strongly emphasized factor is the need to ensure autonomous rights to every individual, which precludes objectification or instrumental treatment, also on the part of public authorities and medical professionals. Accordingly, human dignity is usually guided by the premise of an individual approach based on a sense of empathy and humanitarianism.

Furthermore, when taking specific medical action and making choices regarding genetic testing, it is necessary to refer to generally accepted ethical principles. In difficult situations, ethics can be a particularly useful instrument in facilitating the resolution of conflicts and disputes arising in the medical profession, as it pursues the objectives resulting from ethical reflection and demonstrates a particular hierarchy of values. Professional ethics also interprets the rules of conduct in the spirit of certain ethical doctrines. The most often cited among these include utilitarianism and deontology or personalist ethics, for instance ethical theories that provide a necessary value system, which implies that an ethic must support a certain ethical tradition (Beauchamp and Childress 1996, 329–331). Besides national healthcare legislation, a thus understood professional ethic can form the basis for the recognition of patients' fundamental rights (see: Boratyńska and Konieczniak 2001).

The ethical work of John Gregory and Thomas Percival played a significant role in the formation of medical codes (Szewczyk 2009, 24–25). The latter's *Medical Ethics* is considered the first modern code of ethics for the medical professions (Harris 2001, 2). The subsequent codes for medical professionals combine the Hippocratic tradition with the requirements of the current state of medical science, and highlight the occupational skills of medical staff in order to defend the profession against incompetent practices (Jonsen 2003, 7). This understanding of ethics establishes, on the one hand, a rapport between medical staff, and on the other, determines the attitude of these professionals towards patients (Porter 1997, 287). Percival created a certain canon of perceiving medical ethics as a code of conduct which also determined the value of a doctor's position. This route was then followed by the designers of deontological codes for other occupations. Hence the concept of professional ethics refers primarily to the standards of conduct for members of a given occupation. In this sense, ethics is normative and defines a personality model associated with specific ethically conditioned responsibilities. It formulates the standards for professional practice, and outlines typical ethical dilemmas that may occur in the course of specific jobs (Czarnecki 2008, 16–17.). Professional ethics is, therefore, reflected in the codes of individual occupations and their decorum is, in a sense, governed by these codes.

Medical codes of conduct involve an obligation to respect patients' dignity and their right to self-determination. A patient-doctor relationship is based on the notion of mutual trust and a guarantee that the patient can rely on the doctor. Medical professionals also have a duty to treat their patients equally, and any discrimination is obviously and absolutely forbidden. Their actions should not be affected by their patients' age, gender, marital status, sexual orientation, nationality, religion, political beliefs, genetic endowment, race, skin color or any other preferences or personal characteristics (see: Naczelna Izba Lekarska 2004).

In addition to occupational codes, medical professionals can find important guidance in the so-called "soft laws", i.e. the guidelines and recommendations of the European Union. One such document is Recommendation (2010, 11) of the Committee of Ministers to member states on the impact of genetics on the organization of health care services and training of health professionals (adopted by the Committee of Ministers on 29 September 2010). The Recommendation stipulates that "These services [i.e. genetic services] should include support, care and treatment for those in need and should also include appropriate measures to respond to their *wish to know whether or not they are at risk* [emphasis added] of developing or transmitting a disorder with a genetic component." Furthermore, the criteria that should be considered when deciding on the availability of genetic services clearly specify:

- equity and social solidarity;
- benefit/effectiveness for the individual and family;
- improvement of public health;
- safety of the tests (Jasudowicz and Czepek Kapelańska-Pręgowska 2014: 151–157).

These criteria are directly correlated with the principles of medical ethics where caring for the patient's welfare is the highest standard that involves the principle of respecting autonomy, of nonmaleficence, of beneficence and of justice (see Beauchamp and Childress 1996; Gillon 1997). One should, however, note that the principles of medical ethics function as a *prima facie* principle, which means that a "principle is binding unless it conflicts with another moral principle – if it does, we have to choose between them" (Gillon 2001, 22).

Nonetheless, in order for the principles of medical ethics to function freely, certain conditions must be fulfilled. Autonomy of operation is based on intentional actions taken with understanding and without external pressure. Individuals manage their fates independently, which allows them to achieve their goals, and consequently, deserve unconditional respect. The imposing of another person's will on an individual means that the individual is being objectified. There is also the obligation to respect the views and rights of other individuals, as long as their words or actions do not cause harm to others. This duty correlates with the right to self-determination, which contains the right to confidentiality, personal dignity, freedom and privacy. To make this obligation viable, it is necessary to precisely determine the exceptional situations in which the principle of respect is not feasible. Only people with limited autonomy cannot self-determine. An autonomous individual is free, and acts according to the plan they have independently chosen. However, the principle of respect for autonomy is contingent on circumstances, and may be controlled by various moral arguments. Typical situations that restrict patients' autonomy are decisions that could threaten public health, would require large amounts of money or could cause harm to others. This shows that the principle of respect for autonomy cannot be an end in itself, and that it is possible to waive it when confronted with another principle that is more compelling in a given situation.

Undoubtedly, when applying the principle of autonomy it is important to build good relations with a patient through dialogue in order to obtain the patient's informed consent to any proposed changes to their current or future existence. Furthermore, an unobstructed and clear flow of information between doctor and patient is conducive to obtaining the latter's informed consent regarding the decisions taken in their interest. If the patient understands a situation and does not experience any external pressure, they are capable of making an independent decision. Being able to self-determine in matters concerning their own existence gives a person a sense of self-worth. A particularly important factor in the mutual rapport is how the patient is communicated with, because the stress, suffering or pain they are experiencing may affect how they process information and their ability to make objective judgments. The patient's competences are thus conditioned not only by their age, but also by their physical and mental health or their beliefs. Consequently, this involves not only intentional action, but also the practical aspects of counselling the patient. For example, in diagnostic genetic testing, it should be mandatory to offer every patient genetic counselling, whereby the relevant medical, social and psychological circumstances regarding testing or non-testing are explained in a communication process. This fulfils the primary objective, i.e. it secures the patient's autonomy in decision-making. However, for this to happen, several requirements must be met. One of the basic ones is access to reliable information (about the testing and its safety). Another one is to obtain the tested individual's consent (preferably their independent or cumulative, or ultimately, surrogate consent) with the option to withdraw it at any time. In addition, the patient should be informed about the nature of any risks and the degree of their probability. Information should be personalized to fully take into account the patient's individual situation. Moreover, any communications for pregnant women should concern the impact of the testing on the embryo or fetus. Particularly controversial in this respect is prenatal and preimplantation genetic diagnosis. In this case, the determining factor should be concern for a child's health. Furthermore, testing for non-health-related purposes, e.g. for commercial reasons, is also excluded, as this would be an obvious discrimination and evade the protection of workers' rights. Such practices were used in the past, but have now been abandoned (Fulda and Lykens 2004, 143–147). It is also important that genetic counselling should aim to communicate in a non-suggesting and non-directive way, i.e. it should provide information on the disease probability and on testing and treatment options in a way that does not pressure the patient to make a specific decision, whilst any consent should be expressed in an informed way and be preceded by information adapted to the recipient's perceptive abilities (Oduncu 2002, 53–63; Dryla 2015). This places genetic counselling in opposition to the traditional model of medical communication where doctors often indicate or even suggest procedures to patients. This, in turn, is relevant because the way information about risk is communicated may, by itself, affect the individual's decision to undergo testing or inform others about the risk, or even impact on their marital or childbearing plans. Hence, any decisions on testing, collecting results and the subsequent choices should be made independently by the individual based on their worldview, religious beliefs and values (Oduncu 2002; Dryla 2015). In reality, influence on individual decisions regarding genetic testing is socially and culturally conditioned by factors such as age, religious affiliation, knowledge, marital status and beliefs (Singer et al. 2008).

In addition to the diagnostic aspect, which consists in the estimating of genetic risk, a vital part of genetic counselling is its psychological dimension understood as support for the tested individual, and if necessary, for their family. It should also be emphasized that genetic counselling is an integral part of genetic research and, as such, should be recommended for all types of testing (Clarke 1994).

Assuming that the basic goals of thus understood genetic counselling are:

To provide and explain to the individual the basic facts regarding their diagnosis, the possible development of their disease and the available treatment options;

- To define the possible inheritance model and risk of the disease among the tested individual's family members;
- To explain the risk management methods;
- To explain the benefits and risks associated with testing;
- To support the decision making process regarding testing and family planning while respecting the patient's rights and the fundamental principles of genetic counselling;
- To support the patient in receiving information about the risk or disease they are threatened with;
- To provide continuous support and psychological assistance to the patient.

These goals should be delivered with respect to the main principles of genetic counselling, i.e.:

- The principle of autonomy, which emphasizes that every individual is capable of making their own decisions regarding their health and treatment;
- The principle of privacy, which states that the individual should control who and when should have access to information about their health, and to what extent;
- The principle of confidentiality, which means that genetic information will not be disclosed to third parties and will be used solely for the purpose of testing;
- The principle of beneficence, which assumes that the purpose of a medical procedure is beneficence;
- The principle of nonmaleficence, which requires that a medical procedure does not become a source of suffering for the individual (it should also be emphasized that nonmaleficence is more important than beneficence);
- The principle of equality, which requires medical personnel to perform all services with respect for individual rights and to distribute the services fairly among all who need them;
- The principle of informed consent, which stipulates that the tested individual must not only be informed about but also understand the procedures they are to undergo, including their benefits and risks (Baker et al. 1998; Kapelańska-Pręgowska 2011; Dryla 2014).

13.3   Non-testing Ethics vs. Utilitarianism

In medical practices based on the principles of medical ethics, we can encounter another kind of problem, when a conflict occurs between the right to refrain from testing and social responsibility towards other individuals, and between the right to the so-called "universal happiness".

The term "non-testing ethics" is believed to have been coined by Jacques Testart. In his book *L'Oeuf Transparent* (The Transparent Cell), he postulates the introduction of non-testing ethics as a scientific field which would identify the risks of using scientific research for the wrong (inappropriate) purposes, calling it the "logic of non-discovering" (Testart 1990, 30–31). As a pioneer of in vitro fertilization in

France, Testart eventually renounced this method and further research in the field of assisted procreation. The ultimate message of his book is very clear, and the author emphasizes that not everything that is feasible from the medical perspective should be realized from the public one. The boundary between the purely therapeutic purposes and designing people or changes in human nature is quite blurred, and the "insane prospect of a custom-made child" sounds quite probable (Testart 1990, 27–28). This perception of the problem is very close to the argument of a "downward spiral" in the context of the risks posed by genetic interference (see: Chyrowicz 2002, 225–334). The ethics of non-testing assumes refraining from actions which may potentially cause unpredictable results or even harm to individuals or communities. Therefore, non-testing ethics situates itself in opposition to the common understanding of the trend called utilitarianism, and points out its disadvantage in the form of one of its indicators, namely universality perceived as "the greatest happiness of the greatest number" (see Baum and Antczak 2014). This, however, is an apparent opposition because a wider analysis of utilitarianism reveals its duality.

The determinants of the pursuit of happiness were analyzed by many scholars, including the English utilitarian philosopher Jeremy Bentham (1748–1832), whose main interest was legislation and its improvement. His goal was to create the theoretical grounds for the perfect law and political system, whilst the basic criterion and measure of this perfection was the principle of utility, known as "the greatest happiness principle" (Harrison 1995, 65). Bentham believed that nature had placed humankind under the governance of two masters, namely pain and pleasure, hence by recognizing this governance, the principle of utility "assume[d] it for the foundation of that system, the object of which [was] to rear the fabric of felicity by the hands of reason and law" (Bentham 1958, 18). Bentham's most important purpose was for people to achieve happiness, and the basic tool to ensure this was a law that was consistent with the dictates of reason, and therefore just, because reason dictated to observe the principle of utility (Harrison 1995, 65). Bentham measured the justness of acts by their utility, and utility was measured by the effects of a given act. To him, this was a purely practical matter. He believed that human happiness could be achieved if social orders were improved (through human ingenuity) by ensuring better food, sanitary conditions and education, as well as greater equality of opportunity. Bentham did stress, however, that striving to ensure "the greatest happiness of the greatest number" could, in practice, mean a policy under which some concepts identified with happiness by some individuals would be gradually eliminated. An example of this was the abolition of slavery, which gave freedom to the slaves but made their owners unhappy. Similar dual consequences are also brought about by the introduction of personal or religious freedoms, or the right to decide about one's health.

Even after World War I, this understanding of utilitarianism made it seem a bold and innovative doctrine which "left a positive mark on the history of social critique" (Kymlicka 2009, 67). At the time, Bentham's assumption that one should strive for "the greatest happiness of the greatest number" became a guideline for politicians and governments, who started to believe that wellbeing and mental health should become the new benchmark for the activities of welfare states (Layard 2012). In healthcare, this meant the pursuit of a model which would evolve from a paternalistic approach towards patients to partnerships that respected the patients' informed and autonomous choices. A paternalistic relation model is primarily distinguished by the presence of an absolute authority, a "superior," who indisputably determines how "beneficence" should be understood, whilst the "subordinate" is required to respect their decision. In this case, a treated patient is not regarded as a fully autonomous person but as one who has limited capacity to make their own decisions (see: Gert et al. 1997, 195–216). The partnership model, on the other hand, is based on the assumption of equality in the relationship between the patient and doctor, who trust each other and define mutual goals which they subsequently pursue. The setting of mutual goals makes their relationship take on the form of that between two autonomous individuals and is completely devoid of any authoritative features. The partnership model often refers to the metaphor of friendship or alliance (Szewczyk 2009, 151–169). In the case of the paternalistic model, the doctor's will is imposed upon the patient, while in the partnership model the autonomous patient is the actual source of opinions about the appropriateness of a choice. The basic element of a thus understood alliance is the building of mutual trust. One should also remember that the doctor-patient relationship is not only between a professional and the beneficiary of medical services but also an interpersonal one. Hence, this alliance should be analysed both from an ethical perspective and from that of health psychology. When considering the patient's situation, including their moral dilemmas, it is necessary to thoroughly examine the problem through the prism of one's own beliefs, the code of professional ethics and of the ethics of care (Szewczyk 2009, 157–161). The ethics of care accepts and respects the finite aspect of life. The patient cannot be forced, against their opinion and will, to continue procedures or practices. Moreover, they should not to be forced to know information they would rather avoid, and things should not be implied to them. Usually, it is the patient who intuitively senses what is best for them, and determines their *welfare* based on the fullest possible knowledge of the anticipated effects, both positive and negative. The role of genetic counselling is to facilitate the patient in understanding their individually and subjectively perceived *welfare*.

13.4 Genetic Testing, the Right to (Not) Know and the Duty to Know

Most bioethical codes recognise access to medical information as the patient's fundamental right, and knowledge is, in fact, defined as a condition of empowerment and autonomy. Nonetheless, it is also stressed that in justified cases the doctor is entitled to the so-called "therapeutic privilege" (Edwin 2008; Lajeunesse and Lussier 2010), as if the doctor believes that knowledge about a disease may create a hazard to the patient, e.g. lead to suicide attempts, they may refuse to inform the patient about his/her health status.[1]

However, placing the focus on the partnership model in the healthcare professionals-patient relationship leads to the chief principle of medical ethics being not so much the principle of nonmaleficence but that of the patient's autonomy. It is recognized that, apart from some exceptional cases, it is the individual who is the most competent to make decisions about their own health and life. Therefore, most bioethical codes indicate the right to not know as complementary albeit opposite one to the right to know[2] (Chadwick et al. 1997; Knoppers 2014).

13.4.1　Arguments for the Right to Not Know

The arguments for the right to not know in the context of genetic testing usually refer to the patient's autonomy, their right to privacy and happiness, and to the psychosocial consequences of genetic diagnosis, including stigmatization and discrimination, as well as disturbed social relations (Domaradzki 2015).

By rejecting the claim that freedom of choice presupposes the knowledge of all options, the advocates of the right to not know negate the paternalistic approach toward patient and indicate that it is individuals, not their doctor or family, who should decide whether they want to know (Andorno 2004; Dryla 2012; Helgesson 2014; Laurie 1999; Takala 1999, 2001; Takala and Häyry 2000). Stressing the right to autonomy and self-determination, the advocates maintain that all individuals are sufficiently competent to decide whether they want to know, and any external judgment of their actions is deemed "the new paternalism" (Takala 2001, 490; Wilson 2005). They also reject the assumption that a "rational" individual is one who wants to know, and point out that the individual should balance the benefits and

---

[1] One of the main documents which refers to this right in Poland is the Professions of Doctors and Dentists Act from 1996, which in Art. 31.4 states: "In special situations, when prognosis is unfavourable for a patient, a physician may withhold information about the patient's health and prognosis if, according to the doctor's evaluation, this is in the patient's best interest" (Ustawa o zawodach lekarza i lekarza dentysty 1996). The right is also mentioned in Art. 17 of the Polish Code of Medical Ethics: "[I]nformation about diagnosis and poor prognosis may be withheld from the patient only when a physician is deeply convinced, that its disclosure will cause the patient's serious suffering or other unfavourable health effects" (Naczelna Izba Lekarska 2004).

[2] This right was first formulated in 1981, in Article 7d of the Declaration of Lisbon on the Rights of the Patient where it states: "The patient has the right not to be informed on his/her explicit request, unless required for the protection of another person's life". Currently, the right is founded in the Convention on Human Rights and Biomedicine of 1997 (Art. 8.2), and confirmed by several international documents, including the UNESCO Universal Declaration on the Human Genome and Human Rights of 1997 (Art. 5c) and the World Health Organization Review of Ethical Issues in Medical Genetics of 2003 (Art. 8.2) (Andorno 2004). In Polish legislation, the right is guaranteed in the Act of 2008 on Patients' Rights and Patients' Rights Ombudsman (Art. 9.4), the Act of 1996 on the Professions of Doctors and Dentists (Art. 31.3.) and the Code of Medical Ethics of 1991 (Art. 16.1.) (Kapelańska-Pręgowska 2011).

disadvantages caused by (not) knowing, because any external judgment of the psychosocial consequences of knowing is not scientifically founded (Takala 1999, 292; Takala and Häyry 2000, 109).

Another right that guarantees genetic ignorance is the right to privacy (Gostin 1995). As individuals who are tested cannot fully control who, apart from them, will have access to their information, it is argued that genetic knowledge poses a threat of genetic discrimination by employers or insurers who may perceive the individuals as specific "pre-patients" and deny them the necessary benefits (Ekberg 2005; Soniewicka 2010).

What is more, because genetic knowledge also provides information about the health of third parties, e.g. siblings and offspring, it may affect their lives and family relationships. Therefore, some researchers, for instance Iain Brassington, go as far as claiming that individuals have a duty to not know: that they should waive their right to know, to protect the privacy of others (Brassington 2011).

Since most genetic diseases lack effective therapy, knowledge of a genetic condition or risk does not serve to improve the quality of life. Consequently, it is pointed out that it can be a source of severe stress and pose a threat to the individual's sense of integrity and their right to an open future because the awareness of the risk alone causes a permanent state of tension and anxiety, and becomes the source of a "spoiled" identity (Scambler 2009; Klitzman 2009; Borry et al. 2014). This may prevent individuals from leading a normal life and planning their future regarding marriage, procreation and professional career. In particular, this is exemplified by young caregivers threatened with a genetic disease who live in the shadow of the condition as they care for a sick parent with the knowledge that they cannot avoid it (Sparbel et al. 2008; Williams et al. 2009).

The stigmatizing nature of genetic knowledge is particularly problematic in the case of "genetic predispositions" or risks, where although the information emerging from the tests is purely probabilistic, it is sometimes interpreted in absolute terms, which can cause the phenomenon of "genetic hypochondria" (Pääbo 2001). Despite the fact that a condition may never develop, the individual can take on the role of a "perpetual patient" and anticipate the disease, as in the case of the American actress Angelina Jolie, who learned that she carried a mutation in the BRCA1 which increased the risk of breast cancer, and underwent a prophylactic bilateral mastectomy (Kamenova et al. 2014). Nevertheless, it is argued that such knowledge may even lead to suicide attempts (Bird 1999). Similar examples can be found regarding prenatal genetic testing, which is often not fully conclusive but can negatively impact on the experience of motherhood (Asscher and Koops 2010; Rothman 1993; Rapp 2000; Kelly 2009). Hence, it is stressed that genetic knowledge may have a negative effect on the dynamics of family life and spousal relationships, whereas ignorance protects a family's privacy (Takala 2001, 487; Laurie 1999, 123; Juengst 1999; Featherstone et al. 2006; Røthing et al. 2014).

In this context, the proponents of the right to not know are also referring to the right to happiness. When genetic information is not used to improve the quality of life, knowing about a condition or risk can, at most, lead to the individual's negative appraisal of their life. This, in turn, may endanger their mental health or that of their loved ones, and also impact on spousal, parental and family relationships. The imperative of knowledge is thus deemed immoral, as it impinges on a person's dignity and their individual right to a future unencumbered by knowledge.

13.4.2 Arguments for the Duty to Know

Although most bioethicists assume that the right to information automatically entails the right to ignorance, many also suggests that genetic information holds a contrary status to other medical information, which results from the fact that it may cause consequences to third parties. This so-called "genetic exceptionalism" makes it necessary to treat genetic information on different terms (Soniewicka 2010, 150–158; Dryla 2014).

The hereditary nature of many diseases makes it quite probable that genetic information may have a significant impact on the health and life of the individual's relatives. Hence it is stressed that individuals from risk groups not only have the right but also the duty to know about their potential condition, and to reveal this knowledge to others (Ost 1984; Shaw 1987; Rhodes 1998; Harris and Keywood 2001; Bortolotti and Widdows 2011; Juth 2014). According to some, genetic knowledge is thus not only the individual's private matter, but others are also entitled to it, including spouses, offspring and other relatives, whereas the personal rights to autonomy, privacy and ignorance are not a sufficient argument in the face of other people's potential suffering or danger to their lives.

This argument is especially apparent when there is a possibility of passing a disease to a child. Many accentuate the moral obligation to give birth to a healthy child. This may necessitate genetic testing, because knowledge makes it possible to avoid the child's suffering, whilst genetic ignorance exposes it to suffering and premature death. Thus, according to some, especially in the case of incurable and hereditary genetic diseases, no life can be considered a better situation than living with a severe, fatal disease (Clarkeburn 2000; Chańska 2009; Różyńska 2011). Based on this premise, genetic testing is, in fact, considered a specific form of prevention and a "protective" reproductive practice. Julian Savulescu even writes about "procreative beneficence" and states that parents have a moral duty to take all necessary action to ensure their child's health and quality of life (Savulescu 2001; Savulescu and Kahane 2009). This is of particular significance in the context of medical law, which requires doctors to inform families about the risk of disease (Falk et al. 2003), whilst the failure to do so may lead to litigation based on claims of "bad conception" or "bad life" (Soniewicka 2009; Pelias 1986).

In addition to the risk of passing a genetic mutation to offspring, there are also other consequences which ignorance may cause to family members. Usually, a person who is at risk is eventually cared for by their spouse or children. Hence in this case, the right to know is, in fact, a duty which results from responsibility towards others. Therefore, the patient's right to not know may be considerably restricted by the need to protect other individuals and the quality of their lives. Furthermore, in the case of "radical and irreversible medical procedures", the obligation to inform cannot be completely waived, even at the patient's request" (Michałowska 2003).

13.5   Recapitulation

The doctor-patient relationship should take on the character of an alliance based on dialogue which leads to specific forms of medical care. The essence of good communication is the proper flow of information between the two parties. Its level and form should be adjusted to the recipient's perceptive capabilities. At the same time, ethics is an essential element which facilitates the resolution of many conflicts, taking into account, for instance, the purpose of genetic testing. Therefore, by following the guidelines of occupational codes, of care based on the principles of social justice and of respect for individual autonomy, as well as by using the partnership model in mutual relations, it is possible to build these relationships on the foundation of mutual respect and respect for human dignity. All these elements, in turn, serve the key goal of ensuring the patient's best possible welfare, which was already stressed in ancient times in the maxim: *Salus aegroti suprema lex esto*. Nonetheless, the interpretation of the concept of *welfare* may pose a natural problem, also as an element which connects the past with the future. At present, the past may overlap into and, at times, determine the future. The question is whether this process necessarily leads to more in-depth knowledge and its better application, and if the transparency of people is our right or rather our duty. Will the evolution of the *Homo sapiens* through *Homo faber* and *Homo technologicus* inevitably mean the primacy of the *Homo transparent*?

References

Andorno R (2004) The right not to know: an autonomy based approach. J Med Ethics 30(5):435–440
Arnason V, Hjörleifsson S (2007) Geneticization and bioethics: advancing debate and research. Med Health Care Philos 10(4):417–431
Asscher E, Koops B-J (2010) The right not to know and preimplantation genetic diagnosis for Huntington's disease. J Med Ethics 36(1):30–33
Baker DL, Schuette JL, Uhlmann WR (1998) A guide to genetic counseling. Wiley, New York
Baum E, Antczak S (2014) Szczęście – fenomen niejedno wymiarowy. Konteksty filozoficzne, społeczne i polityczne rozważań nad odczuwaniem szczęścia. In: Kowalska E, Prufer P, Kowalski M (eds) Szczęście w wymiarze pedagogiczno-socjologicznym. Impuls, Kraków
Beauchamp TL, Childress JF (1996) Zasady etyki medycznej. Książka i Wiedza, Warszawa
Bentham J (1958) Wprowadzenie do zasad moralności i prawodawstwa. PWN, Warszawa
Bird TD (1999) Outrageous fortune: the risk of suicide in genetic testing for Huntington disease. Am J Hum Genet 64(5):1289–1292
Boratyńska M, Konieczniak P (2001) Prawa pacjenta. Difin, Warszawa
Borry P, Shabani M, Howard HC (2014) Is there a right time to know? The right not to know and genetic testing in children. J Law Med Ethics 42(1):19–27

Bortolotti L, Widdows H (2011) The right not to know: the case of psychiatric disorders. J Med Ethics 37(11):673–676

Brassington I (2011) Is there a duty to remain in ignorance? Theor Med Bioeth 32(2):101–115

Cebrat S, Cebrat M (2012) Człowiek przejrzysty czyli jego problemy z własną genetyką. Wydawnictwo Kubajak, Krzeszowice

Chadwick R, Levitt M, Shickle D (1997) The right to know and the right not to know: genetic privacy and responsibility. Cambridge University Press, Cambridge

Chańska W (2009) Nieszczęsny dar życia. Filozofia i etyka jakości życia w medycynie współczesnej. Wydawnictwo UW, Wrocław

Chyrowicz B (2002) Bioetyka i ryzyko. Argument równi pochyłej w dyskusji wokół osiągnięć współczesnej genetyki. Towarzystwo Naukowe KUL, Lublin

Clarke A (ed) (1994) Genetic counselling: practice and principles. Routledge, London

Clarkeburn H (2000) Parental duties and untreatable genetic conditions. J Med Ethics 26(5):400–403

Czarnecki P (2008) Dylematy etyczne współczesności. Difin, Warszawa

Dar-Nimrod I, Heine SJ (2011) Genetic essentialism: on the deceptive determinism of DNA. Psychol Bull 137(5):800–818

Domaradzki J (2013) Prawo do niewiedzy a obowiązek wiedzy w opiniach rodzin osób z chorobą Huntingtona. Etyka 47(2):18–33

Domaradzki J (2015) Patient rights, risk and responsibilities in genetic era: a right to know, a right not to know, or a duty to know? Ann Agr Env Med 22(1):156–162

Dryla O (2012) Prawo do niewiedzy a autonomia. Diametros 32:19–36

Dryla O (2014) Informacyjny aspekt testów genetycznych - przegląd zagadnień. Diametros 42:29–56

Dryla O (2015) Niedyrektywność w poradnictwie genetycznym dotyczącym decyzji reprodukcyjnych. Analiza Egzystencja 32:71–91

Edwin AK (2008) Don't lie but don't tell the whole truth. The therapeutic privilege – is it ever justified? Ghana Med J 42(4):156–161

Ekberg M (2005) Governing the risks emerging from the non-medical uses of genetic testing. AJETS 3(1):1–16

Elias N (2000) The civilizing process: sociogenetic and psychogenetic investigations. Blackwell, London

Falk MJ, Dugan RB, O'Riordan MA, Matthews AL, Robin NH (2003) Medical geneticists' duty to warn at-risk relatives for genetic disease. Am J Med Genet 120A(3):374–380

Featherstone K, Atkinson P, Bharadwaj A, Clarke A (2006) Risky relations: family, kinship and the new genetics. Berg, New York

Freeman M (2007) Prawa człowieka. Wydawnictwo Sic! Warszawa

Fulda KG, Lykens K (2004) Ethical issues in predictive genetic testing: a public health perspective. J Med Ethics 32(3):143–147

Gert B, Culver CM, Clouser KD (1997) Bioethics. A return to fundamentals. Oxford University Press, Oxford

Gillon R (1997) Etyka lekarska. Problemy filozoficzne. Wydawnictwo Lekarskie PZWL, Warszawa

Gillon R (2001) Etyka lekarska – cztery zasady plus zakres. In: Alichniewicz A, Szczęsna A (eds) Dylematy bioetyki. Akademia Medyczna w Łodzi, Łódź

Gostin LO (1995) Genetic privacy. J Law Med Ethics 23:320–330

Greco M (1993) Psychosomatic subjects and the 'duty to be well'. Personal agency within medical rationality. Econ Soc 22(3):357–372

Hallowell N (1999) Doing the right thing: genetic risk and responsibility. Sociol Health Ill 21(5):597–621

Harris J (2001) Bioethics. Oxford University Press, Oxford

Harris J, Keywood K (2001) Ignorance, information and autonomy. Theor Med 22(5):415–436

Harrison R (1995) Bentham Jeremy. In: Honderich T (ed) Encyklopedia Filozofii, T. 1. Wydawnictwo Zysk i S-ka, Poznań

Hedgecoe A (1998) Geneticization, medicalisation and polemics. Med Health Care Philos 3(1):235–243

Helgesson G (2014) Autonomy, the right not to know, and the right to know personal research results: what rights are there, and who should decide about exceptions? J Law Med Ethics 42(1):28–37

Hołub G, Duchliński P (eds) (2008) Ku rozumieniu godności człowieka. Wydawnictwo Papieskiej Akademii Teologicznej, Kraków

Jasudowicz T, Czepek J, Kapelańska-Pręgowska J (2014) Międzynarodowe standardy bioetyczne. Dokumenty i orzecznictwo. Wolters Kluwer SA, Warszawa

Jonsen AR (2003) The birth of bioethics. Oxford University Press, Oxford

Juengst ET (1999) Genetic testing and the moral dynamics of family life. Publ Underst Sci 8(3):193–205

Juth N (2014) The right not to know and the duty to tell: the case of relatives. J Law Med Ethics 42(1):38–52

Kamenova K, Reshef A, Caulfield T (2014) Angelina Jolie's faulty gene: newspaper coverage of a celebrity's preventive bilateral mastectomy in Canada, the United States, and the United Kingdom. Genet Med 16(7):522–528

Kapelańska-Pręgowska J (2011) Prawne i bioetyczne aspekty testów genetycznych. Wolters Kluwer, Kraków

Kelly SE (2009) Choosing not to choose: reproductive responses of parents of children with genetic conditions or impairments. Sociol Health Ill 31(1):81–97

Klitzman R (2009) "Am I my genes?": questions of identity among individuals confronting genetic disease. Genet Med 11(12):880–889

Knoppers BM (2014) Introduction: from the right to know to the right not to know. J Law Med Ethics 42(1):6–10

Kymlicka W (2009) Współczesna filozofia polityczna. Aletheia, Warszawa

Lajeunesse RC, Lussier MT (2010) Therapeutic privilege: between the ethics of lying and the practice of truth. J Med Ethics 36(6):353–357

Laurie GT (1999) In defence of ignorance: genetic information and the right not to know. Eur J Health Law 6(2):119–132

Layard R (2012) Making personal happiness and wellbeing a goal of public policy. http://www.lse. ac.uk/economics/research/researchImpact/RichardLayard.aspx. Accessed 21 Jul 2016

Lippman A (1991) Prenatal genetic testing and screening: constructing needs and reinforcing inequities. Am J L Med 17(1–2):15–50

Mauron A (2002) Genomic metaphysics. J Mol Biol 319(4):957–962

Michałowska K (2003) Informowanie pacjenta w polskim prawie medycznym. Prawo Med 13:106

Naczelna Izba Lekarska (2004) Kodeks Etyki Lekarskiej http://www.nil.org.pl/__data/assets/pdf_ file/0003/4764/Kodeks-Etyki-Lekarskiej.pdf. Accessed 21 Jul 2016

Nelkin D, Lindee MS (1999) The DNA mystique. The gene as a cultural icon. W. H. Freeman and Company, New York

Nordgren A (2008) Genetics and identity. Community Genet 11(5):252–266

Nordgren A, Juengst ET (2009) Can genomics tell me who I am? Essentialistic rhetoric in direct-to-consumer DNA testing. New Genet Soc 28(2):157–172

Novas C, Rose N (2000) Genetic risk and the birth of the somatic individual. Econ Soc 29(4):485–513

Oduncu FS (2002) The role of non-directiveness in genetic counselling. Med Health Care Philos 5(1):53–63

Ost DE (1984) The 'right' not to know. J Med Philos 9(3):301–312

Pääbo S (2001) The human genome and our view of ourselves. Science 291(5507):1220

Pelias MZ (1986) Torts of wrongful birth and wrongful life: a review. Am J Med Genet 25(1):7180

Petersen A (2006) The genetic conception of health: is it as radical as claimed? Health 10(4):481–500

Porter R (1997) The greatest benefit to mankind. A medical history of humanity. W.W. Norton & Company, New York

Rapp R (2000) Testing women, testing the fetus. The social impact of amniocentesis in America. Routledge, New York

Rhodes R (1998) Genetic links, family ties, and social bonds: rights and responsibilities in the face of genetic knowledge. J Med Philos 23(1):10–33

Rose S (1995) The rise of neurogenetic determinism. Nature 373(6513):380–382

Rose N (2007) The politics of life itself: biomedicine, power, and subjectivity in the twenty-first century. Princeton University Press, Princeton

Røthing M, Malterud K, Frich JC (2014) Caregiver roles in families affected by Huntington's disease: a qualitative interview study. Scand J Caring Sci 28(4):700–705

Rothman BK (1993) The tentative pregnancy: how amniocentesis changes the experience of motherhood. Norton Company, New York

Różyńska J (2011) Wartość (nie)istnienia. Etyka 44:124–147

Savulescu J (2001) Procreative beneficence: why we should select the best children. Bioethics 15(5–6):413–426

Savulescu J, Kahane G (2009) The moral obligation to create children with the best chance of the best life. Bioethics 23(5):274–290

Scambler G (2009) Health-related stigma. Sociol Health Ill 31(3):441–455

Shaw M (1987) Testing for the Huntington gene: a right to know, a right not to know, or a duty to know. Am J Med Genet 26(2):243–246

Singer E, Couper MP, Raghunathan TE, von Hoewyk J, Antonucci TC (2008) Trends in U.S. attitudes toward genetic testing 1994–2004. Public Opin Q 72(3):446–458

Soniewicka M (2009) Regulacje prawne wobec rozwoju nowoczesnych technik kontroli prokreacji: analiza roszczenia wrongful life. Diametros 19:137–159

Soniewicka M (2010) Dyskryminacja genetyczna. In: Stelmach J, Brożek B, Soniewicka M, Załuski W (eds) Paradoksy bioetyki prawniczej. Wolters Kluwer, Warszawa

Sparbel KJH, Driessnack M, Williams JK, Schutte DL, Tripp-Reimer T, McGonigal-Kenney M, Jarmon L, Paulsen JS (2008) Experiences of teens living in the shadow of Huntington disease. J Genet Couns 17(4):327–335

Stanley M (1981) The technological conscience. Survival and dignity in an age of expertise. The University of Chicago Press, Chicago/London

Stempsey WE (2006) The geneticization of diagnostics. MED Health Care Phil 9(2):193–200

Strohman RC (1997) The coming of Kuhnian revolution in biology. Nat Biotechnol 15(3):194–200

Świątkowski A (2006) Prawo socjalne Rady Europy. Europejska Karta Społeczna. Protokoły. Zrewidowana Europejska Karta Społeczna. Universitas, Kraków

Szawłowski R (1982) Prawa człowieka a Polska. Polonia, Londyn

Szewczyk K (2009) Bioetyka. Medycyna na granicach życia, T. 1. Wydawnictwo Naukowe PWN, Warszawa

Takala T (1999) The tight to genetic ignorance confirmed. Bioethics 13(3–4):288–293

Takala T (2001) Genetic ignorance and reasonable paternalism. Theor Med Bioeth 22(5):485–491

Takala T, Häyry M (2000) Genetic ignorance, moral obligations and social duties. J Med Philos 25(1):107–113

Testart J (1990) Przejrzysta komórka. Państwowy Instytut Wydawniczy, Warszawa

Ustawa o zawodach lekarza i lekarza dentysty (1996) Ustawa z dnia 5 grudnia 1996 r. o zawodach lekarza i lekarza dentysty. Dy. U. 1997 Nr 28 poz. 152. http://isap.sejm.gov.pl/DetailsServlet?id=WDU19970280152. Accessed 21 Jul 2016

Williams JK, Ayers L, Specht JK, Sparbel KJ, Klimek ML (2009) Caregiving by teens for family members with Huntington disease. J Fam Nurs 15(3):273–294

Wilson J (2005) To know or not to know? Genetic ignorance, autonomy and paternalism. Bioethics 19(5–6):492–504

Zeiler K (2007) Who am I? When do "I" become another? An analytic exploration of identities, sameness and difference, genes and genomes. Health Care Anal 15(1):25–32

Chapter 14
Technical and Ethical Limits in Prenatal and Preimplantation Genetic Diagnosis

Małgorzata Karbarz

14.1 Methods in Prenatal and Preimplantation Genetic Diagnosis

Prenatal diagnosis (PND) can be performed using non-invasive techniques: ultrasound, magnetic resonance imaging (MRI); minimally invasive techniques: cell-free fetal DNA (CffDNA), preimplantation genetic diagnosis (PGD); and invasive techniques: chorionic villus sampling (CVS), amniocentesis, fetal blood sampling (Collins and Impey 2012). Ultrasound and MRI will not be discussed here as these methods are not based on genetic analysis. To establish fetal genotype, a risk-free CffDNA may be offered. A maternal blood sample from a pregnancy of approximately 9 weeks is checked for fetal RhD status, sex, several paternally inherited single gene disorders and Down syndrome (Hill et al. 2012). Chorionic villus sampling, amniocentesis and fetal blood sampling procedures are the main choice for couples who wish to have fetal genotyping (Collins and Impey 2012). CVS is an aspiration of trophoblastic tissue under ultrasound guidance. Then for rapid trisomy 13, 18 and 21 and sex chromosome aneuploidy testing techniques like fluorescence in situ hybridization (FISH) and polymerase chain reaction (PCR) are performed. The karyotype analysis is offered later. The limit in the method is the evidence of placental mosaicism in 1% of CVS samples, which is much greater than for amniocentesis samples. Amniocentesis involves taking a small sample of amniotic fluid transabdominally under ultrasound monitoring usually after 15 weeks of pregnancy. The laboratory techniques for genetic analysis are similar to those used for CVS (Collins and Impey 2012). Direct sampling of fetal blood from the umbilical cord, has a much higher

M. Karbarz (✉)
University of Rzeszów, Rzeszów, Poland
e-mail: karbarz.m@gmail.com

© Springer International Publishing AG, part of Springer Nature 2018
M. Soniewicka (ed.), *The Ethics of Reproductive Genetics*,
Philosophy and Medicine 128, https://doi.org/10.1007/978-3-319-60684-2_14

pregnancy loss rate (7.2%) than CVS and amniocentesis (Dugoff and Hobbins 2002). Currently it is only performed for potentially lifesaving therapeutic in utero.

Preimplantation genetic diagnosis, as mentioned above, can be considered as an early form of prenatal diagnosis for couples at high risk of transmitting an inherited disease to their offspring. The purpose of PGD is a diagnosis of a specific disease gene in cells taken from oocytes/zygotes or embryos produced *in vitro* through assisted reproductive technology (ART) and then transferring into the uterus only those embryos that are not affected by this disease (Traeger-Synodinos and Staessen 2014).

A precondition for genetic analysis in PGD is the obtaining of genetic material through a biopsy step. There are basically three developmental stages at which cells suitable for PGD analysis can be biopsied: polar bodies (PBs) from the oocyte/zygote stage, blastomeres from cleavage-stage embryos, or trophectoderm cells from blastocysts. Biopsy at each stage has some limitations. Polar bodies (PBs) are produced in the first and the second meiotic division as oocytes complete their maturation. To prevent misdiagnosis, both PBS should be analyzed 4–12 h after ICSI (intracytoplasmic sperm injection) and 8–16 h for the second polar body. The main limitation of this method is that only genetic disorders of a maternal origin are available for analysis, the paternal genetic information remains unknown. This method may be considered as cost- and time-consuming because it is necessary to employ two samples per run and not all oocytes will be fertilized and form embryos(Traeger-Synodinos and Staessen 2014). Blastomere biopsy is carried out on the third day after fertilization (66–72 h after ICSI), when the embryo is six to ten cells stage. After opening the zona of the embryo (mechanically, enzymatically or laser light) one or two blastomeres are taken. Due to the accuracy of the analysis, collecting two cells was previously recommended, but currently thanks to new technologies of genetic testing it is not necessary. The removal of two cells from an embryo is more harmful to the developing fetus than the biopsy of a single cell. A limitation of this method is the small amount of material to be analyzed and high rates of mosaicism at this early stage of development. Blastocyst – stage biopsy is biopsy on day 5 after fertilization of the ovum. Embryo at the blastocyst stage consists of two layers of cells: trophoectoderm (TE), and the inner cell mass (ICM). From TE placenta is formed and from ICM embryo is formed. A few TE cells are biopsied for analysis. The advantage of TE biopsy is the availability of a large amount of material for testing. The limitations are the survival of only 40–50% of embryos in vitro to the blastocyst stage and a short time to analyze if the fresh embryo transfer is to be performed by day 6 (Traeger-Synodinos and Staessen 2014).

The techniques described above are the methods for collecting material for tests, but the most challenging step in PGD and PND is the stage of genetic analysis. To date there are four techniques and their modifications used for PGD and PND: PCR based methods, FISH, CGH (comparative genome hybridization) and aSNP (single nucleotide polymorphism) array but each of these methods has its limitations.

The method based on PCR allows the amplification of the DNA fragment so we obtain a lot of copies. PCR is used primarily for the diagnosis of monogenic diseases. The major limitation is the risk of contamination with foreign DNA. A method based on fluorescent in situ hybridization uses a fluorescent probe to assess the embryos for chromosome aberrations or selection of male embryos in case of X-linked disease. Comparative genomic hybridization (CGH) is a cytogenetic which simultaneously evaluates all chromosomes from a single cell. Single nucleotide polymorphisms array (aSNP) allows the simultaneous analysis of monogenic diseases and chromosomal aberrations. (Traeger-Synodinos and Staessen 2014).

As the framework of this publication does not allow for a detailed description of all the methods and their limitations, a few cases have been chosen to serve as examples of the technical limits in PND and PGD, but first it is important to answer the question concerning the limits.

14.2   Limits

A limit is a point beyond which it is not possible to go or a point beyond which someone is not allowed to go (Merriam-Webster online Dictionary 2015). This simple definition reflects the two groups of limits that will be considered in this paper. The first part of the definition may concern the technical limit – a point beyond which it is impossible to go. As scientists, owners of in vitro clinics and above all parents, we would like to have 100% certainty of the PGD and prenatal genetic test results. Although current knowledge in human reproductive genetics as well as laboratory techniques are constantly developing, each method has a technical point that limits its utility. We can consider technical limits in two aspects. The first is a purely technical limitation which means that the quality of equipment or the chemistry is not efficient, accurate and generates errors in analysis results so this can be named 'outer' limit. The second, 'inner' limit occurs when for example a stadium of embryo is not advanced enough for more certain analysis results but later analysis would not enable the embryo implantation. Sometimes these inner and outer limits permeate each other in one analysis method.

The second part of definition reflects well ethical limits in prenatal and preimplatation genetic diagnosis. Although there is a technical possibility to perform a procedure there is an ethical limitation, a point beyond which someone is not allowed to go. It is not easy to say where this ethical 'stop sign' is as we do not have a clear 'inner' or 'outer' limits which would designated boundaries. The 'inner' limit could be the status of the embryo and the 'outer' could be the parent's interests. Later in this publication, the selected examples of technical and ethical limits will be presented.

14.2.1   Inner Technical Limit in PND and PGD: Embryo Mosaicism

A good example of inner technical limitation is embryo mosaicism. Anomalies in cleavage-stage embryos which affects 60% embryos generated by in vitro fertilization (IVF), may occur as a result of disturbances during meiosis and then it is present as an uniform abnormality in all cells or may be due to errors in the segregation during the first mitotic division resulting in mosaicism (Baart and Van Opstal 2014). Chromosomal mosaicism is defined as the coexistence of two or more chromosomally different cell lines in an organism which developed from a single zygote and it is a common phenomenon in the early stage of development of the human embryo (Robberecht et al. 2012).

When embryos are composed of a mixture of chromosomally normal and abnormal cells then it is a diploid-aneuploid mosaic and when only of abnormal cells mixture then it is a aneuploid mosaic. Recent analysis using comparative genomic hybridization shows that three quarters of a normally developing human embryo has a chromosomal mosaicism in the cleavage stage and only one fourth is uniformly diploid (Mantzouratou and Delhanty 2011). Previously it was thought that embryo screening would increase the rate of pregnancy after in vitro fertilization. This contributed to the practice of FISH preimplantation genetic screening (PGS) in conjunction with 3-day embryo biopsy, which involves the removal of one or two blastomeres for analysis. PGS was offered in many in vitro centers, but their clinical value has become questionable. The high rate of diploid-aneuploid mosaicism observed during PGS weakens the reliability of the diagnosis, because the blastomere which is biopsied for analysis does not represent the remaining embryo well. If the diploid-aneuploid embryo is biopsied and a diploid cell is taken, the remaining embryo which will be transferred to the uterus contains aneuploid cells and additionally the number of diploid cells decreased after biopsy. In a reverse situation, where aneuploid cells are retrieved during the biopsy, the embryo is not transferred, although in fact the ratio of diploid to aneuploid cells increased.

In Prenatal Diagnosis mosaicism is also a limitation. Shortly after the introduction of chorionic villus sampling it was discovered that the arrangement of fetus and placenta chromosomes can be different, although they originate from the same zygote (Baart and Van Opstal 2014). There are many reports of abnormal karyotypes in chorionic villus that were later not confirmed in fetal cells. CVS can be performed in two ways: the direct method (STC-villi-Short term cultured villi) and the long term preparation method(LTC-villi – Long term cultured villi). These techniques differ in the origin of cells that are analyzed in cytogenetic preparations: the cells in STC-villi come from chorionic cytotrophoblast, and those in LTC-villi come from the mesenchyme. Cytotrofoblast and mesenchymal core of CV have different embryonic origins: cytotrofoblast comes from the trophoblast of the blastocyst, while chorionic mesenchyme derived from inner cell mass (Baart and Van Opstal 2014). Chromosomal analysis in this case is a test for two different compartments of the embryo, which may be chromosomally different as a consequence of the post-zygotic mitotic division errors. The karyotype of mesenchymal core due to its embryonic origin better represents the karyotype of the fetus. Standard cytogenetic prenatal diagnosis by chorionic villus sampling showed mosaicism in 1–2% of the samples. The time and place of a post-zygotic mitotic error during embryonic development, as well as the mitotic or meiotic origin of chromosome aberration determines the pattern of mosaicism. Errors during the early divisions of an originally normal zygote can cause generalized mosaicism covering both the placental and fetal compartment. Later errors affecting specific cell lines lead to a confined placental mosaicism (CPM) and, more rarely, to confined fetal mosaicism (CFM). It is possible to distinguish nine different types of mosaicism: five general and four limited. In the daily practice of prenatal cytogenetic diagnosis, all types can be detected, causing significant cytogenetic variation on the trophoblast-embryo axis.

The example of chromosomal mosaicism well represents the inner limit that has to be taken into consideration before applying a new prenatal or preimplantation diagnostic method commercially, as the wrong test results may be generated. This can happen when laboratory diagnosticians are unaware of the inner limitations or the test developer leaves some margin for a possible mistake. The PGN and PGD are too important to leave this percentage and this kind of technical barrier should be fully respected and accepted as integral feature of human embryo.

Interestingly mosaicism is detected postnatally in 0.4–1% of patients (Ballif et al. 2006; Conlin et al. 2010). However, little is known about the prevalence of mosaicism in various organs in children and adults, as only the blood is routinely tested in postnatal cytogenetic diagnosis. A recent study revealed that mosaic aberrations are present in about 0.8% of phenotypically normal adults (Rodriguez-Santiago et al. 2010). In addition, mosaicism appears to be variable amongst different tissues: chromosomal aneuploidies were detected in approximately 10% of normal human brain cells (Robberecht et al. 2012). This indicates that we still do not know the exact mechanism of mosaicism determination and its role in human life and health and what level of mosaicism is tolerated by an embryo. There have been attempts to set this level. According to the model proposed by Evsikov and Verlinski, there is a phenomenon of self-elimination of the whole embryo if the number of aneuploid cells in the morula stage reaches a predetermined threshold value. Embryos with the number of aneuploid cells below that level develop and enter the blastocyst stage. To date, the model has not been tested directly on human embryos, but the mouse model showed that up to 30% of aneuploid cells are tolerated in apparently healthy animals (Baart and Van Opstal 2014). This model does not take into account the type and number of chromosome aberrations and thus this may affect the threshold value. Some chromosomal abnormalities, such as trisomy 21, can be tolerated in a higher proportion of cells as an extra chromosome 21 is compatible with development. What is interesting here is that the inner mechanism, let's say "nature", treats trisomy 21 less strictly than we as a society do. For example in France, 96% of children with trisomy 21 are not born (Jaranowski 2014). Is it really a good direction to be more precise than nature? Chromosomal instability is an inherent feature of human conception, so there is no way of eliminating this inner technical limit and as the example of mosaicism illustrates there are still weak areas of knowledge in the reproductive genetics. The current trend in PGD is to eliminate every chromosomal instability detected, often by the analysis of one or two cells. As can be seen in the case of mosaicism, one or two cells for analysis is not enough, the threshold level is difficult to set and the misinterpretation of the PGD and PND results are not easy to estimate, therefore the purpose of the diagnosis to keep 'healthy' ones and 'delete' those which are unhealthy is not achieved in some cases.

14.2.2 Outer Technical Limit-Alelle Drop-Out

Single-cell genomics is an important step towards the development of novel clinical methods in prenatal and preimplantation genetic diagnosis. All current single-cell analysis methods not only have limitations in the spectrum of DNA mutations and genetic variants that can be detected in the cell, but also in resolution, accuracy and reliability for detecting genetic variants. The single human diploid cell contains only about 7 pg of DNA, while genomic technologies require hundreds of nano-grams to micrograms of input DNA to perform only one genomic scan of a DNA sample (Kumar et al. 2014). Therefore in all current WGA (whole genome amplifi-cation) methods multiple displacement amplification (MDA) or PCR are performed. There is no WGA approach that produces a linear amplification product of the origi-nal cell's DNA template. The random loss of one allele (called allele drop-out or ADO), preferential amplification (PA) of an allele, or over-amplification or under-amplification of both alleles of a certain locus of the genome occurs during every single-cell WGA, and even varies significantly between different WGA methods (Kumar et al. 2014). Also other less-characterized artifacts, like the production of chimeric DNA molecules and the incorporation of wrong nucleotide, can change the DNA picture of the original cell. It is really difficult to discriminate the WGA artifacts from the cell's true genetic variants. WGA biases over longer distances in the genome can even easily be misinterpreted as genuine copy number changes in the cell.

ADO is a good example of outer technical limitations. Although PCR is consid-ered to be a robust technology that generally provides reliable results, errors during genotyping do occur. One problem is ADO – the occasional amplification failure of one of the two alleles at a given locus (Blais et al. 2015). This may happen due to sequence independent factors or allele-specific sequence variations. In the first case, it might be caused by a variety of sampling and/or molecular events like: variations in DNA extraction quantity or quality, presence of PCR inhibitors, variations in pipetting volumes of reagents or templates, imprecisions in thermocycler tempera-tures occurring unpredictably and independently of the patient's genotype (Pompanon et al. 2005). Such cases are usually not reproducible, and reanalysis of the sample might often provide acceptable results with resolution of the allele drop-out event. The second case may happen when one of the primers used cannot stably hybridize to its specific complementary sequence binding site (Pompanon et al. 2005; Soulsbury et al. 2007).

Alternatively, similar nonrandom amplification failures can also result from polymerase-hindering secondary structures induced by polymorphisms, GC content, or other allele-specific features of the target sequence itself (Lam and Mak 2013). Such allele dropout caused by previously unrecognized polymorphism in a primer binding site or lack of DNA polymerase fidelity caused by secondary structures was previously shown to affect genotyping results of several diagnostic assays such as cystic fibrosis, congenital adrenal hyperplasia, tyrosinemia type 1, multiple endocrine neoplasia I and Wilson disease (Blais et al. 2015).

In prenatal and preimplantation genetic diagnosis, any error in genotyping may generate the wrong genotype result and have important and long term consequences for both the embryo and the parents. Unfortunately, unless a genotyping assay is designed to detect allele dropout events it is impossible for the clinical laboratory to be aware that such an event has occurred. Genotyping errors caused by unpredictable sequence-independent events can be addressed by analyzing each sample multiple times (Pompanon et al. 2005). However, this strategy will not prevent errors from sequence-dependent causes such as polymorphisms that affect primer binding sites or target sequences' secondary structure that cause nonrandom allele dropout.

14.2.3 Inner Ethical Limit: Embryo Status

Some people feel that embryos have the same rights as a full person. Others feel embryos have some value as a potential person and others feel that embryos have no rights. This starting point is crucial in setting the ethical limits in prenatal and preimplantation genetic diagnosis. Moral status of the embryo is constantly subjected to discussion (McGee 2016; Wilkins 2016).

The theories or criteria of humanity can be divided into two basic groups: the first group respects humanity from the beginning and the second group indicates the later moment that is crucial to become human being or human person (Biesaga 2001). The first group may include: the criterion of fertilization, the criterion of the genetic code and the criterion of the continuity of human being development. The second group includes: the criterion of the formed zygote i.e. the theory of 21-h after fertilization, when the formation of the zygote is completed; the implantation criterion that is, the theory of the 14th day after fertilization, which completes the process of implantation, the possibility of twinning division is closed and the formation of the primitive streak begins; neurological theory criterion at day 40 after fertilization, when central nervous system, the brain begins to function; criterion of the ability to independent existence; The criterion of birth; the criterion to establish conscious cognitive-volitional contact with the environment. In this chapter only a few examples of the above criteria would be described.

The criterion of fertilization and the criterion of genetic code will be discussed together as they consider the very beginning of human life and they are both based on biological knowledge. After human copulation, sperm move towards eggs in the female's Fallopian tubes. At the beginning of the process, the sperm undergoes a series of changes, as freshly ejaculated sperm is unable or poorly able to fertilize. The acidic environment of the female reproductive tract causes sperm to become hypermobile and penetrate the outer layer of the egg. A second activation step occurs when, or shortly before, the sperm binds to the zona pellucida (the inner layer that surrounds the egg). During this step, the acrosome (an organelle at the tip of the sperm head) releases digestive enzymes that break down the zona pellucida two cells' membranes fuse and the cells merge (Melcher 2016). This is the fertilization process and it is the beginning of human life. Although scientist are getting closer and closer to a detailed insight into human fertilization, this process is still elusive for them. For example, the most recent discovery is the interaction between two proteins – Izumo1, which is produced by sperm, and Juno, its receptor on eggs – that enables human fertilization. Structural analysis of these proteins separately and in a complex manner provides insight into the recognition process and the subsequent sperm–egg fusion, but the details of this interaction is still unknown (Melcher 2016).

It is also important to look at what happens at the genome level before and directly after fertilization. Mature human oocytes are arrested in metaphase of the second meiotic division. After fertilization, the oocyte completes the second meiotic division and highly condensed chromatin in the sperm nucleus decondenses, resulting in a haploid male pronucleus. Parental pronuclei are physically separated in ooplasm during successive phases: G1, S and G2 during the first cell cycle of the embryo. After the entry into mitosis, maternal and paternal chromatin condenses into chromosomes, which for the first time are at a common metaphase plate. Oocytes and early embryos are transcriptionally inactive. During oogenesis, mammalian oocytes store a large amount of mRNA, proteins and macromolecular structures in the ooplasm to allow the first cell cycles after fertilization and facilitate the transition from maternal to embryonic, a process called embryonic genome activation (EGA). The quality and quantity of this maternal store is likely to be crucial for chromosome segregation regulation during the first embryonic divisions. Until now it was thought that the EGA begins at the stage of 4–8 cell, but recent studies show that this process starts already in the two cell for a selected number of genes, and then comes the main wave of transcription between the sixth and eighth day (Baart and Van Opstal 2014). During fertilization the centrosome is provided from the sperm cell and is responsible for spindle formation in the human zygote and therefore directly related to faithful chromosome segregation. An increased incidence of mosaicism is observed in dispermic human zygotes compared to monospermic or digynic embryos and can be considered evidence of a sperm contribution to abnormal spindle organization (Mantikou et al. 2012).

The above description is just the bare bones of what is known about fertilization and zygote creation. What is important here is that fertilization and new genome creation is a process, not a point. The genome is created when the maternal and paternal chromosome meet for the first time, so it would be proper to say that fertilization is a meeting, not a point or process (of course we can have a meeting point – but in this case it will be the metaphase plate). During the meeting, information is exchanged and something new is created that is exactly what is going on here. The term process is more technical – a series of actions that produce something or that lead to a particular result. The effect of this 'creative meeting' is the new genome from two 'old' parts. The genome is individual, unique, adapted to current times as a gift from the ancestors and it can be very stable and last much longer than a person – for example, the recently discovered assemblage of 28 hominin individuals, found in Sima de los Huesos in the Sierra de Atapuerca in Spain, has been dated to approximately 430,000 years (Meyer et al. 2016). What is worth noting here is that the human being is not a genome and cannot be judged by the genome. We cannot exist without a genome (whole, not a part) whereas the genome can exist without us, even after death, but cannot be expressed. What is also important, as it is in the case of above- mentioned mosaicism, we can have a slightly different genome in different organs or tissues and it does not mean that for example 10% of our brains is not a part of us. The genome is some kind of instruction manual and some editing mistakes may occur. They are impossible to correct after printing, but there can be an erratum- for example some other genes can take function of false gene. The instruction manual can be read with a different interpretation, like silencing or enhancing genes, and changes during our life – epigenetics. The genome as a whole determines our features and is an integrated part of person but we cannot take reproductive decisions on only one small fragment of our DNA as the machinery at the genome level is complicated, do not forget the above-mentioned technical limits. I would say that the criterion of genome is not a criterion of life and, on the other hand, it cannot be the criterion of embryo killing. A person equals genome and surrounding cell/cells plus 'something' whether we name it soul, spirit, mind etc. As long as we don't understand what this "something" is, we don't have the right to choose people based on their genome.

Embryo status is an example of inner ethical limits and human life should be respected from the beginning. It is a continuum and the genetic code just confirms this as genetic code was from the beginning of humanity and will be forever. It goes through different persons in a "constellation" which is specially dedicated for them. The beginning of human life is fertilization, with the meeting of parental chromosomes surrounded by cell/cells. It is the beginning of a continuum so the recognition of some moments in this development as a criterion for humanity is arbitrary and artificial. Scientific knowledge is too weak to tell exactly when life begins. If we still don't know 'what' exactly is going on during human fertilization and development, how can we set the human being threshold on 'when'? There are efforts to determine this moment by changing the status of the embryo and thus allowing PGD and PND with or without some ethical limitations. PND should only be practiced where there is a possibility of helping the embryo.

Another example of a limit or rather no limits in PGD is the criterion of consciousness, which is well expressed by Steinbock (2006):

> Early embryos, indeed early-gestation fetuses, have no consciousness, no awareness, no experiences of any kind, even the most rudimentary (...) It is not wrong to kill embryos because it doesn't matter to an embryo whether it is killed or goes on living. For unlike a fetus, an extracorporeal embryo is not developing into someone with a valuable future

The criterion of consciousness and a valuable future opens the gate for PGD so there is no ethical limitation in this case. In PND, the limit is the second trimester of pregnancy and it is based on Glover and Fisk's (1996) conclusion:

> We do not know for sure when or even if the fetus becomes conscious. However, temporary thalamocortical connections start to form at about 17 weeks and become established from 26 weeks. It seems very likely that a fetus can feel pain from that stage.

What can be easily seen in this quotation is the hesitation with regards the moment of the emergence of consciousness. The criterion of humanity cannot be built on uncertain and constantly changing scientific information, as the status of what is human is not subject to an update.

A valuable future or FLO (future-like-ours) is unpredictable and immeasurable and cannot be a criterion of humanity. In the argumentation of a valuable future there is a distinction between an in vitro preimplantation embryo and embryos in the uterus. The first, when taken for experiments or left-over from in vitro fertilization, have no future whilst the second, if not aborted, have a valuable future. It seems to be true, but this is the consequence of a decision to create embryos outside the maternal body, freeze them or allocate to experiments. This decision determined their 'invaluable' future, except for the "lucky one" that will be implanted and, if it goes through PND with acceptable results, will have FLO. The idea that embryos that are left behind the main stream (implantation) have no valuable future does not answer the question regarding the criteria that were taken into account while saying 'you will have the future' or 'you will not'. As can be seen from above mentioned technical limits in PGD, there is still a lot of genotyping uncertainty and a genetic test result cannot be an indicator of a valuable future.

14.2.4 Outer Ethical Limit: The Interests of Parents

In both PGD and PND the outer limit is the interests of parents and in some cases siblings. The parents want to have healthy baby, which seems to be good and morally accepted. To predict their future child health condition the preimplantation genetic diagnosis or prenatal genetic tests are performed and if the results are abnormal the negative selection for undesired genes (lack of implantation, abortion) is an option. The question is how to measure parental interests and where is the limit? The most common indicator is disease severity, with the harder the disease coming greater acceptance for termination, but what genetic defects are serious enough to prevent implantation? Should abortion be offered for a fetus that carries a breast cancer gene? These are difficult ethical questions. The parents seems to decide about embryo 'valuable genetic future' but is it ethical to make a choice for the fetus regarding something that may happen years in the future?

PGD is morally justified if it avoids the conception of a child affected by serious disease or handicap. To date, PGD has been reported for almost 200 genetic conditions and the number is still increasing (Traeger-Synodinos and Staessen 2014).

There is an idea to prepare a detailed list of all acceptable indications and the most controversial are mentioned below.

When it comes to PGD for an untreatable, middle-onset disorder, Huntington's disease (HD) is a good example. The child will have a long period of good and unimpaired living before the disease will be expressed. Later in life, HD is a highly invalidating disorder and places a sever burden on the family members affected, that is why HD is one of the main and highly accepted indication for PGD practice. The PGD for BRCA1 and BRCA2 is an example of preventable or treatable conditions, because the penetrance of these mutations is incomplete and (future) carriers have preventive or therapeutic options like mastectomy. Is it ethical to decide now what will happen in 30 years' time? After all, gene therapy options may be available for these gene mutations. There is a growing interest in PGD for mitochondrial DNA (mtDNA) disorders like Leigh syndrome and MELAS. A characteristic of these disorders is the coexistence of normal and mutated mtDNA within a single cell. Clinical symptoms depends on the level the mutational load in the cell has to exceed (Traeger-Synodinos and Staessen 2014). The transfer of embryos without a detectable mutant load is the most desirable but sometimes these embryos are not available after PGD and what to do next – undergo another cycle of PGD or chose the mutant with the lowest mutation level?

These are just a few examples of controversial issues in PGD. The limit, the parental wish to have a healthy child is moving toward having a child who will be healthy throughout its entire life, without the danger that disease could occur. There is also a trend to choose the sex and other sometimes dysgenic traits for the future child and it seems that PGD is becoming limitless.

The genetic condition of the embryo is its inner feature, but based on the symptoms of diseases that are manifested in people from outside the uterus (people already born), the elimination indicators for this embryo are set. This is very tricky, because it may be considered as an inner limitation in the embryo, but it is not. It is just the outer estimation of the severity of the disease based on genetic disorders that were detected postnatally in the population. Genetic information is not enough to predict the physical and mental condition of the human. There is no prenatal or preimplantation test, or group of tests, that can detect all types of abnormalities, genetic diseases, and birth defects. Although these tests can diagnose genetic abnormalities, they cannot predict the severity of the resulting disorder. Some children with genetic abnormalities can live a fairly normal life and what if a couple chose the 'genetically perfect' embryo but during delivery there are complications and the child becomes handicapped? After all, they 'ordered' a healthy beautiful child. In vitro clinics are aware of this limitation and they will never give 100% of certainty on any methods. There is nothing like 'zero risk' reproduction, Reproduction is always risky, life is risky and we are not able to control all of its aspects. The most important and sensitive aspects of our life, like conception and death, are beyond our control and should remain as such.

14.3 Summary

Our knowledge, technical possibilities and nature itself put limits on prenatal and preimplantation genetic diagnosis and these limits should be fully respected. There is no simple answer as to why some people with the same genetic disorders (at the genetic level) die and some do not and are well (phenotype level). Nature's internal mechanism is able to choose who is going to live outside the womb or only inside the womb for a while. The molecular machinery in the human body is very sophisticated and we still do not know a lot about the meaning of many of its processes. The elimination criterion that nature uses are not fully available for our brains. Human curiosity is really important in development but respect for who we are and how we are constituted is more important. In current societies there is a cult of genetics and we want to check everything by genetic testing – diseases, diet etc. Of course genetics is to serve the people but it does not have the power to decide on human life. If a genetic test result says you are fine to be implanted or to be born you will be, if not you are not. This is very wrong direction and was well captured by Hadjadj (2014)

> There is a whole side of science that wants to make us believe that what one sees in a microscope is more real than what one sees with a bare eye. Once a probe had been sent inside living beings, man discovered DNA. And science tried to persuade us it had solved the mystery of life.

What is currently being done is the establishment of a genetic disorders list, with the prospect of upgrading, based on the information about genetic changes in the human genome with uncertainly estimated expression and then usage of techniques with inner and outer limitations to decide who should live and who should not.

Human life begins at fertilization and the embryo has the same status as a person after birth and this the only limit in prenatal and preimplantation genetic diagnosis that should be fully respected.

Acknowledgments The writing of this article was funded by the Polish National Science Centre (Dec-2013/10/E/HS5/00157).

References

Baart EB, Van Opstal D (2014) Chromosomes in early human embryo development: incidence of chromosomal abnormalities, underlying mechanism and consequences for development. In: Sermen K, Viville S (eds) Textbook of human reproductive genetics. Cambridge University Press, Cambridge, pp 52–67

Ballif BC, Rorem EA, Sundin K et al (2006) Detection of low-level mosaicism by array CGH in routine diagnostic specimens. Am J Med Genet 140:2757–2767

Biesaga T (2001) Antropologiczny status embrionu ludzkiego. In: Biesaga T (ed) Podstawy i zastosowania bioetyki. Wydawnictwo Naukowe PAT, Kraków, pp 101–113

Blais J, Lavoie SB, Giroux S et al (2015) Risk of misdiagnosis due to allele dropout and false-positive PCR artifacts in molecular diagnostics. Analysis of 30,769 genotypes. J Mol Diagn 17:505–514

Collins SL, Impey L (2012) Prenatal diagnosis: types and techniques. Early Hum Dev 88:3–8. doi:10.1016/j.earlhumdev.2011.11.003

Conlin LK, Thiel BD, Bonnemann CG et al (2010) Mechanisms of mosaicism, chimerism and uniparental disomy identified by single nucleotide polymorphism array analysis. Hum Mol Genet 19:1263–1275

Dugoff L, Hobbins JC (2002) Invasive procedures to evaluate the fetus. Clin Obstet Gynecol 45(4):1039–1053

Glover V, Fisk N (1996) Do fetuses fell the pain. Br Med J 313:796

Hadjadj F (2014) Against the microscope – seeing the smallest one – Tribune of Fabrice Hadjadj Published on 11/17/2014 in Testimonials. http://www.fondationlejeune.org/en/news/testimonials/19-testimonials/976/contre-le-microscope-voir-le-plus-petit-tribune-de-fabrice-hadjadj). Accessed 17 June 2016

Hill M, Barrett AN, White H, Chitty LS (2012) Uses of cell free fetal DNA in maternal circulation. Best Pract Res Clin Obstet Gynaecol 26:639–654. doi:10.1016/j.bpobgyn.2012.03.004

Jaranowski P (2014) Francja: finansowanie badań nad chorobami genetycznymi związanymi z niepełnosprawnością umysłową, Newsletter Bioetyczny 4:4. http://www.poradniabioetyczna.pl/newsletter-bioetyczny/. Accessed 2 June 2016

Kumar P, Estek MZ, Van der Aa N, Voet T (2014) How to analyze a single blastomere? Application of whole genome technologies: microarrays and next generation sequencing. In: Sermen K, Viville S (eds) Textbook of human reproductive genetics. Cambridge University Press, Cambridge, pp 15–31

Lam CW, Mak CM (2013) Allele dropout caused by a non-primer-site SNV affecting PCR amplification a call for next-generation primer design algorithm. Clin Chim Acta 421:208–212

Mantikou E, Wong KM, Repping S, Mastenbroek S (2012) Molecular origin of mitotic aneuploidies in preimplantation embryos. Mol Genet Hum Reprod Fail 12:1921–1230. doi:10.1016/j.bbadis.2012.06.013

Mantzouratou A, Delhanty JDA (2011) Aneuploidy in the human cleavage stage embryo. Cytogenet Genome Res 133:141–148

McGee A (2016) We are human beings. J Med Philos 41(2):148–171. doi:10.1093/jmp/jhv064

Melcher K (2016) Structural biology: when sperm meets egg. Nature 534:484–485. doi:10.1038/nature18448

Merriam-Webster Online Dictionary (2015) http://www.merriam-webster.com/dictionary/limit. Accessed 8 June 2016

Meyer M, Arsuaga JL, de Filippo C et al (2016) Nuclear DNA sequences from the middle Pleistocene Sima de los Huesos hominins. Nature 531:504–507. doi:10.1038/nature17405

Pompanon F, Bonin A, Bellemain E, Taberlet P (2005) Genotyping errors: causes, consequences and solutions. Nat Rev Genet 6:847–859

Robberecht C, Voet T, Utine GE et al (2012) Meiotic errors followed by two parallel postzygotic trisomy rescue events are a frequent cause of constitutional segmental mosaicism. Mol Cytogenet 5:19. doi:10.1186/1755-8166-5-19

Rodriguez-Santiago B, Malats N, Rothman N et al (2010) Mosaic uniparental disomies and aneuploidies as large structural variants of the human genome. Am J Hum Genet 87:129–138

Soulsbury CD, Iossa G, Edwards KJ et al (2007) Allelic dropout from a high-quality DNA source. Conserv Genet 8:733–738

Steinbock B (2006) The morality of killing human embryos. J Law Med Ethics 34(1):26–34. doi:10.1111/j.1748-720x.2006.00005.x

Traeger-Synodinos J, Staessen C (2014) Preimplantation genetic diagnosis. In: Sermen K, Viville S (eds) Textbook of human reproductive genetics. Cambridge University Press, Cambridge, pp 157–170

Wilkins SM (2016) Strange bedfellows? Common ground on the moral status question. J Med Philos 41(2):130–147. doi:10.1093/jmp/jhv066

Chapter 15
From Informed Choice to Distributed Decision-Making: Ethnographic Tales from a Study on Prenatal Testing in Denmark

Nete Schwennesen

15.1 Introduction

In recent decades the ethical principle of informed choice has become an ethical panacea in prenatal testing. Questions of choice (e.g., can decisions on prenatal testing be considered autonomous? Do professionals influence decision-making? Is information non-directive? Can it be?) have become cardinal themes in professional and policy discussions of prenatal testing. This chapter intends to enrich current discussions on choice in prenatal testing, by presenting the insights from a Danish ethnographic study on prenatal decision-making, done as a part of a Ph.D, which I defended in Copenhagen in 2011 (Schwennesen 2011, 2012; Schwennesen and Koch 2009, 2012; Schwennesen et al. 2009, 2010). The Danish case is unique as it was the first country to introduce the technology of first trimester prenatal risk assessment (FTPRA) for Down's syndrome and other chromosomal diseases, free of charge for every individual woman, regardless of age and risk situation. The study takes as a starting point the making of new guidelines on prenatal testing in Denmark (Sundhedsstyrelsen [Danish Board of Health] 2004) which argued for the introduction of the new technology, and it studies ethnographically how they were implemented in practice at an ultrasound clinic in Denmark. In the following I will first briefly describe the route through which the new guidelines were made, how they were framed and what was expected from them. I will then turn to describe the challenges I was faced with when trying to explore processes of prenatal decision-making at the ultrasound clinic. In doing so, I introduce theories on prenatal decision making from the field of genetic counselling and discuss their shortcomings in relation to two aspects: the location of choice and the problem of transformation. I end up – using mainly concepts and theories from the field of STS – by arguing for

N. Schwennesen (✉)
University of Copenhagen, Copenhagen, Denmark
e-mail: ns@anthro.ku.dk

© Springer International Publishing AG, part of Springer Nature 2018
M. Soniewicka (ed.), *The Ethics of Reproductive Genetics*,
Philosophy and Medicine 128, https://doi.org/10.1007/978-3-319-60684-2_15

an approach which conceptualizes decision-making and knowledge production as processes of distributed action. As will become clear throughout the chapter, such an approach articulates other kinds of critical questions for discussion and consideration that are not made visible if problems and solutions about prenatal decision making are continually framed through the lens of choice.

15.2 Towards a New 'Paradigm of Self-Determination'

In 2000 the Danish Board of Health commissioned a medical working group (consisting of doctors and midwifes) to account for new methods in prenatal testing and to provide material for a possible revision of the guidelines of prenatal testing in Denmark that existed at that time (Sundhedsstyrelsen 1994). The work resulted in a report *Prenatal Testing and Risk Assessment* (Sundhedsstyrelsen 2003a) issued in March 2003, which recommended a significant revision of the previous organization of prenatal testing in Danish antenatal care. The working group argued that they considered the previous organization problematic as it was organized around predetermined high risk groups: only pregnant women aged 35 or above, or women who had a known increased risk of giving birth to a child with a chromosomal disease were given the offer to undergo prenatal testing. The working group characterized the program as belonging to a "paradigm of prevention" since the access criteria were established on the basis of economic calculations and a rationale of prevention. They saw the existence of such criteria as an indirect health political request for the women to participate in prenatal testing, signaling an overall health political aim of prevention at a population level. As a solution to what the group considered to be the problematic past of the organization of prenatal testing in Denmark, they suggested a future organization around a new principle of informed choice.

A central element in the suggested new organization was the introduction of a possible offer of first trimester prenatal risk assessment (FTPRA) to every pregnant woman – regardless of age and risk situation. FTPRA is a combined risk assessment for Down syndrome and other chromosomal disorders and is based on a combination of maternal age, nuchal translucency scanning and a biochemical test for serum free beta human chorionic gonadotrophin and pregnancy associated plasma protein A, also called the double test. FTPRA is considered the most effective non-invasive screening technology on the market for Down's syndrome and other chromosomal disorders (Nicolaides 2004), with an estimated detection rate between 85 and 95% (if a cut off at 1:300 is used as access criteria to invasive testing). The working group suggested that on the basis of this assessment women were to be informed about their risk (given as odds, such as 1:250 or 1:10.000) of carrying a fetus with Down's syndrome and that women with a risk above a defined cut off (1:250 was suggested) should be offered an invasive diagnostic test (such as chorionic villus sampling (CVS) or amniocentesis). The main advantages of implementing FTPRA into Danish antenatal care was argued to be its higher predictive value compared to the previous regime and an expected decrease in the number of invasive tests carried out and the number of miscarriages caused by invasive testing (which is about 1% Tabor et al. 1986).

The working group emphasized that information about the possibility of undergoing FTPRA should only be given to the pregnant woman if she expressed a wish to know about the possibilities. As such, information about the possibility of undergoing the test was not to be given automatically to every pregnant woman, but only in response to a request from the pregnant woman herself (Sundhedsstyrelsen 2003a, 57). On this basis, the main criteria of success for the program was formulated as the extent to which women expressing an interest in information about prenatal testing were able to make an informed choice about undergoing prenatal testing. The working group described such a new organization of prenatal testing in Denmark as belonging to a "paradigm of self-determination" which they saw as more in agreement with the intentions expressed in current legislation on patient rights that emphasize the importance of patient autonomy, integrity and self-determination (Act on Patient's Legal Rights and Entitlement in Denmark 1998).

In the report, the working group asked for a clear expression by Parliament about what they considered to be the primary aim of prenatal testing: prevention or choice. In a response, the Danish Parliament discussed the report and in May 2003 issued the following statement:

> The aim of prenatal testing is – within the juridical framework of Danish Law – to assist a pregnant woman, if she wants such assistance, to make her capable of making her own decisions. Neutral and adequate information is a necessary condition to this end.... The aim of prenatal testing is not to prevent the birth of children with serious diseases or handicaps. (Parliamentary Proposal, May 15 2003)

After a formal process of investigation, where patient organizations and central agencies were invited to express their opinions, the Danish Board of Health issued a statement in October 2003 where the central principles of the new guidelines were described (Sundhedsstyrelsen 2003b), and in September 2004 the new guidelines on prenatal testing in Denmark were published (Sundhedsstyrelsen 2004). The guidelines incorporated the working group's advice, and legitimized this incorporation primarily by referring to the official parliamentary statement about the official aim of prenatal testing as facilitating individual choice. This meant in practice that first trimester prenatal risk assessment (FTPRA) was now to be introduced as a possibility to every pregnant woman in Denmark.

In subsequent years, all 15 Danish counties decided to follow the new guidelines and introduce the new offer of FTPRA into antenatal care and by June 2006, the whole of the country was covered (Ekelund et al. 2008). One study covering two counties shows that only 2% of couples who were offered FTPRA in the period from 1 July to 31 December 2005 actively refused it and it is estimated that the overall current uptake is at least 90% (Tørring et al. 2008). In the Copenhagen area – where this study was conducted – the uptake is estimated to be around 95% (Tabor 2006). A survey estimating the overall detection rate for Down's syndrome in the new program, shows that it has increased from 86% in 2005 to 93% in 2006 and that the overall number of infants born with Down's syndrome has decreased by about

50% in the period 2000–2006 (from 55–65 per year in 2000–2004 to 31 in 2005 and 32 in 2006 (Ekelund et al. 2008, 3).

15.3   The Past in the Present

The introduction of informed choice as a new principle in the organization of antenatal care in Denmark expresses a current tendency in western European countries to frame choice as an obvious solution to what is considered to be the problematic past of prenatal testing. Historically, the establishment of "new" medical genetics grew out of the shadow of the eugenics movement and World War II, and had a strong anti-eugenic program. Eugenics was a central element in Nazi ideology and there was a widespread desire after World War II to reject it as a result (Koch 2004a). While eugenics was a concept related to the promotion of public good at the beginning of the century, in the second half of the century it gradually became associated with the execution of state power over marginalized individuals based on faulty scientific proof (Koch 2004a). In an attempt to distance both previous and future organization of prenatal testing from what is considered a problematic past – often associated with coercive state power carried out by an authoritarian and paternalistic doctor – current regimes of prenatal testing commit themselves to liberal values such as free choice, objective information and value neutrality (Rehmann-Sutter 2009, 235; Koch 2004a). In this sense, the considerations of the eugenic experience of the past can be said to have contributed to a transmission of the delegated role of the state in prenatal testing from practices of paternalism, carried out with an aim of protecting the health of the population, to the facilitation of individual, autonomous choice.

In the current regimes of choice, health professionals are key actors. Their main task is to facilitate objective information about the (future) condition of fetal life and the possibility of undergoing prenatal testing. In the last few decades genetic counselling has established itself as a profession, directed to implement this task. Emerging in the shadow of the Eugenic movement and of World War II, the profession has sought to distinguish itself from earlier, more directive forms of interaction in the relationship between a professional and a patient, in order to avoid being accused of any coercion. By doing this, autonomy in decision making has been promoted. As such, the guidelines are in line with a general recognition in genetic counselling that decisions concerning whether or not to have a genetic test should be the counselee's own autonomous decision (Hunt et al. 2005; Weil et al. 2006; Marteau et al. 2001). The premise is that in order to promote an active and autonomous choice, the decision-making process should be freed of any moral judgement or clinical values, so that the woman herself becomes able to make a decision which is consistent with her own values (Hunt et al. 2005; Weil et al. 2006; Marteau et al. 2001).

In the professional literature on prenatal counselling, non-directiveness is described as the ethical gold standard and as a presumption for the realization of a truly autonomous choice. Non-directiveness can be defined as providing complete unbiased information and restraining from giving practical advice (Rehmann-Sutter 2009, 235) and is seen as a safeguard against a potential powerful and authoritarian paternalistic doctor who determines what is right and wrong and dictates the subsequent decision. Authors such as Emery argue, for instance, that "genetic counselling aims to be non-directive, simply providing patients with information regarding their risk of a genetic disorder and options for managing that risk … such a non-directive approach … demonstrate[s] its primary role in offering patients informed choice" (Emery 2001, 81). Likewise, in a highly prominent textbook on genetic counselling, the professor in medical genetics Peter S. Harper points out that "it is not the duty of a doctor to dictate the lives of others, but to ensure that individuals have the facts to enable them to make their own decisions. (…) «non-directiveness» has become a somewhat central tenet of genetic counselling (…) the importance of non-directiveness lies in allowing the decisions to be taken by the individuals involved, not the person giving genetic counselling" (Harper 2004, 16). Even though the ideal of non-directiveness is increasingly being criticized in the professional literature on prenatal counselling for providing an insufficient basis for a profession that addresses moral issues (such as abortion and quality of life), and for failing to address the social and economic context within which individual decision making takes place (Weil et al. 2006), it continues to be the ethical standard against which professional practice is assessed (Williams et al. 2002).

15.4   The Question of Choice Takes Central Stage

In the professional and political debate on the implications of prenatal testing, the question of choice has come to take central stage – for both proponents and critics alike. In the Danish debate, questions have been raised about whether or not the policy of informed decision-making in the context of FTPRA works as intended. Are the users of FTPRA well informed? Are they provided with neutral and non-directive information? What measures may be taken to ensure these preconditions lead to a truly informed and autonomous choice? Critical voices in the Danish debate have pointed out that choices are limited when carried out in a regime shaped by preventive norms and values. Lene Koch expressed the following reservation about the introduction of the guidelines into Danish antenatal care: "It is an increasing problem that the collective establish offers of prenatal testing and then pretend that the individual chooses freely (…) When we talk about free choices, we also have to talk about how such choices are conditioned and maybe are not so free" (Koch 2004b, 116). This viewpoint echoes other scholars critical of genetic counselling, who argue that the very existence of prenatal screening programs represent a powerful recommendation to accept prenatal testing. Referring to prenatal screening programs, Angus Clarke points out: "The very existence of a screening programme amounts in effect to a recommendation that the testing thereby made available is a good thing. Health professionals and society would hardly establish and promote antenatal screening for Down's syndrome unless they wanted people to make use of it – the existence of such a programme is an implicit, but powerful, recommendation to accept any screening made. Screening programmes therefore, simply cannot be non-directive." (Clarke 1997, 401). In a similar vein, Lippman argues that the so-called "need" for prenatal testing is socially and culturally constructed and leads to an increasing control of pregnancy and abortion (Lippman 1991). Anderson argues that the implicit preventive imperative of prenatal testing, where prenatal testing is morally depicted as a social good, might influence the patient's decisions in implicit ways and interrogate the patient's ability to make truly autonomous choices (Anderson 1999, 127). Hunt focuses on the various meanings of risk held by clinicians and patients and argues that a professional's failure to articulate the contrasting meanings of risk held by clinicians and patients may undermine a neutral clinical communication and, thus, the patient's ability to make autonomous informed choices (Hunt et al. 2005).

The question of how much choices are constrained by professional power and the social and cultural surroundings through which choices are framed and made has also been a central point in feminist discussions on reproductive choice. On the one hand, the slogan of a woman's right to choose has often been employed by feminists, in particular in the 1970s, as the slogan for abortion campaigns (Petchesky 1987) and was extremely important for feminists fighting for greater choice for women in the field of reproduction. On the other hand, the slogan has been criticized for endorsing an exclusively liberal, individual approach to the issues of reproduction (McNeil et al. 1990). Critical feminists have pointed out that, under the guise of being given more choice, women are becoming subject to even greater medical and patriarchal control (Corea et al. 1987; Spallone and Steinberg 1987; Arditti et al. 1985). Ironically, the biggest challenge to such a critique comes from the women themselves. In spite of increasing critical feminist attention to new reproductive technologies in the 1980s, women have been very enthusiastic about using such technologies. "Women want it" has often been the argument of supporters.

Other critics use eugenics as a platform for criticizing current regimes of prenatal testing. The American sociologist Troy Duster suggests in his book *Backdoor to eugenics* that although he does not see a current eugenic identifiable with earlier eugenics, converging technologies, interests and disease prevention policies take on an increasing eugenic character (Duster 1990, x). Likewise, disability scholars such as Tom Shakespeare (1998) have raised concerns about the potential eugenic outcome of programs of prenatal testing. He argues that they implicitly frame disability as a medical problem to be avoided through prenatal testing and the termination of pregnancy. Habermas discusses the possibility that contemporary forms of prenatal testing will be a first step towards a liberal eugenic on a free market of choice regulated by supply and demand (Habermas 2003).

15.5   In Between Choice and Coercion

In the following I will not go into a normative discussion about whether or not we are witnessing a current form of liberal eugenics. Rather I will use the debate sketched out above to illustrate the framework through which the impact of prenatal testing has been discussed. The debate depicts the question of choice as the central node, for both proponents and critics alike. The type of choice that should be offered, the right to choose, and the conditions which influence and restrict choices are key to these discussions. On the one hand, the official aim of choice serves as a tool to mark contemporary programs of prenatal testing as distinct from what are seen as problematic quasi- eugenic practices or more paternalistic oriented regimes of the past. The new guideline on prenatal testing in Denmark is a prominent contemporary example of the way in which the aim of choice serves as a rhetorical tool to demarcate any links with past practice (Schwennesen et al. 2009). By linking the new guidelines with values such as choice, self-determination and non-directiveness, prenatal testing is presented as a value-neutral means for enhancing individual reproductive freedom and choice. This move expresses an increased emphasis on liberal values (individual autonomy and rational decision-making) in the context of health and a turn towards a more market- oriented method of health care delivery within programs of prenatal testing in Western-European countries (Helén 2005; Lemke 2005; Kerr 2004; Weir 1996) and is, as such, very much part of a consumerist ethos of modern medicine where health services are depicted as services intended to fulfill users' needs and preferences.

On the other hand, the ideal of autonomous choice has been used as a platform for critical interrogations of choice. The argument is that supposedly free choices are shaped by the social, institutional and cultural surroundings through which they are facilitated and made. Genetic counselling has tried to meet such critiques by committing the profession of genetic counselling to ideals of non-directiveness and choice, thereby seeking to distinguish the present practice of prenatal testing from practices of coercion and eugenic intentions. Ironically, it seems that critical voices of "claims of unfreedom" may sustain and, thus, strengthen the need for more non-directive information and choice. Paradoxically, more choice and more non-directive information thereby become (explicitly or implicitly) re-constituted as obvious means to solve the contemporary "problem" of prenatal testing.

In the following I will present an ethnographic take on my study on prenatal decision making in Denmark, and illustrate how the use of choice as a framing of the complex situations that emerges in the practical use of FTPRA, is problematic as it invokes overly simplified images of the pregnant woman and partners making use of prenatal testing and the process through which decisions are made in practice.

15.6   Tales from a Study on Prenatal Decision Making in Denmark

15.6.1   First Problem: The Location of Choice

I encountered a problem when beginning to investigate processes of prenatal decision making. The problem was related to the question of where to look for choices made in the context of prenatal testing. If we turn to the Danish Board of Health, they come up with a suggestion. The Danish Board of Health has made a model of the various locations of choice which they consider relevant in the context of prenatal testing in Denmark, illustrated in the guidelines (Sundhedsstyrelsen 2004, 26). The model is presented below (Fig. 15.1).

In this model the acts of *giving* information and *making* choices are portrayed as events that replace each other, step by step, throughout a potential trajectory of prenatal testing. After information follows choice, which follows information and so forth. In the model the first location of choice is located at the first pregnancy consultation with the GP. This choice was emphasized as one of the cornerstones in the new regime of prenatal testing in Denmark. The idea was that information should only be given to pregnant women and their partners who expressed an interest in wanting to know about the possibilities of undergoing prenatal testing. The next step in the model is the GPs provision of information about prenatal testing possibilities and then follows the pregnant woman and her partner's choice about whether to undergo FTPRA.

I started the study by interviewing pregnant women and their partners focusing primarily on the experience of their choice of undergoing FTPRA. When I asked them to describe how they would characterize the choice of wanting to have information on prenatal testing (first choice according to the model) and of undergoing FTPRA (second choice according to the model), a characteristic answer was that the act of undergoing FTPRA was not experienced as a choice at all(!) Rather it was experienced as a routine act on their pregnancy trajectory towards giving birth which you undergo in line with other pregnancy checkups during pregnancy. The act of "choosing" FTPRA was rarely experienced as a choice, but may be described as a "default pathway" (Webster 2007, 470) on the normal trajectory towards giving birth. This account of FTPRA choice as routine is also reflected in the overall relatively high uptake of FTPRA, estimated to be more than 90%.

In interviews, the act of undergoing FTPRA was often explained as an outcome of a fundamental trust in the Danish Health care sector and a shared value about using new technology and knowledge in the pursuit of having a healthy baby. Another important aspect was a hope of receiving a visual proof – by undergoing ultrasound and seeing the image of the fetus on the ultrasound screen – that they were expecting a healthy child. This points to the multiple workings of this technology in practice as being, on the one hand, a technology of risk assessment carried out in order to detect disease and possibly to prevent (or prepare oneself for) the birth of a diseased child and, on the other hand, a technology of confirming life, carried out in order to achieve a visual proof that the pregnant woman is expecting a healthy child. This is not to say that the women and their partners who I interviewed were ignorant about the actual content and limitations of FTPRA. In general they were well informed, and many women and their partners had used other media, such as the internet, to obtain more knowledge about FTPRA. However, knowledge about the technical details of FTPRA, and the rationale of prevention underlying it, was consciously placed into the background by many women in favor of a hope and a positive expectation that FTPRA would show itself to become a step on the road towards having a healthy baby. These multiple workings of FTPRA might be seen

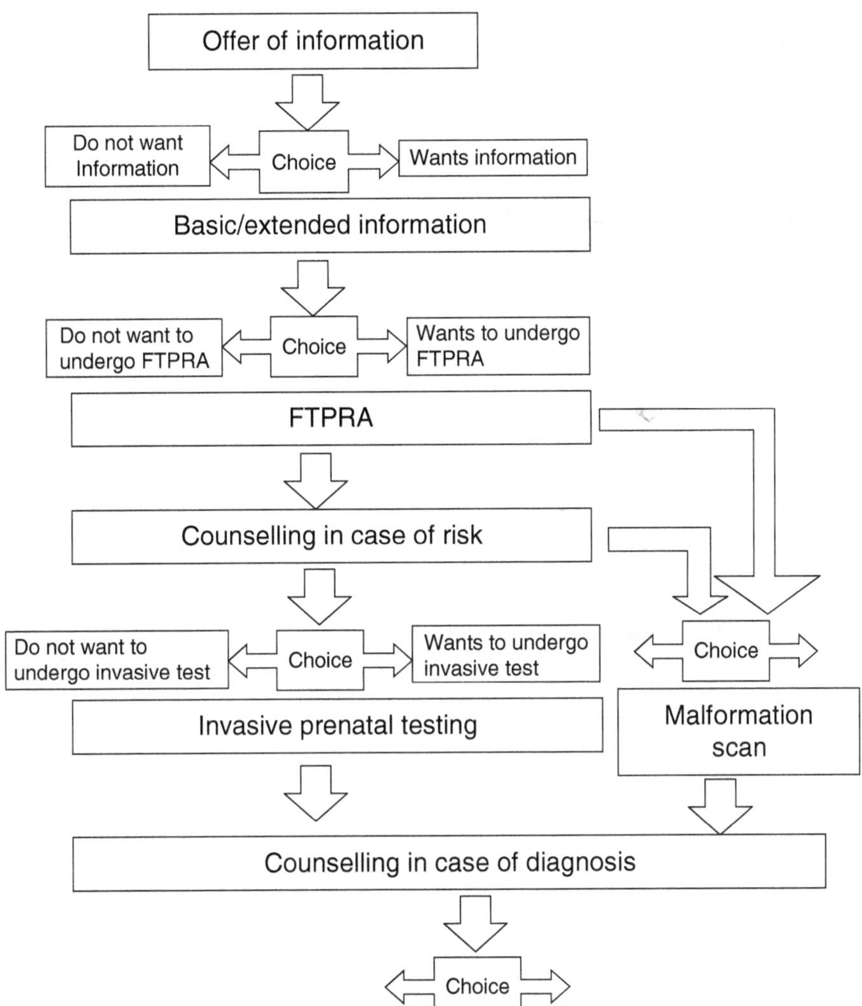

Fig. 15.1 A model of informed choice (Translated from: Sundhedsstyrelsen 2004, 26)

as an example of a point made by STS scholars, that the meanings of technologies – what they are and what they do - are not stable and pre-existing practices, but emerge in ongoing ways through the practices and anticipations involved in their use (Mol 2002; Berg and Mol 1998; Cussins 1998). In trajectories of prenatal testing, FTPRA will, in most cases, emerge as a life confirming technology; the pregnant woman and her partner will achieve a risk figure which is interpreted as "low" and will walk happily away with a visual proof that they are expecting a healthy baby.

These are indeed interesting findings about the ways in which many pregnant women and their partners access and anticipate FTPRA with a hope that it will confirm life. In my endeavor to investigate prenatal decision-making, however, I still wanted to explore situations where pregnant women and their partners experienced themselves as serious decision makers. Therefore I started to look for other situations where decisions on prenatal testing were made and this drew my attention to pregnant women and partners whose risk assessment was somehow considered problematic – either by the sonographers or by themselves. Such processes of problematization of the risk obtained often emerged in situations where a risk figure was calculated as "high" according to the cut off or in situations where the risk obtained was categorized as higher than the woman's age related risk (also categorized as background risk). In such situations, pregnant women and partners suddenly appeared as serious decision makers: the decision they were facing was whether to undergo an invasive test, such as a CVS or an amniocentesis, which involves a risk of inducing a miscarriage at around 1% (Tabor et al. 1986). These decisions were experienced as extremely difficult and demanding by most pregnant women and partners. Being obligated to choose on the basis of uncertain risk knowledge and hereby also take full responsibility of the choice being made, often gave rise to a complicated process of meaning-making emerging through the relationships between the professionals, the clinical socio-material setting, and the social life of the couples. In the face of complex risk knowledge, the pregnant woman and her partner were in many cases reluctant to make choices and transferred authority to the cut off. This meaning making process constituted a double movement in processes of decision making; whereas responsibility for the decision was delegated to the pregnant woman and her partner, authority was delegated to the cut off.

15.6.2   Second Problem: Transformation

In the model of decision making illustrated above, decisions are located as a single event occurring after information is given by a health professional. That is, knowledge and persons are portrayed as stable entities existing independently of each other. This model of the pregnant woman and her partner as individuals with stable preferences reflect an individualistic image of man, which you find in some psychological theories such as planned behavior (Ajzen 1991; Ajzen and Fishbein 1980). Drawing on this notion of human behavior, the psychologist Theresa Marteau has developed a much cited questionnaire designed in order to measure the extent to which informed choices are made (Marteau et al. 2001). The questionnaire is based on the psychological theory of planned behavior of Isek Ajzen (1991) which explains an individual's behavior as a result of knowledge and attitudes. Using this model of health behavior, Marteau classifies choices (as being informed or not) on the basis of three dimensions: knowledge, attitude and uptake. According to this definition an informed choice to undergo prenatal testing "occurs when an individual has a positive attitude towards undergoing a test, has relevant knowledge about the test and undergoes it" (Marteau et al. 2001, 100). An uninformed choice occurs, according to this definition, "when individuals do not have relevant knowledge or when their attitudes are not reflected in their behaviors" (Marteau et al. 2001, 100).

This understanding of choice as a single event and outcome of attitude and knowledge, makes it possible to measure it and then to document as to whether the aim of informed choice is achieved or not. The making of such a questionnaire expresses a need for auditing in contemporary health care sector (Hoeyer 2009; Strathern 2000), and an extensive international literature now reports from such studies (Van Den Berg et al. 2006; Marteau et al. 2001; Crang-Svalenius et al. 1998; see: Dahl et al. 2006a, b for reviews). As such, the principle of informed choice not only serves as a solution to what is considered a problematic past, but has also become a tool to measure and document quality in the current practice of prenatal testing. While such studies are useful in the contemporary culture of audit (Strathern 2000), this model of choice is insufficient if we want to understand the processes through which decisions on prenatal testing are made, and the ways in which they come to have effects in practice.

To make this argument more fully, I will introduce the reader to an anecdote from my field work. During my observations at the ultrasound clinic, I followed women through trajectories of FTPRA; I met them at the reception desk, waited with them in the hall, I accompanied them into the ultrasound scan, went out again and sat with them and waited with them for the result of the test. And I was present when the sonographer informed them about the resulting risk figure. While initially these observations were only considered to be background data for my study, I came to think of FTPRA as a process through which the pregnant women and their partners and entities such as life, risk and responsible behavior underwent a transformation. The period of pregnancy might be thought of as a transformative process of separation, through which two entities (a pregnant woman and her partner) are gradually transformed to become a family consisting of three entities (a mother, a partner and a child). Undergoing ultrasound and seeing the image on the screen, however, may potentially escalate this process of separation by detaching fetal life from the woman's body. During ultrasound, the image of the scan was often interpreted as an image of life, as if it existed outside and independent of the woman's body. In the context of risk communication, however, the fetus was attached to the woman and her partner again, now in the form of a life being at risk. Through this simultaneous process of detachment and attachment new identities of motherhood and fatherhood were created and new relationships of responsibility (between parents and fetus and sonographer and pregnant woman) evolved.

Understanding FTPRA as a process through which transformation takes place and new relationships emerge, made me realize that looking only at the location where choices were formally taking place, would miss out important knowledge about the ways in which knowledge about fetal life where creatively produced and managed in the socio-material encounter of FTPRA (involving various nonhuman actors such as visual images, risk figures and the organization of space) and would miss the generative effects of such knowledge production and, subsequently, the ways in which situations of choice were shaped, experienced and acted upon.

If we understand FTPRA as a process of transformation, the limitations of the model of informed choice becomes visible. This model only measures the *outcome* of choice and thereby makes invisible the process through which situations of choice are made and the ways in which technologies such as the ultrasound scan and the risk assessment are involved in this process of knowledge production. On this basis, I started thinking about decision making as a transformative process of knowledge production that evolves over time and is shaped and reshaped by both human and non-human actors.

Recent attempts have been made in the field of genetic counselling, to develop models of counselling which take into account the dynamic and interactive relationship between counsellor and counselee, most notably the model of shared decision-making (SDM), developed by Elwyn and colleagues (1999, 2000). In the model of SDM "the clinician/counsellor and the client share information on the basis of which a decision is to be made. They then discuss their views and come to an agreed decision for which they share the responsibility" (Elwyn et al. 2000, 136). The model of SDM acknowledges a more active role of the counsellor than the model of non-directiveness, by allowing the principle to "flourish" (Elwyn et al. 2000, 136) in cases where it is considered relevant. It involves, for instance, that the counsellor can contribute with his professional opinion in the decision making process. While this model of counselling acknowledge that decisions are reached through relationships between the professional and the patients, it does not take into account how knowledge is shaped over time and involves socio-material encounters beyond the immediate space of the consultation room. I argue for an approach that studies decision making as processes of knowledge production that develop over time, and which may potentially involve a transformation of the entities involved, such as the pregnant woman and her partner and notions of risk, life and responsibility, and can, as such, be seen as a suggested extension of the model of shared decision making.

15.7   Accounting for Transformation and Non-human Agency

In order to develop an understanding of decision making as a process of transformation and knowledge production which develops over time and involves both human and non-human forms of agency, I turned to the field of Science and Technology Studies (STS) and actor-network theory (ANT). STS/ANT is not a coherent theory which provides a general frame of explanation, but can be described as an attitude (Gad and Bruun Jensen 2009) that takes its point of departure in an interest in how knowledge, agency, subjects and objects in ongoing ways emerge in particular socio-material settings. Scholars from these fields share a common interest in combining ethnographic work with investigations of philosophical questions, something which Mol also calls "empirical philosophy" (Mol 2002, 4). It is, therefore, also an analytical resource which invites the researcher to ask other forms of questions than those relating to correspondence (how to produce and communicate true and value-neutral facts) towards questions related to their becoming (how are entities coming into being in socio-material practice and what are the (unintended) effects) (Mol 2002, 4).

One of the basic premises in STS/ANT is a constructivist ontology, which states that entities (such as humans, technologies, facts, life, risk) are not distinct and definable, but are fundamentally intertwined with each other in socio-material networks and gain their specific qualities through these relations (Barad 2007; Mol 2002; Latour 1999). That is, no a priori assumptions about the existence of certain entities are made from the outset. Rather, the analytical focus is on the heterogeneous ways in which hybrid associations are made between people and things. A controversial tenet therefore, is that agency is not to be considered a property of human beings but an effect of relational associations made up by both human and non-human actors (Barad 2007; Latour 1999). As such, everything that can be observed to cause an effect on the course of action can be conceived as an actor (Latour 1999, 124).

To illustrate the ways in which humans and technology are considered co-produced, Latour uses the well-known example of a gun and a man, and ask: who or what is responsible for the act of killing? One answer would be that it depends on who or what is seen as the actor. If one takes at the outset a psychological understanding of man, which explains human behavior as the outcome of individual intentions and beliefs, the gun is rendered a neutral tool to make predefined and fixed intentions possible. According to this view it is insignificant if (or which) technology is used. A murderer is a murderer with or without a gun. This individualistic perspective on human behavior is similar to the one found in the FTPRA guidelines. On the other hand, a technological deterministic account would argue that human action is determined by the gun. In such a view "the gun acts by virtue of material components irreducible to the social qualities of the gunman" (Latour 1999, 176). Latour argues against both views and suggests the starting point is not from discrete entities (man or gun, human or technology) but it is from the association of the hybrid human-and- gun. This new hybrid actor (or actant) comprising of human-and-gun translates the identities of both the human and the gun into new ones. This translation is symmetrical, both the gun and the human change through the association made between the two:

> A good citizen becomes a criminal, a bad guy becomes a worse guy; a silent gun becomes a fired gun, a new gun becomes a used gun, a sporting gun becomes a weapon. (Latour 1999, 180)

From this outset neither the human being nor technology can be understood as the actor, rather it is the association of the human and the technology that acts. He concludes, that "it is neither people nor the gun that kills. Responsibility for action must be shared among the various actants" (Latour 1999, 180). Latour uses this simple two-actor scenario to illustrate the point that action is an effect of human and non-human associations. This is not to say that action is a result of a certain subset of actors. Rather, Latour's point is, that all (human and non-human) actors participating in bringing about action, together constitutes the action.

In a parallel argument about the fundamental entanglement of the material and the human, the feminist science studies scholar Karen Barad develops her theory on agential realism (Barad 2007). From Bohr's theory of Quantum Physics, Barad develops the concept of *phenomenon* to capture the ways in which entities are constituted and reconstituted out of specific configurations of apparatuses of observation. Barad's point is that the specific configuration of apparatuses of observation we use to make entities knowable should not be framed as passive or innocent (Barad 2007, 33). Rather, they should be conceived as productive and part of the phenomena which emerge through processes of knowledge production. If we apply this framework to the practices of FTPRA, we should understand the fetus as a phenomenon which "includes" the whole configuration of apparatuses of observation which is brought about in the production of knowledge about what the fetus "is" (such as visual images, risk figures, the pregnant woman, her partner, the sonographer, discursive interpretations etc.). As such, who or what the fetus becomes (human/nonhuman, low risk/high risk) in FTPRA processes of knowledge production is not given in the fetus itself, but involves the whole configuration of apparatuses of observation used in bringing about knowledge of the fetus. This framework has serious implications for how to think about agency and responsibility in FTPRA. If agency is distributed to involve the whole apparatuses through which knowledge about the fetus are produced, responsibility for the knowledge produced and the decision made cannot be attributed to a single actor (the woman and her partner) but must be conceived as distributed among the various apparatuses and persons. Barad explicitly draws out the ontological and ethical implications of this framework. She argues that if we acknowledge that the basic units of reality are phenomena, then we are in part responsible for what there "is" in the world: "we are responsible for the world in which we live, not because it is an arbitrary construction of our choosing, but because agential reality is sedimented out of particular practices that we have a role in shaping" (Barad 2007, 247). Following this, Barad suggests that ethical concerns must not simply be considered supplemental to the practice of knowledge production (which has been the case in FTPRA where ethical concerns are centered on questions of correspondence, such as how to produce objective and non-directive communication). Rather, ethical concerns must be considered as an integral part of knowledge production and involve also concerns about the effect of knowledge production for those being involved.

If we apply this framework to the analysis of prenatal decision making we are able to think of prenatal decision making as a process of knowledge production, through which new forms of associations between human and technology are made possible and new phenomena are made to matter: Who the pregnant woman (and her partner) "is" (mother/pregnant woman, responsible/irresponsible, empowered/victim) and what the fetus becomes (human/nonhuman, low risk/high risk) might potentially change through the number of associations the pregnant woman and her partner engage with when they undergo FTPRA. This translation process is symmetrical; the technologies and entities involved in FTPRA (the ultrasound scan, the risk assessment, the policy of informed choice) what they are, and what they do, might also undergo a change. What kind of associations are made and what their effects are, however, is an empirical question.

15.8   (Bio)Power as an Effect of Action

This framework has important implications for how we may understand power. While the FTPRA guidelines articulate a view on power as an external force, exemplified by the paternalistic or authoritarian doctor, STS/ANT scholars work with a notion of power as an *effect* of action (Latour 1986). Within this framework, what constitutes power is the ways in which an entity is translated and comes to have effects in practice. As Latour explains:

> Those who are powerful are not those who "hold" power in principle, but those who practically define or redefine what "holds" everyone together. This shift from principle to practice allows us to treat the vague notion of power not as a cause of people's behaviour but as the consequence of an intense activity of enrolling, convincing and enlisting (...) power is not something you can hold or posses, it has to be made. (Latour 1986, 273-274)

This form of power is productive in the sense that it works through actors' engagement in processes of translation. Building on a notion of power as working through productive relationships of knowledge production, Foucault develops the concept of bio-power to characterize the ways in which the modern state internalizes concerns about being healthy and informed in the citizen through the production of normative expert claims in the form of statistics and surveys (Foucault 1976). Such normative claims work by problematizing a given state of affairs (Foucault uses the example of sexual behavior) and simultaneously produce new spaces of possible action, through which people can act according to the norm (Foucault 1976). This implies a radical rephrasing of the notion of freedom and choice. Whereas the guidelines takes information and choice to lead to an increase in individual freedom, this form of power works exactly through the choices and the new spaces of possible action made available to its citizens (Rose 1999).

The sociologist Nikolas Rose argues in the article *The politics of life itself* (Rose 2001), that bio-power has taken a new form in contemporary liberal society which reconfigures the relationship between the state and the individual. Rose explains:

> [W]ithin the political rationalities that I have termed "advanced liberal" the contemporary relation between the biological life of the individual and the well-being of the collective is (...) no longer a question of seeking to classify, identify, and eliminate or constrain those individuals bearing a defective constitution, or to promote the reproduction of those whose biological characteristics are most desirable, in the name of the overall fitness of the population. Rather, it consist in a variety of strategies that try to identify, treat, manage or administer those individuals, groups or localities where risk is seen to be high. The binary distinctions of normal and pathological, which were central to earlier bio-political analyses, are now organised within these strategies for the government of risk. (Rose 2001, 6–7)

It is obvious that FTPRA may be considered as an instance of bio-power because of its concern with classification of fetuses being at risk. Through the practices that constitute FTPRA, knowledge about risk is created and managed and new spaces of action evolve where knowledge is acted upon (the decision to undergo invasive diagnostic testing).

Knowledge about risk in the context of prenatal testing expresses the probability of a future event (having a child with Down's syndrome) and the possibility of acting in the present (by undergoing an invasive, diagnostic test), in order to control that future (Hacking 1990). In this new space of action, the cut off, defined by the Danish Board of Health, take central stage, by organizing access to invasive testing. The cut off is settled on the basis of complex cost-benefit calculations made with the aim of defining a limit of access to invasive testing that most effectively distributes risk into the categories of high (requiring an invasive test) and low (not requiring test). This calculation is based on a large sample of epidemiological data (Nicolaides 2004) and expresses a relationship between detection rate, false positive rate and economic costs. There is a trade-off between these factors, and where the exact limit is set is not objectively evident, but is based on normative and political decisions balancing the different interests. Latour developed the concept *immutable mobile* (Latour 1987) to capture the ways in which standards are capable of translating interests across time and space. I argue that we may understand the cut off as such a standard, which transports the rationality of effective prevention into the space of possibility that emerges in the context of prenatal decision-making and comes to have effects on the ways in which risk is understood and acted upon. The cut off may be conceived as the result of a translation of a long chain of heterogeneous associations made up of people (epidemiologists, doctors, nurses, politicians, pregnant women), bodily substances (blood, fetuses), things (computers, software,) and expresses normative, political and economic intentions and values about the desired outcome of the program on a population level. When the cut off emerges as a proportional risk figure in the context of prenatal decision making, however, it is detached from the heterogeneous chain of associations from which it is created and thereby crafted as objective and value-free. In Theodore Porter's terms, statistics are a basic technology for crafting objectivity and stabilizing facts (Porter 1995).

A number of social science scholars have studied the implications of an increasing discourse of risk in the context of pregnancy and prenatal testing, from a Foucauldian bio-power perspective (Meskus 2009; Helén 2004, 2005; Ettorre 2002; Lupton 1999; Weir 1996; Lippman 1991). While such scholars seem to emphasize the disciplining effects of risk knowledge on pregnant women and the control and constraints on action derived from such knowledge, this study adds to them, by illustrating the complex ways in which knowledge about risk is managed and made meaningful in a concrete socio-material clinical practice. The point is that how standards such as risk and cut off points are translated in practice, and what their effects are, is not given in the nature of the standard itself, but is an empirical question which has to be investigated.

If we apply an STS/ANT framework to the study of decision-making we have to dismiss the idea of power as an external force constraining our actions, and the image of decision making as (ideally) an outcome of the stable preferences and wishes of single actors. Rather, understanding decision making within such a framework allows us to consider decision making as a result of distributed action emerging through the dispersed socio-material practices of FTPRA. This de-centered understanding of agency and decision making directs the analytical focus away from questions such as whether or not the pregnant woman and her partner has, in fact, received or achieved non-directive information or whether or not their choice was actually an outcome of an autonomous act, towards questions related to how new associations of pregnant women and their partners and technologies evolve in FTPRA and their (unintended) effects. On this basis, I argue that if we want to investigate how decisions are made on prenatal testing, we shall not define entities such as humans, life and risk a priori or assess whether or not knowledge is communicated in a non-directive way, but follow empirically the way in which such entities are linked in practice and look for the effects arising from these linkages.

This approach allows one to study processes of decision making as an effect of an *assemblage* of multiple forms of actors involved in the process of FTPRA, such as human actors (the professionals, the pregnant woman and her partner), technologies (such as the ultrasound scan), standards (such as the cut off and other risk figures), policies (the policy of informed choice) and nonverbal aspects of knowledge production (such as the organization of space).

15.9   Concluding Remarks: From Informed Choice to Distributed Decision Making

The translation of the principle of informed choice into the heterogeneous socio-material practices of prenatal decision making is not as straightforward as the Danish Board of Health imagined or argued in the new guidelines. In this process, actors are associated in new ways, and an unintended redistribution of entities such as humans, life, risk and responsibility unfolds. What was intended by introducing the principle of informed choice into Danish antenatal care was to increase empowerment and the self-determination of pregnant women and their partners. However, many pregnant women and partners experience the exact opposite. Having to make serious decisions on the basis of uncertain risk knowledge – which potentially involves a miscarriage of a healthy fetus – was not experienced as empowering, but as frustrating and as giving rise to a sense of losing control. Decisions did not emerge as a result of prudent acts and rational considerations, but as the result of a complex process of meaning making where a wide range of socio-material actors were involved, such as the ultrasound scan, risk figures, health professionals and the pregnant women and their partners. In complex processes of decision making, the cut off often came to work as the dominant norm through which specific risk figures became meaningful in practice. As such, the norm of effective prevention was translated into the practice of prenatal decision making and became an important actor, in the process of making risk figures meaningful. Even so, responsibility for the decision being made was delegated only to the pregnant woman and her partner, which was experienced more as a burden than something desired.

A report from the Ethical Council in Denmark discusses what they see as the ethical dilemmas arising from the use of prenatal testing in Denmark. The Ethical

Council intends to contribute to the debate about how to organize and practice current and future prenatal testing in an "ethically acceptable way" (Ethical Council 2009, 7). One of their main concerns is how to make sure that informed choices are really made in the context of prenatal testing. The Ethical Council states: "It is important that the woman can make decisions on an informed basis, that she understands the details of the situation and is able to cope with the consequences the different decisions have for her and other people involved. It is also important that she is not pressured to make particular decisions by specific circumstances or health professionals (...) A recurrent theme in the Ethical Council's discussions has been, if there are specific circumstances about prenatal testing, which can make it difficult for the pregnant woman to make independent and well informed decisions" (Ethical Council 2009, 13–14). In this statement and in the report as such, the question of choice comes to take central stage in the discussion of how to practice prenatal testing in a moral and ethical way. It is emphasized that the woman has to "understand the situation", so that she will be able to "cope with the consequences" and that it is important not to "pressure" the woman in decision-making, so that "independent" and "well informed" decisions can be made. Thereby, the discussion about how to practice FTPRA well is reduced to being primarily a question about how to inform the pregnant women and their partners, in order to make them capable of making an autonomous choice and be able to cope with the consequences.

I argue that framing the problems and solutions of prenatal testing and decision making through the lens of choice might blind us from the ways in which the principle of informed choice actually comes to have effects in practices of prenatal decision making. Although the principle is designed to bring about certain effects, the specific socio-material relations it enters determines its actual capability in practice. While the Ethical Council acknowledges that the realization of the principle is perhaps not as straightforward as imagined, the discussion on how to make the organization of prenatal testing in Denmark work well do not take into account what the principle of informed choice "does", when translated into complex practices of prenatal decision making.

In discussing the issue of how to organize prenatal testing in an "ethical acceptable way", the Danish Board of Health starts from the premise that knowledge is an entity which can be handed to the pregnant woman and her partner in a non-directive and value-neutral way, so that she is able to make a decision which is in accordance with her values and wishes. I suggest that we have to start from the opposite premise, that prenatal knowledge production and decision making is an intervention – regardless of whether or not it is deemed non-directive – which redistributes entities of life and risk and relationships of responsibility. If we acknowledge that such entities and relationships are fluid and not given in the nature of things, then we also have to think about prenatal decision making and knowledge production as an intervention into the very categories through which pregnant women and their partners experience pregnancy and come to understand themselves and their relationship to others. I argue, that we have to consider ethical concerns in prenatal decision making not simply as supplemental to the practice of knowledge production (which has been the case in FTPRA where ethical concerns are centered on questions of correspondence, such as how to produce objective and non-directive communication), but also involve concerns about the concrete production and effects of such knowledge production.

If we take this as a starting point other (ethical) questions comes to the fore, such as how to be accountable for the entities enacted in the process of prenatal decision making and knowledge production. What forms of agencies are being excluded or left out in processes of decision-making? Are women and their partners who do not act according to the norm of effective prevention allowed to articulate alternative notions of life and risk, without being condemned as wrong? How to interfere in knowledge production with care? In opposition to an ethics aiming at non-interference (non-directiveness and value neutrality), such questions express an ethics of being locally accountable for the ways in which programs of prenatal testing intervene in pregnant women's lives and of taking responsibility for the entities and phenomena that emerge through such knowledge production.

References

Act on patient's legal rights and entitlement in Denmark (1998) Act. No. 482 of July

Ajzen I (1991) The theory of planned behavior. Organ Behav Hum Decis Process 50:179–211

Ajzen I, Fishbein M (1980) Understanding attitudes and predicting social behavior. Prentice Hall, Englewood Cliffs

Anderson G (1999) Nondirectiveness in prenatal genetics: patients read between the lines. Nurs Ethics 6(2):126–136

Arditti R, Duelli-Klein R, Minden S (1985) Test-tube women: what future for motherhood? Pandora Press, Ontario

Barad K (2007) Meeting the universe halfway: quantum physics and the entanglement of matter and meaning. Duke University Press, Durham

Berg M, Mol A (eds) (1998) Differences in medicine. Duke University Press, London

Clarke A (1997) The process of genetic counselling. In: Harper P, Clarke A (eds) Genetics, society and clinical practice. BIOS Scientific Publishers, Oxford, pp 179–200

Corea G, Duelli RK, Hanmer J, Holmes HB, Hoskins B, Kishwar M, Raymond JR, Rowland R (1987) Man-made women: how new reproductive technologies affect women. Hutchinson, London

Crang-Svalenius E, Dykes A, Jörgensen C (1998) Factors influencing informed choice of prenatal diagnosis: women's feelings and attitudes. FetalDiagn Ther 13:53–61

Cussins C (1998) Ontological choreography: agency for women patients in infertility clinics. In: Berg M, Mol A (eds) Differences in medicine. Duke University Press, London, pp 166–201

Dahl K, Kesmodel U, Hvidman L, Olesen F (2006a) Informed consent: attitudes, knowledge and information concerning prenatal examinations. Acta Obstet Gynecol Scand 85(12):1414–1419

Dahl K, Kesmodel U, Hvidman L, Olesen F (2006b) Informed consent: providing information about prenatal examinations. Acta Obstet Gynecol Scand 85(12):1420–1425

Duster T (1990) Eugenics through the back door. Routledge, London

Ekelund CH, Jørgensen FS, Petersen OB, Sundberg K, Tabor A (2008) Impact of a new national screening policy for sown's syndrome in Denmark: population based cohort study. BMJ 337:a2547

Elwyn G, Edwards A, Kinnersley P (1999) Shared decision making in primary care: the neglected second half of the consultation. Brit J of Gen Pract 49:477–482

Elwyn G, Gray J, Clarke A (2000) Shared decision making and non-directiveness in genetic counselling. J Med Genet 37:135–138

Emery J (2001) Is informed choice in genetic testing a different breed of informed decision-making? A discussion paper. Health Expect 4:81–86

Etisk råd [Ethical Council] (2009) Fremtidens fosterdiagnostik. Etisk råd, København

Ettorre E (2002) Reproductive genetics, gender and the body. Routledge, London

Foucault M (1976) The history of sexuality: an introduction, vol 1. Pantheon Books, New York

Gad C, Bruun Jensen C (2009) On the consequences of post-ANT. Sci Technol Hum Values 34(6):55–80

Habermas J (2003) The future of human nature. Polity, London

Hacking I (1990) The taming of chance. Cambridge University Press, Cambridge

Harper PS (2004) Practical genetic counseling. Arnold, London

Helén I (2004) Technics over life: risk, ethics and the existential condition in high-tech antenatal care. Econ Soc 33(1):28–51

Helén I (2005) Risk management and ethics in antenatal care. In: Bunton R, Petersen A (eds) Genetic governance. Health, risk and ethics in the biotech era. Routledge, London, pp 47–63

Hoeyer K (2009) Informed consent: the making of a ubiquitous rule in medical practice. Organization 16(2):267–288

Hunt L, de Voogd KB, Castañeda H (2005) The routine and the traumatic in prenatal genetic diagnosis: does clinical information inform patient decision-making? Patient Educ Couns 56(3):302–312

Kerr A (2004) Genetics and society. A sociology of disease. Routledge, London

Koch L (2004a) The meaning of eugenics: reflections on the government of genetic knowledge in the past and the present. Sci Context 17(3):315–331

Koch L (2004b) Hvad er der galt med formynderi? In: Meyer G (ed) Ekspert og samfund. En interviewbog om offentlig diskussion og videnskab. Samfundslitteratur, Copenhagen, pp 108–116

Latour B (1986) The powers of association. In: Law J (ed) Power, action and belief. A new sociology of knowledge. Routledge, London, pp 264–280

Latour B (1987) Science in action. Open University Press, Milton Keynes

Latour B (1999) Pandora's hope: essays on the reality of science studies. Harvard University Press, Cambridge, MA

Lemke T (2005) From eugenics to the government of genetic risks. In: Bunton R, Petersen R (eds) Genetic Governance. Health, risk and ethics in the biotech era. Routledge, London, pp 95–106

Lippman A (1991) Prenatal genetic testing and screening: constructing needs and reinforcing inequities. Am J Law Med 17(1–2):15–50

Lupton D (1999) Risk and the ontology of pregnant embodiment. In: Lupton D (ed) Risk and sociocultural theory: new directions and perspectives. Cambridge University Press, New York, pp 59–85

Marteau TM, Dormandy E, Michie S (2001) A measure of informed choice. Health Expect 4(2):99–108

McNeil M, Varcoe S, Yearley S (1990) The new reproductive technologies. Macmillian, London

Meskus M (2009) Governing risk through informed choice: prenatal testing in welfarist maternity care. In: Bauer S, Wahlberg A (eds) Contested categories. Life science in society. Ashgate, Surrey, pp 49–68

Mol A (2002) The body multiple. Ontology in medical practice. Duke University Press, Durham

Nicolaides KH (2004) The 11–13 week scan. Fetal Medicine Foundation, London

Parliamentary Proposal (2003) V 105, under interpellation 61 on prenatal diagnosis, May 15

Petchesky R (1987) Foetal images. The power of visual culture in the politics of reproduction. In: Stanworth M (ed) Reproductive technologies: gender, motherhood and medicine. University of Minnesota Press, Minneapolis, pp 57–80

Porter T (1995) Trust in numbers: the pursuit of objectivity in science and public life. Princeton University Press, Princeton

Rehmann-Sutter C (2009) Allowing agency. An ethical model of communicating personal genetic information. In: Rehmann-Sutter C, Müller H (eds) Disclosure dilemmas. Ethics of genetic prognosis after the "right to know/not to know" debate. Ashgate, Surrey, pp 231–260

Rose N (1999) Power of freedom. Cambridge University Press, Cambridge

Rose N (2001) The politics of life itself. Theory Cult Soc 18(6):1–30

Schwennesen N (2011) Practicing informed choice: inquiries into the redristributing of life, risk and relations of responsibility in prenatal decision making and knowledge production. Center for Medical Science and Technology Studies, Copenhagen University

Schwennesen N (2012) At the magins of life: making fetal life matter in trajectories of first trimester prenatal risk assessment. In: Vermeulen N, Tamminen S, Webster A (eds) Bio-objects: life in 21th century. Ashgate, Surrey, pp 117–131

Schwennesen N, Koch L (2009) Calculating and visualizing life: matters of fact in the context of prenatal risk assessment. In: Bauer S, Wahlberg A (eds) Contested categories. Life science in society. Ashgate, Surrey, pp 69–87

Schwennesen N, Koch L (2012) Representing and intervening: doing good care in first trimester prenatal risk assessment (FTPRA). Soc Health Illn 34(2):283–298

Schwennesen N, Koch L, Svendsen MN (2009) Practicing informed choice: decision making and prenatal risk assessment – the Danish experience. In: Rehmann-Sutter C, Müller H (eds) Disclosure dilemmas. Ethics of genetic prognosis after the "right to know/not to know" debate. Ashgate, Surrey, pp 191–204

Schwennesen N, Svendsen MN, Koch L (2010) Beyond informed choice: prenatal risk assessment, decision making and trust. Clin Ethics 5(4):207–216

Shakespeare T (1998) Choices and rights: eugenics, genetics and disability equality. Disabil Soc 13(5):655–681

Spallone P, Steinberg DL (1987) Made to order: the myth of reproductive and genetic progress. Pergamon, Oxford

Strathern M (2000) Audit cultures. Anthropological studies in accountability, ethics and the academy. Routledge, London

Sundhedsstyrelsen [Danish Board of Health] (1994) Vejledning om prænatal information, rådgivning og undersøgelse. Sundhedsstyrelsen, Copenhagen

Sundhedsstyrelsen [Danish Board of Health] (2003a) Fosterdiagnostik og risikovurdering [Prenatal testing and risk assessment]. Schultz Information, Copenhagen

Sundhedsstyrelsen [Danish Board of Health] (2003b) Notat vedrørende nye retningslinjer for fosterdiagnostik [Note regarding new guidelines of prenatal testing]. Sundhedsstyrelsen, Copenhagen

Sundhedsstyrelsen [Danish Board of Health] (2004) Nye retningslinjer for fosterdiagnostik [New guidelines for prenatal testing]. Sundhedsstyrelsen, Copenhagen

Tabor A (2006) Personal communication. Professor in foetal medicine. Rigshospitalet, Copenhagen

Tabor A, Philip J, Madsen M, Bang J, Obel EB, Noergaard-Pedersen B (1986) Randomised controlled trial of genetic amniocentesis in 4,606 low-risk women. Lancet 1:1287–1296

Tørring N, Jølving LR, Petersen OBB, Holmskov A, Hertz JM, Uldbjerg N (2008) Prænatal diagnostik i Århus og Viborg Amter efter implementering af første trimester-risikovurdering. Ugeskr Laeger 70:50–54

Van den Berg M, Timmermans DRM, ten Kate LP, van Vugt JMG, van der Wal G (2006) Informed decision making in the context of prenatal screening. Patient Educ Couns 63(1–2):110–117

Webster A (2007) Crossing boundaries. Social science in the policy room. Sci Technol Hum Values 32(4):458–478

Weil J et al (2006) The relationship of nondirectiveness to genetic counseling: report of a workshop at the 2003NSGC annual education conference. J Genet Couns 15:85–93

Weir L (1996) Recent developments in the government of pregnancy. Econ Soc 25(3):373–392

Williams C, Alderson P, Farsides B (2002) Is nondirectiveness possible within the context of antenatal screening and testing? Soc Sci Med 54:17–25

Chapter 16
Legal and Cross-Cultural Issues Regarding the Termination of Pregnancy: African Perspectives

Sylvester C. Chima

16.1 Introduction

In recent times the concept of reproductive health as a basic human right has been afforded international recognition following two United Nations' conferences in the mid-1990s, namely (a) The International Conference on Population and Development (ICPD), held in Cairo, Egypt in 1994 (ICPD 1994), and (b) The International Conference on Women, held in Beijing, China in 1995 (RFWCW 1995). The Programme of Action adopted at the Cairo conference defined reproductive health as "a state of complete physical, mental, and social wellbeing (…) in all matters relating to the reproductive system and to its functions and processes" (ICPD 1994). The 1995 Beijing Platform for Action (Beijing Declaration) reiterated this definition, listing as a human right the right of a woman to control her own sexuality and reproduction (RFWCW 1995; Ngwena and Cook 2005). Further international judicial institutions such as the ECtHR have recognized reproductive rights as falling within an individual's rights to privacy, right to liberty and security, and right to equality and non-discrimination as recognized by Articles 5, 8 and 14 of the European Convention for the Protection of Human Rights and Fundamental Freedoms (ECHR) (Mullaly 2005; ECPHRFF 1950). It has been argued that UN Human Rights instruments such as the International Convention on Civil and Political Rights (1966) which guarantee fundamental rights such as the right to life; freedom from cruel and inhuman punishment; liberty and security of the person; privacy, dignity; health equality and non-discrimination are similarly consistent with the right to reproductive health guaranteed to all persons especially women (Ngwena 2010a). In the African regional context it has been argued that the African Charter on Human and Peoples Rights (hereinafter African Charter 1981) contains

S.C. Chima (✉)
University of KwaZulu-Natal, Durban, Republic of South Africa
e-mail: chima@ukzn.ac.za

© Springer International Publishing AG, part of Springer Nature 2018
M. Soniewicka (ed.), *The Ethics of Reproductive Genetics*,
Philosophy and Medicine 128, https://doi.org/10.1007/978-3-319-60684-2_16

rights which are supportive of reproductive rights such as the right to equality, life, human dignity and health (Ngwena and Cook 2005; Ngwena 2010a; African Charter 1981). As well as the rights to protection of the family and self-determination (Mbazira 2006). Further, it has been argued that within the ambience of Article 16 of the African Charter- the right to health which states that:

> (1) Every individual shall have the right to enjoy the best attainable state of physical and mental health. (2). States parties to the present Charter shall take the necessary measures to protect the health of their people and to ensure that they receive medical attention when they are sick.

In seeking to enhance and protect the human rights of women in Africa including reproductive rights; the African Union passed a supplemental protocol, Protocol to the African Charter on Human and Peoples' Rights on the Rights of Women in Africa, (hereinafter Maputo Protocol 2003) consistent with the United Nations Convention on the Rights of women (CEDAW 1979). The Maputo Protocol apart from emphasizing women's rights to equality, human dignity, life, integrity and security of the person, explicitly emphasized the right to reproductive health in Article 14 where it states that "States Parties shall ensure that the right to health of women, including sexual and reproductive health is respected and promoted" (Maputo Protocol 2003). Further, the Maputo Protocol requires in Section 14 (2) that States Parties shall take all appropriate measures to:

(a) Provide adequate, affordable and accessible health services, including information, education and communication programmes to women especially those in rural areas;
(b) Establish and strengthen existing pre-natal, delivery and post-natal health and nutritional services for women during pregnancy and while they are breast-feeding;
(c) Protect the reproductive rights of women by authorizing medical abortion in cases of sexual assault, rape, incest, and where the continued pregnancy endangers the mental and physical health of the mother or the life of the mother or the foetus.

The provision of section 14 (2) (c) of the Maputo protocol therefore obliges African countries as signatories to the African Charter and Protocol, to liberalize abortion laws and provide mechanisms for easy access, consistent with the requirements of Article 16 of the African Charter; so that women can exercise the right of choice to abortion or TOP (Ngwena and Cook 2005; Ngwena 2010a, b; African Charter 1981; Mbazira 2006; Maputo Protocol 2003). Despite the international and regional human rights instruments which require that state parties pay greater attention to the right to health particularly reproductive health rights, and despite the requirement for non-discrimination against women in international conventions such as the ICCPR (1966), CEDAW (1979), and International Convention on Social Economic and Cultural Rights (1966) there is abundant evidence to show that the majority of countries in the African continent, Latin America and in some developed countries such as Ireland and Poland have retained or promulgated restrictive abortion laws which have hindered many women, especially vulnerable women such as minors and adolescents, from enjoying the rights to autonomy and self-determination with regards to full reproductive health rights, particularly the right to abortion. This chapter will explore some of the legal, ethical, and cross-cultural issues with regards to the failure of full implementation of the right to abortion/TOP with particular reference to the rights of women in African countries. I will briefly analyze the spectrum of clinical, ethical and cross-cultural factors and elucidate some moral dilemmas surrounding TOP and its impact on SSA and the general impact of unsafe abortion on women's reproductive health and human rights in Africa (cf. Chima and Mamdoo 2015).

16.2 The Legal Status of Abortion or Termination of Pregnancy in African Countries

Despite the achievements of the CEDAW with regards to recognizing the equal rights for women (1979), some authors have argued that the UN Committee on CEDAW, while recognizing the role of restrictive abortion laws in the genesis and prevalence of unsafe abortion, with consequent increase in maternal morbidity and mortality, has historically shown a reluctance in adopting safe and legal abortion as a solution to this aspect of women's right to health (Ngwena 2010a, 2013). It was reported that the ICPD in Cairo (1994) showed a reluctance to call for abortion law reforms despite the obvious impact of restrictive abortion laws on maternal morbidity and mortality probably due to political expediency (ICPD 1994; RFWCW 1995; Ngwena 2010a). Based on the background of these failures, and with the knowledge that maternal mortality due to unsafe abortions is a major contributor to maternal morbidity in African countries (Okonofua 1997; Brookman-Amissah and Moyo 2004; Adinma et al. 2011), the African Union took the bold step of enacting the Maputo Protocol (2003) on women's rights as a regional platform for the reduction of maternal deaths in Africa (Ngwena 2010b). However, despite this laudable objective, it is evident that the majority of African countries still maintain restrictive abortion laws derived from colonial laws (Okonofua 1997; Brookman-Amissah and Moyo 2004; Adinma et al. 2011; Centre for Reproductive Health 2014). Though the Maputo protocol was enacted in 2003, till date only about 36 countries have fully signed and ratified the instrument (Maputo Protocol 2003). Recent reports indicate that 49 of 54 African countries who are members of the African Union, still maintain restrictive abortion laws which generally allow abortion only to preserve the life of the woman (Centre for Reproductive Health 2014). So far, only three African countries, namely; Cape Verde, South Africa and Tunisia, have fully liberalized their national abortion laws to allow TOP for any reason (Centre for Reproductive Health 2014). It has been argued that while Zambian laws theoretically allow abortion for liberal reasons including socio-economic factors (Ngwena 2010a; Centre for Reproductive Health 2014), implementation in practice has remained problematic due to resource constraints such as the requirement for the concurrence of three medical doctors prior to abortion, rendering the law almost ineffective due to the onerous requirements for patients seeking abortion, especially in rural communities (Ngwena and Cook 2005; Ngwena 2010a; TOPA 1972; Koster-Oyekan 1998; Coast and Murray 2016). In one study it was reported that up to 69% of Zambian women knew of another woman who had died from an unsafe or illegal abortion, despite the liberalization of Zambian abortion laws (Koster-Oyekan 1998; Coast and Murray 2016; Grimes et al. 2006). Globally, most developed countries in the northern hemisphere and East Asia have generally liberalized their abortion laws (Centre for Reproductive Health 2014), leading to easy access to TOP and contraceptive services with associated decrease in maternal morbidity and mortality (Centre for Reproductive Health 2014; Grimes et al. 2006). By contrast, countries in the global 'South' including most developing countries in SSA, still maintain restrictive abortion laws (Centre for Reproductive Health 2014) with a comparative increase in the incidence of maternal mortality and morbidity (Grimes et al. 2006); with a few exceptions in countries like South Africa, Colombia, Cambodia, Zambia and Cape Verde (Centre for Reproductive Health 2014; Grimes et al. 2006). However, it has been argued that merely liberalizing abortion laws without providing effective mechanisms for women to access abortion or contraceptive services will not improve the outcome for unintended pregnancies or reduce the incidence of unsafe abortions with its associated maternal morbidities (Grimes et al. 2006; Leke 2014). This has been demonstrated in countries like Zambia where the liberalized abortion laws still requires onerous conditions such as the signature of three medical practitioners thereby making it difficult for women in the rural areas to access abortion on demand (TOPA 1972; Koster-Oyekan 1998; Coast and Murray 2016; Grimes et al. 2006). Similarly in Nepal it was reported that the cost of abortion was too exorbitant for ordinary women, prompting the Nepal Supreme court to urge the government to subsidize the cost of abortion for women who cannot afford such services (Centre for Reproductive Health 2014). In the case of developed countries like Poland and Ireland, the UN Human Rights Committee (HRC) has determined that even where abortion is allowed by law to safeguard the health of women, the absence of just administrative action to enable enjoyment of these rights, was contrary to the fundamental rights of women as well as being discriminatory (Human Rights Committee 2016; Enright 2016; Gentleman 2016; McDonald 2013; IFPA 2014; Tysiąc v. Poland 2007; P. and S. v. Poland 2008; R. R. v. Poland 2011). Similar decisions were rendered by the HRC in the case of *KL v Peru* (Human Rights Committee 2005), where the HRC determined that the government of Peru was in violation of several aspects of the ICCPR including Article 2 (right to effective remedy); Article 7 (right to be free from inhuman and degrading treatment); Article 17 (right to privacy) and Article 24 (right to special protection as a minor) (Human Rights Committee 2005). Similar findings were elucidated against Argentina in the case of *LMR v Argentina* (Human Rights Committee 2011) in addition to Article 3 (right to equal enjoyment of rights) (Bates 2013). As shown in *LC v Peru* (Bates 2013; Committee on the Elimination of Discrimination against Women 2009), refusal of abortion services to minor women even where the law allows such services to preserve the health of a woman (LC) or due to severe fetal anomalies (KL) (Human Rights Committee 2005), would be regarded as cruel and inhuman punishment in contravention of the ICCPR (1966) and was also discriminatory against women, especially vulnerable adolescents in contravention of CEDAW (Committee on the Elimination of Discrimination against Women 1979, 2009; Bates 2013).

It has been argued that despite recent attempts at reform of abortion laws in African countries based on regional and international human rights injunctions such as the Maputo Protocol (2003), African Charter (1981), and CEDAW (1979), opinions by the HRC as well as the recent General Comment 2 from the African Commission on Human and Peoples Rights (ACHPR 2014), which provided further clarifications on the application of Article 14 of the Maputo Protocol and the African charter (2014) (Ngwena et al. 2015); the majority of African countries still persist with restrictive abortion laws till today (Centre for Reproductive Health 2014). Historically, it has been shown that the majority of abortion laws in Africa are a legacy of colonialism, wherein most national abortion laws in African countries, were adopted into local laws by the colonial authorities (Ngwena 2010a, 2012, 2013; Brookman-Amissah and Moyo 2004; Ngwena et al. 2015). Originally, such colonial laws were derived from ancient religious injunctions which abhorred the so-called 'mortal sin' (Ngwena 2010a, 2012); while others have argued that it was introduced to curb maternal mortality from primitive and illegal methods of procuring abortion (Brookman-Amissah and Moyo 2004). It must be emphasized that such abortion laws which are a legacy of colonialism and foreign religions are among the spectrum of factors which were introduced by colonialism which disrupted traditional African values and belief systems, and arguably still holds African communities captive till today, leading to the disruption of the social fabric of African societies and persistent underdevelopment (Chima 2015a). Some authors have lamented this state of affairs, suggesting that while colonialism was supposed to have ended with the independence of African states, colonial practices still hold African women captive (Brookman-Amissah and Moyo 2004). In former Anglophone colonies and latter independent countries like Nigeria and Kenya, the legacy of colonial laws with regards to abortion still persist to this day (Ngwena 2010a, 2013). For example in Nigeria, current abortion laws are amongst the most restrictive in the world, with abortion only allowed to save the life of a pregnant woman (Okonofua 1997; Adinma et al. 2011). Currently, Nigerian abortion laws are encoded in the portions of the criminal and penal codes related to pregnancy miscarriage (Brookman-Amissah and Moyo 2004; Adinma et al. 2011; OAPA 1861). These laws were adopted from the ancient English 'Offences against the Person Act of 1861' which is applied in prosecuting cases of violence against individuals short of murder (1861). Based on this law, section 228 of the Nigerian criminal code applicable to Southern Nigeria stipulates as follows:

> Any person who, with intent to procure miscarriage of a woman whether she is or is not with child, unlawfully administers to her or causes her to take any poison or other noxious thing, or uses any force of any kind, or uses any other means whatever, is guilty of a felony, and is liable to imprisonment for fourteen years.

Similarly, section 232 of the penal code applicable in Northern Nigeria states as follows:

> Whoever voluntarily causes a woman with child to miscarry shall, if such miscarriage be not caused in good faith for the purpose of saving the life of the woman, be punished with imprisonment for a term which may extend to fourteen years or with fine or with both.

From the above it is obvious that Nigerian abortion laws are highly restrictive prescribing penalties of up to 14 years or life imprisonment for late termination, against guilty offenders including pregnant women and HCPs, while allowing lawful abortion only to save the life of a pregnant woman (Adinma et al. 2011; Mullaly 2005). The Kenyan abortion law which was equally restrictive and similarly derived from the 'Offences against the Person Act' (Brookman-Amissah and Moyo 2004; OAPA 1861) was recently reformed via constitutional means to allow TOP when needed in an emergency, when the life or health of a pregnant woman is threatened or if permitted by any other written law (Ngwena 2010a, 2012, Ngwena 2013; Constitution of Kenya 2010). However it has been argued that implementation in practice has been problematic because constitutional provisions also assert that life begins at conception- thereby putting a chilling effect on HCPs who attempts an abortion for any reason (Ngwena 2010a, 2012; Constitution of Kenya 2010). Therefore even in jurisdictions like Kenya, Swaziland, Zambia and Ghana which have attempted to reform their colonial abortion laws to become more consistent with current international norms, problems with implementation including human and financial resource constraints and lack of political will have hindered the implementation of such reforms in practice (Ngwena 2010a, 2012; Ngwena et al. 2015; TOPA 1972; Koster-Oyekan 1998; Coast and Murray 2016; Constitution of Kenya 2010). It has equally been observed that countries with such restrictive abortion laws have the highest rates of maternal mortality and morbidity in SSA including those with attempted reform of local laws like Zambia and Kenya (see e.g. Sedgh et al. 2012, 2016; Grimes 2003; Singh 2006; Okonta et al. 2010; Okonofua et al. 2009). It has been reported that other factors contributing to the persistence of unsafe abortions in many African countries include lack of knowledge of the rudimentary local laws by HCPs and policy makers (Okonta et al. 2010; Okonofua et al. 2009), and patients are not fully aware of local abortion laws and circumstances where the law allows or restricts abortion (Adinma et al. 2011), thereby most stakeholders still mistakenly assume that abortion is totally proscribed or illegal even in countries where the law allows TOP in extreme circumstances (Adinma et al. 2011; Koster-Oyekan 1998; Okonta et al. 2010; Okonofua et al. 2009). Therefore, one can conclude that despite the Maputo protocol which was designed to assist African countries in liberalizing restrictive abortion laws, the majority of countries in SSA where unsafe abortion and maternal mortality is prevalent, have not sufficiently liberalized their abortion laws due to lack of political will and other factors such as religious and cultural belief systems. Furthermore, in those SSA countries like Zambia and Ghana where such laws have been relatively liberalized on paper, human resource and financial constraints have not allowed for the effective implementation leading to persistence of high maternal morbidity and mortality similar to countries with restrictive laws (Centre for Reproductive Health 2014; Koster-Oyekan 1998; Coast and Murray 2016; Sedgh et al. 2012, 2016; Grimes 2003; Singh 2006; Okonta et al. 2010; Okonofua et al. 2009). In view of this, it has been suggested that there should be a systematic framework for the study and management of abortion and its impact on African countries and communities (Okonofua 1997; Leke 2014). Further it has been suggested that the problems relating to abortion and unintended pregnancies in Africa ought to subjected to preventive measures ranging from primary prevention strategies such as contraception to quaternary prevention which would in include post-abortion counselling as part of a package of comprehensive public health measures for managing unintended pregnancies and TOP in African countries (Leke 2014; Sedgh et al. 2014; Dickens and Cook 2007).

16.3 The Importance of Abortion Law Reforms for African Communities

Unsafe abortion has been described by some authorities as a silent scourge and preventable global pandemic (Grimes et al. 2006; Grimes 2003). TOP according to the South African Choice Act (Republic of South Africa 1996) "means the separation and expulsion, by medical or surgical means, of the contents of the uterus of a pregnant woman" (Republic of South Africa 1996). Abortion may also be classified as 'safe' or 'unsafe' based on the methods employed in achieving it. According to the World Health Organization (WHO) 'unsafe abortion' is defined as a procedure for terminating an unintended pregnancy either by individuals without the necessary skills or in an environment that does not confirm to minimum medical standards or both (Leke 2014). While some have argued that this definition in is wrong because it generally assumes that TOP done in resource-poor countries or settings is automatically unsafe thereby increasing the number of unsafe abortions reported from developing countries, especially those in Africa. Nonetheless, abortion has a high prevalence in countries of the global South, especially in SSA countries (Centre for Reproductive Health 2014) with restrictive abortion laws, poor healthcare infrastructure, and human resource constraints (Ngwena 2010a; Adinma et al. 2011; Koster-Oyekan 1998; Singh 2006). Recent reports indicate that the number of abortions performed globally has increased annually from 50.4 million during the period 1990–1994 to 56.3 million per annum during the period 2010–2014 (Sedgh et al. 2016). The authors of this report also estimated that there were 35 abortions per every 1000 pregnancies in women aged 15–44 years in the years 2010–2014 (Sedgh et al. 2016), which showed a slight reduction in rates, when compared the rate of 40/1000 reported between1990 and 1994 (Sedgh et al. 2016). In developed countries there was significant decline of abortion rates from 46/1000 pregnancies to 27/1000 pregnancies when compared from 1990–1994 to the 2010–2014 (Sedgh et al. 2016). However in developing countries the decline in numbers was insignificant from 39 to 37 per 1000 pregnancies (Sedgh et al. 2016). Overall the authors suggested that 25% of all pregnancies result in abortions or TOP of which 73% occurred with married women while 27% were obtained by unmarried women (Sedgh et al. 2016). Based on these studies it was concluded that abortion rates have declined significantly in developed countries since 1990, but not in developing countries especially SSA (Sedgh et al. 2016). However, the authors found no association between abortion rates and the grounds on which abortion is permitted (Sedgh et al. 2016). Generally abortion or TOP occurs as a consequence of unintended pregnancies (Sedgh et al. 2014). According to recent reports, the global pregnancy rate decreased marginally from 2008 to 2012. During this period there were 85 million unintended pregnancies representing about 40% of all pregnancies globally in 2012, with about 50% resulting in abortion or TOP, 13% ending in a miscarriage, while 38% resulted in an unplanned birth (Sedgh et al. 2014). It has been reported by several authors that unintended pregnancies are usually associated with many negative health, economic, social and psychological outcomes for women and children (Chima and Mamdoo 2015; Dickens and Cook 2007). In Africa and other parts of the developing world, the consequences of having an unintended pregnancy usually forces women to seek TOP, due to the social economic burden associated with raising an unwanted child (Chima and Mamdoo 2015; Grimes 2003; Singh 2006; Okonta et al. 2010; Okonofua et al. 2009; Sedgh et al. 2014, 2016). Furthermore, due to the paucity and absence of legal healthcare facilities which allow safe abortion, most of these women may end up resorting to unsafe methods of abortion by unqualified providers in unhealthy healthcare facilities leading to the high incidence of maternal deaths and morbidity due to unsafe abortion (Okonofua 1997; Coast and Murray 2016; Grimes 2003; Singh 2006; Okonta et al. 2010; Okonofua et al. 2009) (Fig. 16.1).

As shown in the above figure, the causes of unsafe abortions, with increased maternal mortality and morbidity in African countries and elsewhere are multifactorial, and would require a comprehensive approach for its resolution (Okonofua 1997; Leke 2014). According to the final UN Millennium Development Goals (MDG) report (MDGR 2015); the problems of reducing maternal mortality (MDG 5) and infant mortality (MDG 4), could not be achieved in African countries, although significant improvements were made in these areas (MDGR 2015). Subsequently these two areas are also being targeted in future sustainable development goals (SDGs). Despite many achievements in women's rights during the past 14 years, most women remain suppressed by and excluded from the society. While the Millennium Development Goals' (MDGs) contribution to their advancement were notable, the Sustainable Development Goals (SDGs) are expected to preserve and further enhance them (Shariq et al. 2015).

Based on the final MDG report (MDGR 2015), there was a 49% decline in maternal mortality from 830/100,000 live births to 510/100,000 live births in SSA women aged 15–49 years; while the infant mortality rate also decreased by 52% from 179 to 86 per 1000 live births between 1990 and 2015 (MDGR 2015). When these numbers are compared to the rest of the world especially developed countries, it becomes obvious that a lot more work needs to be done (MDGR 2015). In recent times, there has been global recognition that it is critically important not only to prevent death and morbidity from pregnancy-related causes, but also to reduce infant mortality and lay a good foundation for sustainable economic development especially in developing countries in SSA (MDGR 2015; Shariq et al. 2015). Some have suggested that the failure to achieve MDG may be related to discriminatory practices against women, while the HRC has recognized in many cases that

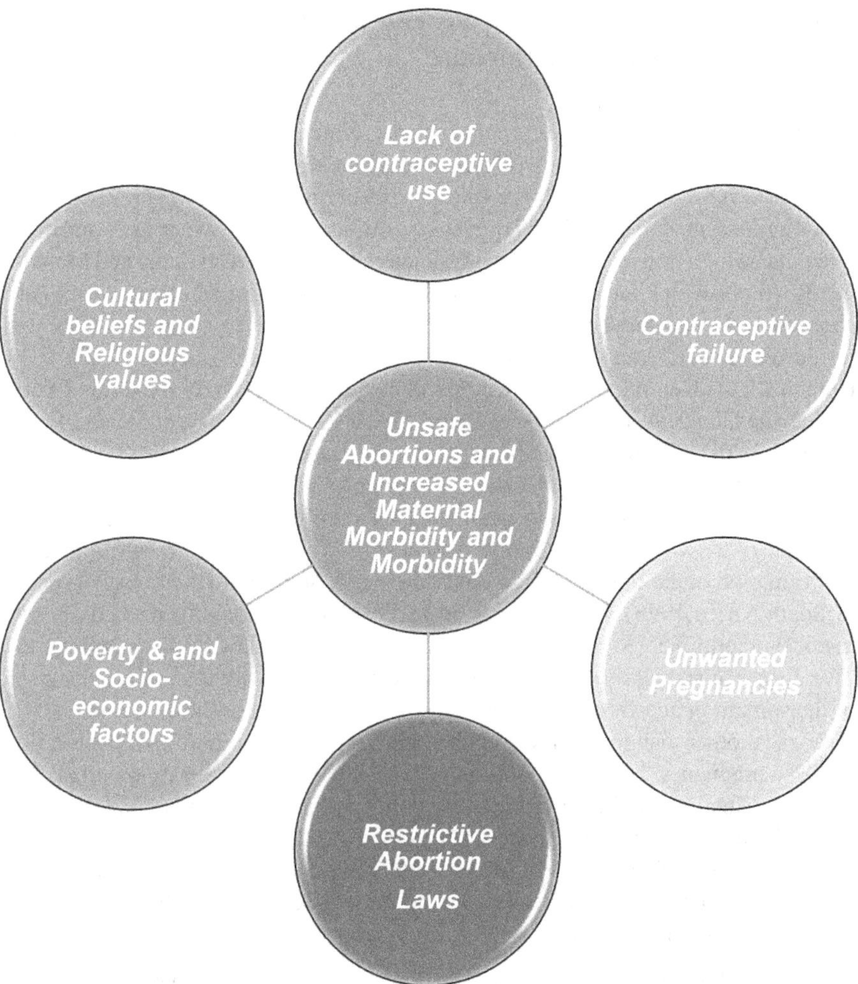

Fig. 16.1 Conceptual model for factors associated with unsafe abortions and maternal morbidity and mortality in African countries

preventable maternal mortality or morbidity could be considered a human rights violation (Ngwena 2010a; Dickens and Cook 2007; Shaw and Cook 2012). In light of its impact on human development, some authors have concluded that maternal mortality could devastate families especially in resource-poor settings such as SSA, impacting on infants and the elderly, thereby weakening families and vulnerable communities (Leke 2014; Shaw and Cook 2012). Based on these observations, the Cairo Programme of Action and the World Conference on women in Beijing urged countries to "deal with the health impact of unsafe abortion as a major public health concern" (ICPD 1994; RFWCW 1995).

16.4   A Brief Analysis of Some Moral and Ethical Dilemmas Associated with Abortion

TOP for severe fetal anomalies is controversial, ethically and emotionally challenging for both HCPs and parents (Chima and Mamdoo 2015). Ideally, the expected outcome of most pregnancies is the delivery of a normal baby, however, when severe abnormalities are detected during pregnancy, this leads to ethical conflicts and moral dilemmas which impact on both the physicians and the parents (Chima and Mamdoo 2015). Abortion for fetal anomalies is allowed in many jurisdictions with liberal abortion laws and maybe grounds for allowing abortion in some countries with restrictive laws (Chima and Mamdoo 2015; Centre for Reproductive Health 2014; Human Rights Committee 2005; Codigo Penal de Peru 2008). In the case of South Africa, the Choice Act (Republic of South Africa 1996) allows termination of pregnancy from the 13th to the 20th week of gestation where "there exists a substantial risk that the fetus would suffer from a severe physical or mental abnormality" and after the 20th week, where two qualified HCPs are of the opinion that continued pregnancy: "(i) would endanger the woman's life; or (ii) would result in a severe malformation of the fetus; or (iii) would pose a risk of injury to the fetus" (Republic of South Africa 1996). In such cases however, TOP must only be carried out by a medical practitioner (Republic of South Africa 1996). The law further stipulates that TOP for these and all other cases can only be performed with the informed consent of the woman in question. "Notwithstanding any other law or the common law (…) no consent other than that of the pregnant woman, shall be required for the termination of a pregnancy" (Republic of South Africa 1996). As previously reported, TOP for severe fetal anomalies is associated with many moral conflicts and dilemmas even where the woman has made an informed decision in favor or against TOP (Chima and Mamdoo 2015). Moral and ethical dilemmas which may arise in the cases of TOP for fetal anomalies or other abortion cases could be classified into five major categories as follows:

1. The first dilemma that arises in late TOP is the issue of maternal-fetal conflict i.e. balancing the rights of the fetus versus the rights of the mother
2. The second dilemma is balancing the rights of the individual (mother or both parents) versus cultural, religious and societal values etc.
3. The third area of conflict would be legal consequences of medical negligence leading to 'wrongful birth' or wrongful life' litigation (14)
4. The fourth area of conflict would be balancing the rights of the physician v societal expectations, the law, and the rights of the parents to medical treatment e.g. problems of 'the physician as a conscientious objector'
5. The fifth area of conflict are the issues of distributive justice and resource allocation

While space constraints will not allow a full discussion of the moral dilemmas outlined above in this chapter. I will briefly discuss the impact of cultural issues, distributive justice and resource allocation in the management of abortion cases.

16.5 Cross-Cultural Issues, Distributive Justice and Resource Allocation

Cultural issues, especially religion and other belief systems, have a very important role to play on the issue of abortion. Some religions, especially the Roman Catholic Church, actively discourage their members from contraceptive use and TOP based on the general belief that children are a gift from God and all potential outcomes must be accepted. Similarly, the Islamic religion also discourages its adherents from abortion especially after the period of ensoulment which is variably interpreted as 40, 90 or 120 days after conception (Chima and Mamdoo 2015; Al-Matary and Ali 2014; Hessini 2007); leading to various injunctions either allowing abortion under certain circumstances or TOP being totally forbidden (Al-Matary and Ali 2014; Hessini 2007). Africans are generally very religious, accepting either imported foreign religions such as Christianity or Islam, or practicing African Traditional Religions (ATR) (Chima 2015a; Mbiti 1969). Following the importation of Christianity and Islam into Africa, many individual Africans and communities have adopted the guiding principles of both religions in addition to ATR and cultural belief systems (Chima 2015a; Mbiti 1969). This has created a strong pressure system in some countries e.g. Nigeria, which has a large population of both Christians and Muslims, leading to the rejection of all attempts at abortion laws reforms, whereby abortion laws still remain very restrictive despite efforts by community activists and physician groups to change such laws (Ngwena 2013; Okonofua 1997; Okonofua et al. 2009; Brookman-Amissah and Moyo 2004; Adinma et al. 2011; Leke 2014; Okonta et al. 2010) and therefore negating attempts to change Nigerian abortion laws to become more consistent with international and regional human rights instruments to which Nigeria is a signatory (ICCPR 1966; African Charter 1981; Maputo Protocol 2003; CEDAW 1979; ISECR 1966). Overall religion has a very strong impact on the persistence of restrictive laws in African countries and this has had a chilling effect on the impetus to change laws and allow safe abortion in many African jurisdictions (Ngwena 2010a, 2013; Okonofua 1997; Brookman-Amissah and Moyo 2004; Adinma et al. 2011), which would ultimately lead to the decrease in maternal morbidity and mortality and the achievement of future SDGs (MDGR 2015; Shariq et al. 2015). Therefore any attempt to remove restrictive abortion laws must also confront the issue of religious belief systems before responsible authorities such as politicians could muster the political will and necessary support to change restrictive abortion laws. Like most traditional societies, another cultural influence on the issues of abortion, pregnancy and contraception is the traditional African belief in the subservient role of the woman which requires women to fulfill the cultural role of child-bearing regardless of their personal circumstances, or autonomous beliefs (Chima 2011, 2015a). In such situations, an unequal power relationship exists between a woman, her husband and the extended family; whereby women can be put under pressure by the husband or extended family to either become pregnant or carry a pregnancy to term, otherwise they would be confronted with stigmatization and sometimes exclusion from communal rights and activities

(Chima 2011; Chima and Mamdoo 2015; Famoroti et al. 2013). This creates emotional and psychological pressure forcing women to either succumb to the wishes of the husband/family or become objects of ridicule. This may impact on the decision to seek contraceptive methods to control pregnancy or on decisions whether to carry a pregnancy to term and other reproductive health choices (Chima 2011; Chima and Mamdoo 2015; Famoroti et al. 2013). With regards to distributive justice and resource allocation; one can question whether it is morally justifiable to allocate scarce health care resources, especially in resource-poor countries and communities like Africa; to the care of children with severe congenital anomalies? It could be argued that in terms of just resource allocation and distributive justice, it may be morally justifiable to encourage women pregnant with fetuses diagnosed with severe lethal anomalies to seek termination of such pregnancies for their own personal good, and in the best interests of their families and society (Chima and Mamdoo 2015). Although this should be strictly based on the autonomous choice of the woman, as demonstrated in the two cases previously reported from South Africa (Chima and Mamdoo 2015; Republic of South Africa 1996). In support of this point of view, the Health Professions Council of South Africa (HPCSA) has opined that:

> There are circumstances when withholding treatment, even if it is not in the best interest of the patient is permissible. This will apply to continued care in special units such as critical care (including neonatal units) and chronic dialysis units for end stage kidney failure. A health care institution has the right to limit life-sustaining interventions without the consent of a patient or surrogate by restricting admission to these units. However, such restriction must be based on national admission criteria agreed upon by the expert professional bodies in the relevant specialty, as well as the HPCSA. A health care institution is, obliged to provide the appropriate palliative care and follow up when specialized care is withheld (….) The HPCSA considers it unethical to continue with life-prolonging treatment for the sole purpose of financial gain (or where treatment is futile) Moreover, it is unacceptable that patients are transferred to state institutions after all their funding has been exhausted as a result of prolonging futile treatments. (HPSCA 2007)

In light of the above opinions and observations, one can suggest that in cases where women are pregnant with fetuses diagnosed with lethal congenital anomalies such as anencephaly as in the case of *KL v Peru* (Human Rights Committee 2005); it could be ethical and morally justifiable to terminate such pregnancies to preserve scarce healthcare resources as well as preventing inhuman and degrading treatment as established by the HRC (Human Rights Committee 2005). On the other hand, in the cases where a patient is diagnosed with non-lethal anomalies with variable presentations such as Downs syndrome (Rydval and Lynöe 2008; Pueschel 1991) or a fetus with holoprosencephaly as previously reported (Chima and Mamdoo 2015), then it could be morally justifiable to take a more reticent approach towards TOP (Rydval and Lynöe 2008; Pueschel 1991; Flieschman et al. 1998), strictly subject to the informed consent of the pregnant woman (Republic of South Africa 1996; Chima 2015b).

16.6 Summary and Conclusions

From the above analysis, it is obvious that issues relating to abortion have global implications for public health, reproductive ethics, and human rights. The subject of abortion is ethically and morally challenging, and controversial. Requests for TOP usually occurs as a consequence of unintended pregnancies of which about 85 million occur annually representing about 40% of all pregnancies globally (Sedgh et al. 2016). Due to the impact of unintended pregnancies on maternal health and the wellbeing of families and communities globally, a large proportion of such unintended pregnancies usually end in induced abortion with a global incidence of 35 abortions per 1000 women aged 15–44 years; estimated at 25% of all pregnancies; of which 73% or the majority of cases were procured by married women (Sedgh et al. 2014). Because of its large impact on human life, the issue of pregnancy and TOP has implications for human religious beliefs and practices as well as cultural values, morals and human rights, especially women's rights (Ngwena and Cook 2005; Ngwena 2010a; Chima 2011, 2015a; Al-Matary and Ali 2014; Hessini 2007; Mbiti 1969). Due to the ongoing recognition of women's rights, and the need to enhance women's autonomy and dignity, and empower women globally, the right to abortion or termination of unintended pregnancy is increasingly being recognized as a fundamental human right in accordance with international human rights instruments such as ICCPR (1966) and CEDAW (1979). Similarly regional human rights regulations such as the Maputo Protocol (2003), as well as the ECHR (ECPHRFF 1950) have been used as instruments for recognition and enforcement of women's right to autonomy, including the right to abortion (discussed in e.g. Enright 2016; Gentleman 2016; McDonald 2013; Bates 2013). Despite the high incidence of unintended pregnancy and the subsequent prevalence of induced abortions, there is no universal agreement on the importance of TOP as an instrument for controlling population growth and for reduction in maternal mortality and morbidity and promoting global reproductive health rights (Centre for Reproductive Health 2014). Different jurisdictions have devised their own laws for either allowing or restricting the performance of abortion based on cultural or religious belief systems. In general more developed countries in the global 'North', with a few exceptions such as Ireland and Poland, have established legal mechanisms for liberalized abortion (Ngwena 2013; Centre for Reproductive Health 2014). On the other hand, most developing countries in the global 'South' including many countries in Africa, with the exception of South Africa, still maintain rigid and restrictive abortion laws (Ngwena 2010a; Okonofua 1997; Brookman-Amissah and Moyo 2004; Adinma et al. 2011; Centre for Reproductive Health 2014). The irony of this legal dichotomy is that in countries with restrictive abortion laws, women have had to resort to unsafe methods of TOP leading to unsafe abortion being described as a 'preventable global pandemic' with its associated increase in maternal mortality and morbidity (Okonofua 1997; Brookman-Amissah and Moyo 2004; Adinma et al. 2011; Leke 2014, Koster-Oyekan 1998; Okonta et al. 2010; Okonofua et al. 2009). In addition, such countries especially those in SSA, have been unable to meet the targets of the global MDG such as reduction in maternal morbidity and infant mortality (MDGR 2015; Shariq et al. 2015). Considering the high incidence of unsafe abortion in African countries and subsequent impact on maternal morbidity and mortality secondary to post-abortion hemorrhage, sepsis, uterine rupture, infertility, maternal deaths, psychological distress and stigmatization associated with clandestine abortions in Africa (Okonofua 1997; Adinma et al. 2011; Koster-Oyekan 1998; Coast and Murray 2016; Leke 2014; Grimes 2003; Singh 2006). The African Union introduced the Maputo protocol in 2004 (Maputo Protocol 2003) as a corollary instrument to the African Charter (1981), which amongst other recommendations obliges African countries to liberalize abortion laws and:

> Protect the reproductive rights of women by authorizing medical abortion in cases of sexual assault, rape, incest, and where the continued pregnancy endangers the mental and physical health of the mother or the life of the mother or the fetus. (Maputo Protocol 2003)

Despite the obligations established by the Maputo protocol, many African countries such as Kenya (Ngwena 2013; Constitution of Kenya 2010) and Nigeria (Adinma et al. 2011) still maintain restrictive abortion laws (Okonofua 1997). Furthermore, even in African countries with liberalized abortion laws such as Zambia, implementation in practice has been hindered by the lack of human and financial resources (Koster-Oyekan 1998; Coast and Murray 2016) with a continued negative impact on mostly rural and indigent African women (Adinma et al. 2011; Koster-Oyekan 1998; Coast and Murray 2016; Leke 2014; Okonta et al. 2010; Okonofua et al. 2009). Therefore maternal mortality and morbidity secondary to unsafe abortions and unintended pregnancies may have a huge and negative impact on families and communities in resource-poor settings such as SSA; prejudicing the health of victims' families with a negative impact on population and reproductive health (Leke 2014; Sedgh et al. 2014, 2016; Dickens and Cook 2007). Overall, abortion raises many ethical and moral issues including women's autonomy rights versus a fetus' rights to personhood (Chima and Mamdoo 2015; Flieschman et al. 1998). Cross-cultural issues, including people's religious beliefs and traditional cultural practices, as well issues of distributive justice and resource allocation (Chima 2011, 2015a; Mbiti 1969; Famoroti et al. 2013; Rydval and Lynöe 2008) are also factors to consider. Other moral dilemmas elicited by the problem of TOP include liability of HCPs due to medical negligence, which may follow misdiagnoses in cases of fetal anomalies which could lead to liability for 'wrongful birth' or 'wrongful life' (Chima and Mamdoo 2015) and issues of informed consent (Republic of South Africa 1996; Chima 2015b). In conclusion one can advocate that seeing as global evidence points to the fact that unsafe abortions are most prevalent and devastating in countries with restrictive abortion laws (Okonofua 1997; Brookman-Amissah and Moyo 2004; Adinma et al. 2011; Coast and Murray 2016; Leke 2014; Okonta et al. 2010; Okonofua et al. 2009). Hence, to accomplish a rapid and effective liberalization of national abortion laws, African countries may want to learn from the South African Choice Act (Republic of South Africa 1996) which has led to direct reduction in the incidence of maternal mortality and morbidity associated with unsafe abortion following a liberalization of South African abortion laws in

1996 (Mhlanga 2003; Buchmann et al. 2007; Faundes and Shah 2015). South African law allows TOP for socio-economic reasons in the first trimester of pregnancy and also allows any woman of any age to request TOP on demand during the first 12 weeks of gestation subject to the informed consent of the woman alone (Republic of South Africa 1996). The consequence of this has been removal of stigma and discrimination related to abortion (Chima and Mamdoo 2015; Republic of South Africa 1996; Mhlanga 2003; Buchmann et al. 2007; Faundes and Shah 2015), as well as other positive impacts on the healthcare system, thereby generally enhancing the dignity and human rights of women.

References

ACHPR (2014) African Commission on Human and Peoples Rights General Comment No.2 on Article 14 (1) (a) (b) (c) and (f) and Article 14 (2) (a) and (c) to the Protocol on the African Charter on Human and Peoples Rights on the Rights of Women in Africa. Adopted by the African Commission on November 28, 2014

Adinma ED, Adinma JIB, Ugboaja J et al (2011) Knowledge and perception of the Nigerian abortion law by abortion seekers in south-eastern Nigeria. J Obstet Gynaecol 31(8):763–766. doi:1 0.3109/01443615.2011.593645

African Charter (1981) African Union African Charter on Human and Peoples Rights Adopted in Nairobi June 27, 1981 Entered into Force October 21, 1986

Al-Matary A, Ali J (2014) Controversies and considerations regarding the termination of pregnancy for foetal anomalies in Islam. BMC Med Ethics 15:10

Bates C (2013) Abortion and a right to health in international Law: LC v Peru. Cambridge J Int Comp Law 2(3):640–656

Brookman-Amissah E, Moyo JB (2004) Abortion law reform in Sub-Saharan Africa: no turning back. Reprod Health Matters 12(Supp 24):227–234

Buchmann E, Kunene B, Pattinson R (2007) Legalized pregnancy termination and septic abortion mortality in South Africa. Int J Gynecol Obstet. doi:10.1016/j.ijgo.2007.10.017

CEDAW (1979) UN convention on the elimination of all forms of discrimination against women. 18 November 1979, 1249 UNTS 13

CEDAW (2009) Views of the Committee under Article 7, paragraph 3 of the Optional Protocol to the Convention on the Elimination of All Forms of Discrimination against Women at its fiftieth session concerning Communication No 22/2009, L.C. v Peru- CEDAW/C/50/D/222009

Centre for Reproductive Health (2014) The world's abortion laws. New York, USA www.world-abortionlaws.com. Accessed 10 Aug 2016

Chima SC (2011) A Primer on medical law bioethics and human rights for African scholars. Chimason Educational Books, Durban

Chima SC (2015a) Religion politics and ethics: moral and ethical dilemmas facing faith-based organizations and Africa in the 21st century-implications for Nigeria in a season of anomie. Niger J Clin Pract 18(Suppl 1):S1–S7

Chima SC (2015b) 'Because I want to be informed – to be part of the decision making': patients' insights on informed consent practices by healthcare professionals in South Africa. Niger J Clin Pract 18(Suppl 1):S45–S56

Chima SC, Mamdoo F (2015) Ethical and legal dilemmas around termination of pregnancy for severe fetal anomalies: a review of two African neonates presenting with ventriculomegaly and holoprosencephaly. Niger J Clin Pract 18(S1):S31–S39

Coast E, Murray SF (2016) "These things are dangerous": understanding induced abortion trajectories in urban Zambia. Soc Sci Med 153:201–209

Codigo Penal de Peru (2008) Legislative No. 635 Art 119

Constitution of Kenya (2010) The Constitution of the Republic of Kenya, Published by the National Council for Law Reporting with the Authority of the Attorney General

Dickens BM, Cook RJ (2007) Reproductive health and public health ethics. Int J Gynecol Obstet 99:75–79

ECPHRFF (1950) European convention for the protection of human rights and fundamental freedoms opened for signature 4 November 1950, 213 UNTS 221, Europ TS. No 5 (entered into force 3 September 1953)

Enright M (2016) Amanda Jane Mellet v Ireland-The Key Points- Human Rights in Ireland. http://humanrights.ie/uncategorized/amanda-jane-mellet-v-ireland-the-key-points/. Accessed 29 Aug 2016

Famoroti TO, Fernandes L, Chima SC (2013) Stigmatization of people living with HIV/AIDS by healthcare workers at a tertiary hospital in KwaZulu-Natal, South Africa: a cross-sectional descriptive study. BMC Med Ethics 14(Suppl 1):S6

Faundes A, Shah IH (2015) Evidence supporting broader access to safe legal abortion. Int J Gynecol Obstet 131:556–559

Flieschman AR, Chervenak FA, McCullough LB (1998) The physicians moral obligations to the pregnant woman, the fetus, and the child. Semin Perinatol 22(3):184–188

Gentleman A (2016) UN calls on Ireland to reform abortion laws after landmark ruling' The Guardian, 9 June 2016

Grimes DA (2003) Unsafe abortion: the silent scourge. Br Med Bull 67:99–113

Grimes DA, Benson J, Singh S et al (2006) Unsafe abortion: the preventable pandemic. Lancet 368(25):1908–1919

Hessini L (2007) Abortion and Islam: policies and practice in the Middle East and North Africa. Reprod Health Matters 15(29):75–84

HPCSA (2007) Guidelines for the withholding and withdrawing of treatment, 2nd edn. Health Professions Council of South Africa, Pretoria

Human Rights Committee (2005) K.L. v Peru ICCPR/C/85/D/1153/2003, United Nations HRC, 3 November, 2005

Human Rights Committee (2011) LMR v Argentina Communication No. 1608/2007, CCPR/C/101/D/168/2007

Human Rights Committee (2016) Views adopted by the Committee under article 5(4) of the Optional Protocol, concerning communication No. 2324/2013-Amanda Jane Mellet v Ireland CCPR/C/116/D/2324/2013, 9 June 2016

ICCPR (1966) United Nations International Convention on Civil and Political Rights

ICPD (1994) Programme of action of the International Conference on Population and Development, adopted 18 Oct 1994. U.N. GAOR, Ch. VII, 7.2, U.N. Doc. A/CONF.171/13

IFPA (2014) UN human rights committee and Irelands abortion laws. www.ifpa.ie/node/603. Accessed 10 Aug 2016

ISECR (1966) United Nations International Convention on Social Economic and Cultural Rights.

Koster-Oyekan W (1998) Why resort to illegal abortion in Zambia? Findings of a community-based study in Western Province. Soc Sci Med 46:1303–1312

Leke RJI (2014) Contribution of obstetrics and gynecology societies in West and Central African countries to prevention of unsafe abortion. Int J Gynecol Obstet 126:S17–S19

Maputo Protocol (2003) African union protocol to the African charter on human and peoples' rights on the rights of women in Africa, Adopted by the 2nd Ordinary Session of the Assembly of the Union Maputo, 11 July 2003. Available via http://www.banfgm.org/IT/IT/La_Campagna_files/Protocol_to_the_African_Charter_on_Human_and_People_s_Rights_on_the_rights_of_women_in_Africa_simplified_pdf%20(1).pdf. Accessed 10 Nov 2015

Mbazira C (2006) Enforcing the economic, social and cultural rights in the African Charter on Human and Peoples' Rights: twenty years of redundancy, progression and significant strides. Afr Hum R Law J 6(2):358–381

Mbiti JS (1969) African religions and philosophy. Heinemann, London

McDonald H (2013) Irish abortion ban case taken to UN human rights committee. The Guardian 21 November 2013

MDGR (2015) United Nations Millennium Development Goals Report 2015

Mhlanga RE (2003) Abortion: developments and impact in South Africa. Br Med Bull 67:115–126

Mullaly S (2005) Debating abortion rights in Ireland. Hum Rights Q 27(1):78–104

Ngwena CG (2010a) Inscribing abortion as a human right: significance of the protocol on the rights of women in Africa. Hum Rights Q 32(4):783–864

Ngwena CG (2010b) Protocol to the African Charter on the Rights of Women: implications for access to abortion at the regional level. Int J Gynecol Obstet 110:163–166

Ngwena CG (2012) State obligations to implement African abortion laws: employing human rights in a changing legal landscape. Int J Gynecol Obstet 119:198–202

Ngwena CG (2013) Developing regional abortion jurisprudence: comparative lessons for African Charter organs. Netherlands Q Hum Rights 31(1):9–40

Ngwena CG, Cook RJ (2005) Rights concerning health. In: Brand D, Heyns CH (eds) Socio-economic rights in South Africa. Pretoria University Law Press, Pretoria, pp 107–151

Ngwena CG, Brookman-Amissah E, Skuster P (2015) Human rights advances in women's reproductive rights in Africa. Int J Gynecol Obstet 129:184–187

OAPA (1861) The offences against the Person Act 1861 24 & 25 Vict c 100

Okonofua F (1997) Preventing unsafe abortion in Nigeria. Afr J Reprod Health 1(1):25–36

Okonofua FE, Hammed A, Nzeribe E et al (2009) Perceptions of policy makers in Nigeria toward unsafe abortion and maternal mortality. Int Perspect Sex Reprod Health 35(4):194–202

Okonta PI, Ebeigbe PN, Sunday-Adeoye I (2010) Liberalization of abortion and reduction of abortion related morbidity and mortality in Nigeria. Acta Obstet Gynecol 89:1087–1090

P. and S. v. Poland (2008) No. 57375/08 Eur. Ct. H.R

Pueschel SM (1991) Ethical considerations relating to prenatal diagnosis of fetuses with Down syndrome. Ment Retard 29(4):185–190

R. R. v. Poland (2011) No. 27617/04, Eur. Ct. H.R., para. 267

Republic of South Africa (1996) Choice on Termination of Pregnancy Act 92 of 1996. Government Gazette No. 17602, Notice No. 1891, dated 22 November 1996, with current amendments. http://www.saflii.org/za/legis/consol_act/cotopa1996325/. Accessed 10 Nov 2015

RFWCW (1995) Report of the fourth world conference on women – Beijing Declaration and Platform for Action, adopted 17 Oct., U.N. GAOR, 50th Sess., 94–96, U.N. Doc. A/CONF.177/20

Rydval A, Lynöe N (2008) Withholding and withdrawing life-sustaining treatment: a comparative study of the ethical reasoning of physicians and the general public. Crit Care 12:R13. doi:10.1186/cc6786

Sedgh G, Singh S, Henshaw SK et al (2012) Induced abortion: incidence and trends worldwide. Lancet 379(9816):625–632

Sedgh G, Singh S, Hussain R (2014) Intended and unintended pregnancies worldwide in 2012 and recent trends. Stud Fam Plan 45(3):301–314

Sedgh G, Bearak J, Singh S et al (2016) Abortion incidence between 1990 and 2014: global, regional and subregional levels and trends. Lancet 388:258–267

Shariq A, Sirakaya A, Mumbi B, et al. (2015) Opinions by young analysts on the UN Sustainable development goals In: Hampel M (ed) Special report on the UN sustainable development goals. Politheor: European Policy Network

Shaw R, Cook RJ (2012) Applying human rights to improve access to reproductive health services. Int J Gynecol Obstet 119:S55–S59

Singh S (2006) Hospital admissions resulting from unsafe abortion: estimates from 13 developing countries. Lancet 368:1887–1892

TOPA (1972) Termination of Pregnancy Act. Republic of Zambia. Chapter 304 of the laws of Zambia. 13 Oct 1972

Tysiac v Poland (2007) European Court of Human Rights Application 5410/03 Strasbourg, 20 March 2007

CPSIA information can be obtained
at www.ICGtesting.com
Printed in the USA
BVHW040227081118
532517BV00005B/22/P